CSR, Sustainability, Ethics & Governance

Series Editors

Samuel O. Idowu, London Metropolitan University, Calcutta House
London, UK

René Schmidpeter, Cologne Business School
Cologne, Germany

In recent years the discussion concerning the relation between business and society has made immense strides. This has in turn led to a broad academic and practical discussion on innovative management concepts, such as Corporate Social Responsibility, Corporate Governance and Sustainability Management. This series offers a comprehensive overview of the latest theoretical and empirical research and provides sound concepts for sustainable business strategies. In order to do so, it combines the insights of leading researchers and thinkers in the fields of management theory and the social sciences – and from all over the world, thus contributing to the interdisciplinary and intercultural discussion on the role of business in society. The underlying intention of this series is to help solve the world's most challenging problems by developing new management concepts that create value for business and society alike. In order to support those managers, researchers and students who are pursuing sustainable business approaches for our common future, the series offers them access to cutting-edge management approaches.
CSR, Sustainability, Ethics & Governance is accepted by the Norwegian Register for Scientific Journals, Series and Publishers, maintained and operated by the Norwegian Social Science Data Services (NSD).

Marisa Mühlböck
Editor

Sustainable Transformation and Well-being

The Role of Business for Individual and Collective Flourishing

Springer

Editor
Marisa Mühlböck
Independent Researcher
Vienna, Austria

ISSN 2196-7075 ISSN 2196-7083 (electronic)
CSR, Sustainability, Ethics & Governance
ISBN 978-3-031-75568-2 ISBN 978-3-031-75566-8 (eBook)
https://doi.org/10.1007/978-3-031-75566-8

This Springer imprint is published by the registered company Springer Nature Switzerland AG
The registered company address is: Gewerbestrasse 11, 6330 Cham, Switzerland

Foreword

The essays in this volume break some important new ground. Sustainable transformation is a very difficult idea in itself, and when aimed at nothing less than transforming the way we conduct value creation and trade (business), it becomes a persistent, wicked, and extremely difficult issue to address. We will not be successful with "ideas as usual." The traditional disciplines of business need reinvention, and we need to address the whole of human knowledge and understanding to gain more and deeper insights.

The editors and authors of these papers have pulled out all the stops and shown us how to begin to broaden the conversation about business and the good life. They have suggested that we need to enroll the humanities, from philosophy to music in recasting an understanding of what it means to live a good life, a life of well-being and meaning. And, how we need to reconstruct our business organizations to support such a transformation.

This process must begin with understanding our purposes and values, in individual and organizational terms. However, there is a problem with the idea that we are the best judges of our values. We should be humbled by Freud, Darwin, Einstein, and all of their followers as they have decentered the world over the last two centuries. We are often mistaken about what lies in our hearts, and we need strong relationships with others to help us figure out how we want to live. We need to leave no stone unturned as we take on the project of living in an authentic way. In short we need to create our lives, much in the way that artists create works of art. A life created has meaning, is based on a sense of values and purpose, and the resulting sense of well-being will benefit both individuals and society. The editors and authors of these essays have done all of us a favor by giving us a wonderful catalog of ideas that have the potential for real sustainable business transformation.

University Professor
University of Virginia
Charlottesville, VA, USA

R. Edward Freeman

Acknowledgment

Editing this book was a fascinating and at the same time challenging process. To me, holding the outcome of many working hours in my hands does not only mean an intellectual accomplishment but a result of personal development. Thus, I would like to express my deepest gratitude to all those who have contributed to the creation of this book.

Firstly, I'd like to thank the researchers and experts who have shared their knowledge and insights from different perspectives. In times of overfilled working schedules, fulfilling a commitment on top of that should never be taken for granted. This book deals with how to promote a good life for everyone, and as such, it was important to me that none of the contributors had to compromise their individual state of well-being when writing the respective chapters. For that reason, I would like to express my respect and extend my gratitude also to those who were initially interested in contributing but in the end decided not to participate after taking their well-being into account. This book represents only a starting point for the investigation into and debate on the topics of sustainable business transformation and well-being, and I hope to be able to integrate many more relevant perspectives in future projects.

A big thank-you goes out to all contributors who are featured in Part III of this publication, which focuses on real-life cases. It is a brave endeavor to share personal experiences openly, in particular when that includes challenging phases in one's life. For that, they deserve special recognition.

I am grateful for René Schmidpeter's encouragement, first in his role as a series editor in which he motivated me to publish this book. At the same time, he has been a congenial co-author who has repeatedly pushed me to develop my thinking further. Salvatore Lavecchia and Mathias Schüz have been a tremendous support with valuable advice throughout the editing process. Furthermore, I would like to thank Ed Freeman. As a pioneer in his field, he has reshaped society's perception of business, which has inspired me in many ways throughout my career and in particular when writing my own chapters of this book. I am therefore very much honored that he took the time to write a foreword to this book.

I would like to extend my appreciation to all publishing team members at Springer who have been involved in bringing this publication to life, to everyone I had the pleasure to communicate with in person, but also to those "hidden figures" without whom I would not have been able to hold this work in my hands today. A special thanks goes out to my proofreader Trevor Simpson, who has edited and thus improved each text with great diligence. And last but not least, I would like to express my gratitude to my family and friends for their moral support during the entire process. They have provided me with spaces to recharge my batteries in times of frustration.

Let's now celebrate the "birth" of this book for an instant. It (only) represents a picture of the moment. Thus, I am already looking forward to developing the topic further—together with the brilliant expert network which has emerged while working on this book and with all those who will join the journey in the future.

Vienna, Austria Marisa Mühlböck

Introduction

Even though the "Anthropocene" may not yet have been officially recognized as an era, the influence of humans on the Earth's biological, geological, and atmospheric processes cannot be denied. At the same time, human thoughts, feelings, and actions not only have an impact on an individual's surrounding or their "outside" but also on their inner sphere or "inside." Inner and outer or individual and collective well-being is therefore inextricably linked. This makes the exploration of human behavior in connection with the concept of well-being increasingly significant. Can personal well-being become a driver to create more well-being for everyone? And what kind of well-being are we talking about anyway?

Despite increased discussion on the notion of well-being, both among the general public and in academia, the concept remains vague. Definitions depend on the specific discipline and context. In other words, it makes a significant difference, if you ask a philosopher, a biologist, a psychologist, or an anthropologist what well-being is for them. Similarly, at the practical level, one might receive different answers from a business manager in Norway, a single mother in Indonesia, or a farmer in Brazil. Thus, it is of relevance to clarify whom or what we are looking at: Is the well-being of a person, society, or the planet in the focus? What is the cultural context? Are we talking about well-being for today's or future generations?

The business sector is characterized by a particularly broad scope of action and impact. As employers and providers of products and services, companies interact with all other sectors of society at the national and sometimes even international level. That multifaceted role requires that decision-makers in business take on the significant responsibility, which emerges from that context. If the right framework conditions are in place, the business sector thus has a great deal of transformative power. With this in mind, this book addresses the following key questions: What could a zeitgeisty role description for an economy that focuses on the well-being of everyone look like? What interplay and interrelations of individual and collective well-being can we expect in this context? And where can we find points of contact in theory and practice to support and promote such a sustainable transformation in the economy?

While the topics of sustainable business and corporate social responsibility have already been the subject of research and debate for some time, linking the concepts of sustainability, and particularly sustainable business, and well-being is still a relatively new approach. Therefore, the field that is emerging still holds many questions, but also intriguing opportunities that need to be explored. As this rather complex playground calls for holistic and multidimensional system thinking, it has been a central concern of this publication to take an inter- or rather transdisciplinary approach into account. This book thus deliberately offers impulses from different scientific disciplines as well as expert knowledge from theory and practice, under the consideration of different cultural backgrounds. The result is a presentation of different perspectives of and approaches to the concept of sustainable business transformation and well-being. The book therefore does not claim to provide a universal definition, but rather aims to understand which approaches exist in which disciplines, fields, and sectors in order to identify further questions that need to be investigated to ultimately contribute to greater individual and collective well-being.

Structure of the Book

In line with the motto "entering a new paradigm," the book launches its investigation by exploring conceptual thoughts and concepts for a new role of business in the context of sustainable transformation and well-being. In Part 1 questions are addressed, such as: Why is mastering the "art of transformation" essential when aiming for more sustainability and well-being? Why do we need to question current narratives and images of the self, and which image aligns best with the given context? And what do Taoism and Japanese philosophy teach us about that?

In the chapter "The Art of Transformation: How We Can Simultaneously Adapt to and Change the World," Stefan Brunnhuber illustrates why a "deficient" human species must master the art of transformation in a new paradigm in which we recognize not only the complexity of the outside world but also what is inherent in us humans, and what needs to be done in the process.

In the chapter "The Present of an Ancient Notion: Eudaimonic Well-Being as a Concept for Sustainable Entrepreneurs?," Salvatore Lavecchia takes us back to ancient Greece and, drawing on the original meaning of the concept of eudaimonía, which assumes that human beings only manifest their true selves in their encounter with the world, examines its suitability as a basis for a sustainable transformation of the economy. He shows that sustainable eudaimonic well-being must be inextricably linked to an appropriate image of the self that goes beyond an understanding of individuals as atomistically separated from the outside world.

The chapters "How Purpose Drives Sustainable Business Transformation and Well-Being: Practical Wisdom of Taoism" and "Centennial Companies in Japan: Ethical Strategies for Sustainability and Well-Being," in turn, take us on a journey to the Far East. Liangrong Zu immerses us in the philosophy of Taoism, demonstrates its inseparable connection with the concept of "purpose," and from that derives recommendations for a sustainable transformation of the economy. Maki Sato, on the other hand, takes a keen look at *shinise*—companies that have existed for at least a hundred years and which are particularly numerous in Japan, and thus familiarizes us with ethical strategies for greater sustainability and well-being that have their origins in the various philosophical schools of thought and concepts prevalent in Japanese society.

The chapter "Paying Tribute to the Complexity of Human Beings: A Business Narrative Harmonizing the Well-Being of Individuals and Society" serves as a bridge from the first to the second part of the book. Marisa Mühlböck and René Schmidpeter ask why the prevailing narrative in and about the economy is significantly characterized by skepticism and reservations. They explain why a new narrative is essential for sustainable development and which elements could serve as cornerstones for this. In doing so, they incorporate current scientific findings as well as developments in business practice and show that such a new narrative also requires a further development of economics in the direction of increased inter- and transdisciplinarity.

The second part of the book therefore focuses on specific fields and contexts of application for sustainable business transformation and well-being. To kick things off, in the chapter "Philosophy as a Resource for Sustainable Transformation and Entrepreneurs' Well-Being" Yusuf Mehmet Örnek uses practical experience to showcase that entrepreneurs are not only willing and able to engage in philosophical reflection, but that such a debate contributes greatly to individual and collective well-being. Gustav Mahler's Symphony No. 8, in turn, serves Oliver Graber in the chapter "The Zen of Music: On the Energetics of Music as a Source of Personal Well-Being" as the basis for an evidence-based illustration of the well-being-enhancing effects of art and music in particular.

The chapters "The Workplace as a Challenge and a Resource for Adult Development" and "Cornerstones of the School of the Twenty-First Century in the Context of Sustainability and Well-Being" are dedicated to specific places that, given the right framework conditions, are suitable as "breeding grounds" for greater sustainability and well-being. Rainer von Leoprechting first takes a close look at the workplace and demonstrates its potential based on findings from adult development. Andreas Salcher focuses on schools as central places of teaching and learning and, accompanied by concrete best practices, presents principles, forms of work, and methods that offer fertile ground for teaching skills that benefit students, our society, and our planet in equal measure.

The chapters "Impactful Ventures, Burned-Out Founders? Insights from Two Studies of Social Entrepreneurs," "Understanding the Nuances of Women-Owned Businesses in Emerging and Developing Economies," and "The Phenomenon of Family Businesses: A Cross-Generational Commitment to Sustainability—Attitudes and Values of the Successive Generation in Family Businesses in the Case of Austria" take a look at specific economic sectors in the context of sustainable business and well-being. Peter Vandor and Michael Meyer start by focusing on social enterprises and tackle the issue of burnout as a form of "negative well-being" (or mental ill-being). Based on the results of two international studies, they show the degree of burnout prevalence, antecedents of burnout at the individual and organizational level, as well as coping strategies. Yanire Brana, Nayan Mitra, and Mohammad Mostak Al Farhad then turn their attention to women-led companies and take us on a journey through various emerging countries around the world. They point out that psychological or eudaimonic well-being has hardly been researched in these regions, but that the focus has so far mainly been on financial, institutional, and infrastructural, regulatory and socio-cultural factors that stand in the way of women in their role as entrepreneurs. Tien Than Nguyen and Christoph Robinson focus on family businesses and show that the special entanglement of family and business favors long-term and sustainable thinking and action and at the same time has an impact on the well-being of the individual family members.

The last topic cluster in Part 2 is dedicated to external factors and sectors that influence the microeconomic level and thus the decisions of companies. In the chapter "How to Achieve a Wellbeing Economy from Macro to Micro," Fritz Hinterberger points out the necessity of intertwining macro, meso, micro, and individual levels of action in policy, business and personal fields and discusses framework conditions for more well-being for everyone by bringing together theoretical insights and practical learnings. In the chapter "Eudaimonic Well-Being: A Foundation for Public Administration Systems Fit to Build a Sustainable Future," Fredrik Lindencrona then moves on to the administrative level, outlining a framework concept for an improved system of public administration based on eudaimonic principles and concrete empirical values. Finally, in the chapter "The Philanthropic Sector Creating Momentum for a Sustainable Transformation and the Well-Being of Changemakers," Tomas Björkman, Claudia Garuti, Susanne Hillebrand, and Marisa Mühlböck highlight the importance of the philanthropic sector in the context of the book. They make it clear that the sector itself must undergo a transformation in order to continue to serve society. Based on practical experience, they argue in favor of taking up new topics and fields of activity and questioning current framework conditions, structures and processes in philanthropic institutions.

The chapter "Teaching and Implementing Ethics for Well-Being in Business: A Field Report" paves the way from Part 2 to Part 3 of the book. Mathias Schüz combines his theoretical knowledge as a business ethicist with a detailed report on his more than 30 years of experience in both university teaching and the implementation of business ethics concepts in entrepreneurial practice. In this way, he builds the perfect bridge to the practical section, which, taking a transdisciplinary approach into account, concludes with a portrait of four entrepreneurs who openly report on their personal experiences with sustainable business and well-being. They share private insights that, for them personally, have proven to be individual and entrepreneurial success factors.

Bart Weetjens, who is not only the founder of several companies but also a trained Zen monk, shares a particularly challenging personal experience to illustrate what entrepreneurs can learn from the practice of Zen in the chapter "The Skull in Our Suitcase: What We Can Learn from Zen in Challenging Moments". In the chapter "From Paragliding in the Himalayas to Finding Purpose in Unleashing Women's Full Potential: A Personal Report," Veroniek Vermeulen describes her transformation from a rational manager in a consumer goods company to a passionate entrepreneur who finds and creates meaning by helping other companies to establish a workplace environment that promotes women's well-being. The chapter "If I Can Really Make an Impact, Then My Life Has Been a Success: A Personal Report on the Transformation of a Company and an Individual" is dedicated to the life story of Thomas Welser, which is closely linked to the family business that has existed since the seventeenth century. He describes a personal and entrepreneurial transformation process and emphasizes that for him, a successful life is always linked to the opportunity to achieve maximum impact and why sustainability and social responsibility are a logical consequence of that. Finally, in the chapter "Lessons from a Wasp Sting: A Plea for Inner Development in Order to Change the World for the Better," Hannah Lux shares her experience. Her report also looks at so-called "negative" drivers for social entrepreneurship based on a personal traumatic experience and emphasizes the importance of inner development for sustainable transformation on the outside.

The final chapter "A Transdisciplinary Economy for Individual and Collective Well-Being: Starting a Scientific Journey in Economics and Management Science" attempts to integrate the wealth of knowledge and insights from all previous contributions in this book and highlights the need for further research activities in the context of sustainable business transformation and well-being. It is a call for a "new scientific journey" in economics and management science by embracing transdisciplinarity as an approach to better depict reality, or in other words the complexity of our inner and outer world. On that mission, Marisa Mühlböck and René Schmidpeter sketch cornerstones for a transdisciplinary context-specific research program in the future that puts well-being at its heart. They discuss what a suitable research philosophy might look like, play with first ideas about a research framework and introduce potential research questions that result from the various inputs in this book.

Sustainability and well-being are both concepts that will probably always be challenging to grasp, in the real world and in particular from a scientific perspective. The multidimensionality that accompanies them can sometimes be frustrating, yet at the same time fascinating and intriguing. Clearly, they both invite us to embracing complexity rather than striving to reduce it. The 27 contributors in this publication provide food for thought to further investigate the symbiotic effects of sustainability and well-being from a business perspective. As such, I hope that this book will help to inspire the emergence of better questions, and with that ultimately better answers to today's various social, ecological and economic challenges.

Vienna, Austria Marisa Mühlböck

Contents

About the Editor

Dr. Marisa Mühlböck holds a doctorate in social and economic science. Over the course of her career she has gained extensive insight into different fields of work, which has fostered her understanding of diverse organizational structures, thinking patterns, and social context. As a former social entrepreneur, she also knows the entrepreneurial life from practice and incorporates this experience into her research. As a Research Country Co-Lead, she is involved in the Inner Development Goals Initiative (IDGs), which is dedicated to researching and promoting inner development to achieve the UN Sustainable Development Goals. She also served as a founding board member of Verband für gemeinnütziges Stiften, Austria's first association representing and advocating for charitable foundations. Her current research interests are based on her theoretical background as well as on her practical experience and focus on sustainability in different areas, sustainable entrepreneurship, and well-being.

Part I
Entering an New Paradigm: Exploring Thoughts and Concepts for a New Role of Business in the Context of Sustainable Transformation and Well-Being

The Art of Transformation: How We Can Simultaneously Adapt to and Change the World

Stefan Brunnhuber

Abstract Transformation is a choice and an art. It is no linear process but a spiral going up and down within fixed boundaries, widening and opening up on the way to the top. The author of this chapter illustrates why the 'deficient' human species is dependent on specific skills and capacities to ascend that spiral staircase and that this requires a juggling act to keep five balls in the air simultaneously. He showcases that this path involves dealing with complex mental processes and that mastery can only be achieved through hard and disciplined work. This chapter is a call to take on this challenge and a warning that the ability to adapt, which has represented mankind's 'selection advantage', is at stake in a world where our cultural achievements have enabled us to coevolve and create a culture beyond nature. At the same time, it reminds us that transformation is the very process by which we can best deal with uncertainties, contradictions and disruptions. As such, mastering the art of transformation is relevant for both sustainability transformation at the collective level and more well-being for individuals, determined by a mindset that is capable of staying open to what the future may hold and that favours a 'both-and' thinking over the 'either-or'.

Keywords Transformation · Anthropology · Adaptation · Mindset

1 Introduction

It's not five to midnight, as we've been told for the last 50 years (Crutzen, 2002). But we are indeed in the midst of a transformation process. This raises the question as to what we actually mean when we talk about transformation. Transformation isn't something that just happens, or an inevitable evolution. It is not a revolution

S. Brunnhuber (✉)
Colditz, Germany
e-mail: brunnhuber.cor@gmxpro.de; http://www.stefan-brunnhuber.de

M. Mühlböck (ed.), *Sustainable Transformation and Well-being*, CSR, Sustainability, Ethics & Governance,
https://doi.org/10.1007/978-3-031-75566-8_1

that destroys the existing order, or a reform that adapts and optimises it. It is not a furious protest in which we yell at each other, break things and then go home again.

Transformation is something more, something completely different. It is a conscious, intentional process and therefore a choice. We can call a process transformative if it respects how things are and appreciates what has been achieved but also critically points out problems, resolves contradictions, incorporates ambivalences and at the same time transcends them all (Brunnhuber, 2023).

It seems we are more anxious about transforming each other and our society than we are about the negative consequences of all the regular crises we face. For over 50 years, research has shown that it is not objective stressors like pandemics, climate change and species loss that increase stress and dysfunctional decision-making but rather subjective and collective loss of control over the process. That eventually leads to tunnel vision where we see only the perils, not the potential. In short: in situations characterised by pressure and increased complexity, we tend to do whatever will restore our feeling of control, even though it is often insufficient, ineffective and inefficient.[1] If we were honest with each other, we might admit that we don't really know what to do and what form the necessary transformation should take (Šlaus, 2020; Lovelock & Margulis, 1974).

2 The Spiral Staircase

Transformation is a spiral staircase. It goes up and down within fixed boundaries, widening and opening up as it approaches the top. With each level we climb, the focus of our consciousness, our thinking, our feeling, our actions and our chosen rules changes. This process is largely determined by our mindset and by changes in our practices and is vertical rather than horizontal. With each step we take, everything is at stake. Understood this way, transformation is not a circle in which we meet ourselves over and over again[2] nor a matter of numbers and algorithms or technology. Rather, as we ascend the spiral staircase, we improve and become able to see more, to integrate opposites. Our consciousness becomes more inclusive, comprehensive and integral.[3] Arguably, it is our intellect or mindset that enables a certain form of cultural achievement, which in turn creates a new consciousness. At first glance, this seems somewhat circular. And indeed it is. But that doesn't matter,

[1]Another reflex in the face of complexity and adversity is escapism. Instead of facing the actual challenges, we tend to escape into digital worlds or our personal lives. Both tendencies shy away from true transformation (Brunnhuber, 2023).

[2]Describing transformation as a circle excludes the possibility of genuine creativity and novelty, since the circle metaphor suggests that we keep going round again and again without changing.

[3]Our perceptual baseline is thus shifting towards higher standards and a better quality of life. Whereas 200 years ago 90% of the total population was poor, at the beginning of the twenty-first century we now consider a 10% poverty level unacceptable. In short, things have never been better, yet at the same time have never felt worse (Acemoglu & Johnson, 2023).

because a spiral staircase also has a partly circular structure. The art of transformation cannot be reduced to an algorithm or a linear decision tree, nor is it a chaotic play-dead reflex in response to a shock or exponential development. Rather, it is something else: a spiral. The world is breaking down or unravelling at its edges, creating more diversity. Those inner and outer limits are gradually widening and opening up as we progress upwards, and that is what helps the art of transformation to succeed. Our economy, our agriculture, our social life and nature as a whole must follow this law of the spiral to flourish. Even regional economic cycles, regenerative agriculture and all the experiences that are handed down from one generation to the next follow the same principle. Each time, something is added and something is removed or left out. It is never just a circle that revolves around itself but always a circular movement up or down. The fact that we keep turning it into a narrative of linear progress is probably a flaw in our thinking rather than something amiss in the world itself (Harari, 2018).

As is well known, the model is not the reality it represents, the menu is not the food it describes and the map is not the place it depicts. But the metaphor of the spiral staircase does capture the process of transformation and the reality we are living in best. The path we are following can go up (progress) or down (regression), we can fall over the railings (limits) of the spiral staircase and we can stop transforming. Transformation is the process by which we can best deal with uncertainties, contradictions and disruptions. It gives us a direction. We now know which way is up. And the spiral staircase opens out at the top into a terra incognita, an unknown land that lies ahead of us. This is nothing new, of course. Throughout history, people have had to transform themselves time and again. But they have not had to do so constantly and everywhere, like we need to today. And as we climb the spiral staircase, we simultaneously have to perform a juggling act.

3 Juggling Five Balls in the Air

There are five balls we have to keep in the air as we climb the staircase of transformation.

First, there is an acute reflex in response to adverse events and shocks, such as the effects of rising temperatures, viral pandemics or biodiversity loss. Such events force us to rethink and demand a 'creative latency'. But that's only the first step. If we stopped there and just called for 'degrowth' or 'haircuts' and thus simply repeated the mantra 'less is more', we would hit pause on our thinking at the very moment when our critical faculties are needed more than ever.

As we stop business as usual and put 'less is more' into practice, transformation also requires us, *secondly*, to *reappraise* all areas of life: to reflect on the collateral damage we have caused and set priorities. We have to work out what is important, what can be postponed and what we can do without. For instance, building dams, hospitals and nurseries, protecting nature reserves and retention basins and reforming the rules for the financial markets are vital. It's become apparent that 'doing less' during the pandemic—characterised by degrowth and haircuts—plunged an additional 100 million people into poverty and hunger. The UN's Human Development Index (HDI) rolled back 10 years, so that after the global pandemic of 2019–2022 we find ourselves set back by a decade. We can't accept that.[4]

A *third* ball then comes into play: a complex decision-making process, which no longer focuses on optimising and adjusting the existing rules of the game but rather on setting new rules, regulations and incentives. That includes introducing national citizens' assemblies, disruptive technologies like AI or changed mandates for central banks. Other examples are the regionalisation of our economy, new rules for our commons, increasing our resilience and building more regenerative agriculture. But none of that is enough by itself.

Fourth, there is the consideration that our decisions must also benefit future generations. It is important to separate situational risks from systemic uncertainties. Situational risks can be hedged or even avoided, but systemic uncertainties will remain; they can only be mitigated by sensible precautions. We are gradually moving towards the right, intergenerational approach, where sustainability equals long-termism.

Fifth, we must learn how to coordinate the first four processes. A bird's eye view, also known as metacognition or a witness perspective, becomes key. This is a mindset that is tolerant of failure that is always open to revision and adaptation.

We might say we're keeping five balls in the air all at the same time—*responsiveness, rebalancing, new rules, prevention and metacognition*—while at the same time ascending a spiral staircase.[5] All five processes can be collected under the general heading of successful adaptation. This adaptation takes place not just externally but also in our minds. Indeed, it is first and foremost a mental process and only secondarily a material one. The changes in our thinking, mindset or consciousness come first, and changes in technology, governance and our way of life follow. The final ball, the ability to keep an eye on all the other balls, is probably the most important and at the same time the most difficult to manage. It requires a different focus of consciousness: role-taking, anticipation, renunciation, networked thinking and consideration of subsequent effects. And it requires new emotional states and attitudes, such as humility, compassion and responsibility. It's a complex mental process, in which the transformation proceeds not linearly and sequentially but more or less simultaneously. Given constraints of time, geographical borders, scarce resources and incomplete information, it is important to make smart decisions

[4] The transformation process will likely need to remain agnostic regarding GDP. What is important is not how high the figure is but what exactly grows; hospitals, nurseries and nature reserves are some of the things we might think really matter.

[5] To be more precise, each of the areas represented by the five balls can be further divided into six subdomains: demographics (birth control, education), technology (resource efficiency, new energy carriers), governance (open societies, autocratic regimes), lifestyle changes (cycling, meditating, going vegetarian), culture (third sector, science, media) and finance (new financial engineering instruments).

today—in the midst of this transformation process. And while doing all that, we are keeping the five balls in the air as we climb a spiral staircase.

4 Transformation Is an Art

All these factors turn the process of transformation into an art. The term ‘art’ is meant here in the sense of the ancient Greek *techne* (τέχνη), that is to say, a learned, practical rule-guided craft of creating something new. We have to be careful: the gap between how the world is and how it needs to be has never been wider. Transformation is a craft that we can only practise, perfect and master through hard, disciplined work. And we can also unlearn it again.

As is well known, it takes over 10,000 hours of supervised practice to master a craft. Bearing that in mind, we should take all the time we can spare, which is available to us to patiently master the ‘art of transformation’.

At its core, transformation confronts us with our individual and collective finitude.[6] The number of possibilities we can act on is inversely proportional to our finitude. The more opportunities that are open to us, the more we must learn to renounce and reject. We must simultaneously let go of the familiar and work towards something new.

The task we must learn is not just to optimise and make more efficient what we already have. It is not just a matter of innovation, research and development. It is about doing things completely differently, for we can otherwise quickly slip into doing the wrong thing without noticing it. At every step, we can ask what we give up if we decide to hold on to something and what we miss out on if we keep clinging to an internal and external logic of expansion. If we focus merely on growth and expansion, we will lose valuable experiences such as silence, solitude, simplicity, slowness, mindfulness, having time and living in the analogue world. In the twenty-first century, we need a transformation that shifts us away from quantitative growth, constant extraction and expansion and external spectacle and instead focuses on internal development.

5 Where It All Began[7]

When human beings invented the needle around 50,000 years ago, something amazing happened. They began sewing furs that allowed them to survive in colder climates and explore new habitats. This complex adaptation process involved not just

[6] We can identify two general forms of limitations: external limitations (e.g., planetary boundaries) and internal limitations (e.g., our mental frames and biases). Both confront us with our inadequacy to deal with the transformation that lies ahead (Byers, 2014).

[7] This section refers to Darwin (1872), Elias (2000) and Jaspers (1949).

tool use and the executive functions required to make a needle. It also involved a new distribution of temporal and material resources. The division of labour began, with some people hunting buffalo, some sewing furs and some building huts. Groups of people made a collective decision to change their lives. At the same time, the decision to devote time to making needles meant spending less time on other activities. People had to learn to weigh things up and prioritise. Perhaps they could not collect as many berries as before or their children received less or different education. Ultimately, the decision to make needles came out of the judgement that it was worthwhile, as it would provide a valuable skill for the future (Hare, 2016).

To get straight to the point: the 'culture of enough', which goes hand in hand with a complete rejection of any economic growth, is often the starting point for critical reassessment of the social status quo. But it has never been the end point of a successful adjustment.[8] It is always possible to give things up without replacement, to renounce excess and to do less. And doing so comes, in a sense, for free. It has no institutional preconditions and does not violate any law. Above all, it requires self-discipline and the ability to say no. That is all true and very important. Yet, a successful adaptation involves far more.

What the needle was 50,000 years ago is what the sustainability agenda is today. In both cases, a complex process of adjustment is associated with opportunities and sacrifices, with curiosity, courage and risk. And in both cases, this process is transformative and involves multiple stages.

Transformation will always remain an 'art' and one where it is possible to fail as well as to succeed: causing fatalities, disease and suffering, creating winners and losers and bringing massive restrictions with it. Transformation always involves wrong decisions that have to be corrected, as well as risks and uncertainties that have to be borne. Edison is said to have performed over 6000 failed experiments before finally getting the light bulb to shine. *Homo sapiens* has successfully transformed time and again over the past 50,000 years. It should be possible to repeat the success of the needle today with the transition to a sustainable way of living. Rigorous scientific findings and age-old life wisdom should determine in equal measure the areas in which the principles of 'business as usual', 'less is more' and 'more but different' should be applied. We will have to try new things under conditions of uncertainty and great time pressure. We will have to keep self-correcting and learning from our mistakes and failures while holding on to the hope that the whole system will not fail. Because this process of change and adjustment will happen anyway, whether we want it or not. The only question is whether we will succeed in mastering an 'art of transformation' or instead become 'masters of disaster', stumbling from one panic-and-neglect or delay-and-denial reaction to the next.

[8] We move more sediments than all the world's rivers and seas combined. Two-fifths of our land area is used for agriculture, and 50% of that area has been transformed by humans (cities, roads). The populations of two-thirds of all species have declined, and since the start of the twentieth century we have used the equivalent of 50 kg of cement per square meter. Atmospheric concentrations of microplastics, plutonium, phosphorus and CO_2 have soared. That is why the present is referred to as the Anthropocene—an era in which the world is shaped by human beings.

6 Adaptation Is (Almost) Everything

How does the human species fit into this complex world? At our core, we are not truly and fully part of nature. Unlike other animals, which are fully adapted to their niche, humans are a deficient species and therefore need a crutch to fill the gap between themselves and the natural world. That gap is filled by technology, governance, culture, language, science and so on.[9] On the one hand, this crutch represents a form of self-organisation or autopoiesis. On the other hand, it is always self-made and so incomplete and open, a terra incognita. The Darwinian framing that places us at the top of evolution is simply the wrong mental framework to explain humans in the twenty-first century. We are not standing at the top of a ladder and looking down at evolution. Rather, we are marginal figures, string players within the great orchestra of living beings, who can participate in the countless parallel worlds of other creatures, who can resonate and communicate with those worlds without fully mastering or understanding them (Brunnhuber, 2024; Prigogine & Stengers, 1984; see Fig. 1).

Processes of adapting to nature are nothing new. What is new is that in the course of 10,000 years of civilisation, we have forgotten how to properly adapt. Instead, our cultural achievements have enabled us to coevolve and create a culture beyond nature. We have become more fragile, more prone to interference and deception and more vulnerable but at the same time more co-operative and friendlier ('survival of

Fig. 1 The ladder and the string player

[9]While nature is entropic, constantly creating disorder, human activities are in part negentropic. We create a (fragile, brittle, transient) order that goes against the tide of thermodynamic heat death and the laws of natural selection.

the friendliest', we might say) (Hare, 2016). It sounds paradoxical: we have a selection advantage due to our lack of adaptation to nature. Externally, we invent technologies and institutions, while at the same time our internal psychological space changes, expands inwards and becomes more complex. We start to tell ourselves fictional stories that we all believe in and that coordinate the behaviour of large cohorts of people. Narratives about God, about a monetary or legal system that only exists because we all believe in it. This isn't how things are for other species. On the other hand, this 'civilising process' (Elias, 2000) has led to an ongoing internalisation of norms and imperatives that has made us more disciplined and rule-governed. These narratives and internalised rules interact with external cultural achievements such as tool use, legal systems and language and serve to reinforce and correct each other. They are very important to humans' selection advantage over other species.[10]

However, over thousands of years of history, this process of adaptation led to Malthus cycles. Crop failures, wars, catastrophes and crises reduced the population, only for it then to increase again further down the line. These cycles alternated over centuries, with only a small, privileged minority benefiting from advances. For many centuries, life was a circle. However, in the nineteenth century, things changed abruptly. For the first time in human history, large swathes of the population began to enjoy greater life expectancy, prosperity and well-being. Liberal rule-of-law mechanisms guaranteed the protection of minorities, child labour was banned and social security systems and access to education and a public health system were granted to a broader population. Unprecedented capital accumulation enabled more wealth for more people. Mankind didn't have to worry as much anymore, because food, housing, technological advances, education and health were available and the 'social quest' of the nineteenth and twentieth centuries was vigorously pursued. The loss of adaptability to nature primarily affects the 20% of the world's population who have benefited directly from the massive accumulation of capital and increase of prosperity since around 1820. Now, in the twenty-first century, the 'social quest' to redistribute goods and services is once again being overshadowed by an 'ecological quest'. We will have to learn to adapt again, as we cannot negotiate with nature: it follows biological and physical laws, not the social laws of labour disputes, strikes, changing ownership and liability or the adapted mechanisms of rule of law. We must be careful to ensure this upcoming transformation process doesn't fall back into the cycles we knew for centuries of human history. However, again, truly transformative processes are not circles but more like a spiral staircase (Jackson, 2005).

Humans are the most adaptable creatures on this planet and the least genetically determined.[11] We can't help but constantly adapt, correct and optimise ourselves. We can live as hermits or in large cohorts, can become meat eaters or vegetarians,

[10] The paradox is that the web of life is complex and we humans cannot fully understand and control it. But we do constantly strive to reduce that complexity, which in turn adds another layer of complexity. We might call it hypercomplexity.

[11] We share 60% of our genome with bananas and have the same number of genes as the roundworm. The genetic difference between individual humans is less than 0.1%. Thus, genetics alone

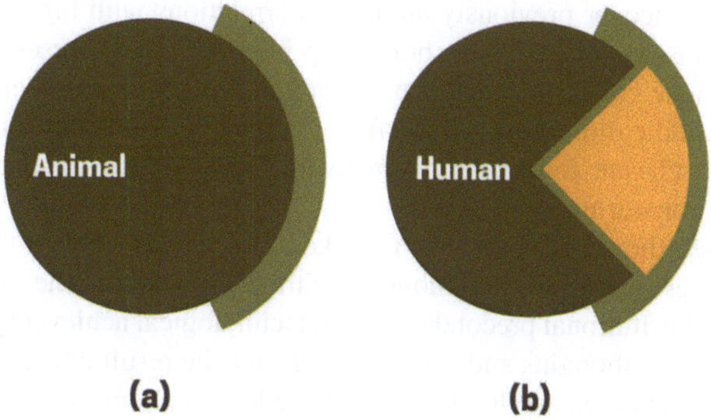

Fig. 2 The gap and the crutch. Note: (**a**) Animals are fully adapted; (**b**) Humans are not

can fly or stay grounded, can eat or fast and can speak or be silent. A lion, for example, will not become a vegan, nor can it really be held responsible for its actions. Cooking and gardening are two special adaptations unique to humans. The ability to adapt sometimes goes so far that we even adapt, without realising it, to practices or conditions that are unhealthy and irrational, inefficient and addictive (Fig. 2).

In Fig. 2, Graph (a) represents animals as fully adapted to their ecological niche, while Graph (b) represents humans who are never fully adapted to nature and therefore require a crutch to fill the gap, which is symbolised by the yellow wedge.[12]

7 The Middle Dimension[13]

Even if we reach down into the nanoworld, overcome unimaginable cosmic distances or exceed the speed of light, all our new knowledge will still in some way or another be bound to our 'middle dimension'. The world humans inhabit is one of metres and minutes, hammers and nails and keyboards and steering wheels. Our senses of sight and hearing are restricted to certain wavelengths and our senses of smell, taste and touch to certain domains. Our emotional and cognitive capacities are limited, too. We have a base set of six to ten innate primary affects, which are then refined over our life cycle. Our thinking is determined more by frames (i.e., mental expectations and social norms) than by objective facts. And our actions are guided by convincing narratives that we all believe in, even if they're factually incorrect, rather than statistically supported risk assessments or utility calculi. And

cannot explain the variations between people, culture and societies throughout history (Droc et al., 2013; Wang et al., 1999).

[12] See Blumenberg, 2014; Bowlby, 1995 [1950]; Gehlen, 2014 [1940]; Plessner, 1975; Plessner, 1983; Scheler, 2007.

[13] This section refers to Schumacher (1973) as well as to Tomasello (2019).

even if we can uncover previously unknown correlations with big data algorithms and treat Parkinson's disease with brain-chip interfaces and viruses or cancers by modifying genetic codes, such adaptations are only really comprehensible to us as part of that 'middle dimension'. If we fail to achieve self-efficacy and self-control in the specific relevant forms, our distinctively human ability to adapt will be lost and the art of transformation will fail.

However, we should not make the mistake of focusing solely on the contents of our consciousness, emotions and thoughts without considering the diverse cultural achievements, institutional preconditions and technological achievements that make them possible. Our thoughts and actions are always the result of a complex historical process of adaptation and learning that precedes our attempts to critically reflect on them. This is all attributable to our status as a 'deficient species' that can only successfully adapt to our environment through constant acquisition of cultural achievements and technologies. Humans, unlike animals, constantly need crutches to survive. There is no fundamental difference between the needle example discussed above and the imperative to create a sustainable way of life today. We now need drones, dams, drugs and data to successfully adapt to a new, complex environment.

Such adjustments are always linked to a process of critical revision and an increase in our experience, information and wisdom. Arguably, successful adjustment is the same as successful problem-solving, which is accompanied by greater knowledge and learning. That in turn leads to an expanded consciousness, which then provides the necessary basis for successful adaptation, and so on.

8 Mindset Determines our Course[14]

It is our mindset or consciousness that will determine our course up the staircase—a consciousness characterised by courage and humility, renunciation and generosity, patience and determination, control and devotion, knowledge and indecision and necessity and chance (Kahneman, 2011; McGilchrist, 2021). The great traditions of Western and Eastern wisdom can guide us here, teaching us to ask the right questions and find the right answers so that we can successfully adapt to and change the world. That means thinking in terms of 'both-and' rather than 'either-or', not resolving opposites but accepting them and thus creating potential for change. It is like the poles of a battery: we always need both to achieve the correct voltage. In psychology, this is referred to as 'tension and ambivalence tolerance'. It is considered one of the key cognitive capacities for successful development. Thus understood, a change process does not seek to explain the world in terms of simple correlations and causal relationships but is always open to what the future may bring. Transformation is a vertical process, not a horizontal one. It requires more depth and

[14] This section refers to Aurobindo (1997) and Maturana and Varela (1987) as well as to Wilber (1997).

more breadth and will allow us to see more, embrace more and do better. In terms of the staircase metaphor, we are caught between two steps, existing in a state of limbo. The familiar common ground is coming to an end, yet the new, open space is not yet within reach. That is why transformation should be considered an art and why we must keep on climbing the staircase.

References

Acemoglu, D., & Johnson, S. (2023). *Power and progress: Our thousand-year struggle over technology and prosperity*. Basic Books.

Aurobindo, S. (1997). *Primary works* (12 vols). Lotus Press.

Blumenberg, H. (2014). *Beschreibung des Menschen*. Suhrkamp.

Bowlby, J. (1995 [1950]). *Maternal care and mental health* (2nd ed.). Jason Aronson.

Brunnhuber, S. (2023). *Die Kunst der Transformation* (2nd ed.). Oekom.

Brunnhuber, S. (2024). *The third culture, the impact of AI on knowledge, society and consciousness in the 21st century*. Springer Nature.

Byers, W. (2014). *Deep thinking: What mathematics can tell us about the mind*. World Scientific Publishing.

Crutzen, P. (2002). Geology of mankind. *Nature, 415*(23), 23.

Darwin, C. (1872). *The origin of species by means of natural selection* (6th ed.). John Murray.

Droc, G., Larivière, D., Guignon, V., Yahiaoui, N., This, D., Garsmeur, O., Dereeper, A., Hamelin, C., Argout, X., Dufayard, J. F., Lengelle, J., Baurens, F. C., Cenci, A., Pitollat, B., D'Hont, A., Ruiz, M., Rouard, M., & Bocs, S. (2013). The banana genome hub. *Database: The Journal of Biological Databases and Curation, 2013*, bat035. https://doi.org/10.1093/database/bat035. PMID: 23707967; PMCID: PMC3662865.

Elias, N. (2000). *The civilizing process: Sociogenetic and psychogenetic investigations* (revised ed.). Blackwell.

Gehlen, A. (2014 [1940]). *Der Mensch, seine Natur und seine Stellung in der Welt* (16th ed.). AULA.

Harari, Y. N. (2018). *21 Lessons For The 21st century*. Jonathan Cape.

Hare, B. (2016). Survival of the friendliest: Homo sapiens evolved via selection for prosociality. *Annual Review of Psychology, 68*(1), 155–186.

Jackson, M. D. (2005). *Existential anthropology: Events, exigencies and effects*. Berghahn Books.

Jaspers, K. (1949). *Vom Ursprung und Ziel der Geschichte*. Piper.

Kahneman, D. (2011). *Thinking, fast and slow*. Farrar, Straus and Giroux.

Lovelock, J. E., & Margulis, L. (1974). Atmospheric homeostasis by and for the biosphere: The Gaia hypothesis. *Tellus, 26*(1–2), 2–10.

Maturana, H. R., & Varela, F. J. (1987). *The tree of knowledge: The biological roots of human understanding*. Shambhala.

McGilchrist, I. (2021). *The matter with things: Our brains, our delusions, and the unmaking of the world*. Perspectiva Press.

Plessner, H. (1975). *Die Stufen des Organischen und der Mensch: Einleitung in die philosophische Anthropologie* (3rd ed.). De Gruyter.

Plessner, H. (1983). Die Frage nach der Conditio humana. In *Gesammelte Schriften: Conditio humana* (Vol. 8, pp. 136–217). Suhrkamp.

Prigogine, I., & Stengers, I. (1984). *Order out of chaos: Man's new dialogue with nature*. Flamingo.

Scheler, M. (2007). *Die Stellung des Menschen im Kosmos* (16th ed.). Bouvier.

Schumacher, E. F. (1973). *Small is beautiful: A study of economics as if people mattered*. Blond & Briggs.

Šlaus, I. (2020). *Transforming our world: Necessary, urgent and still possible*. Cambridge Scholars Publishing.
Tomasello, M. (2019). *Becoming human: A theory of ontogeny*. Cambridge, MA.
Wang, D. Y., Kumar, S., & Hedges, S. B. (1999). Divergence time estimates for the early history of animal phyla and the origin of plants, animals and fungi. *Proceedings of the Biological Sciences, 266*, 163–171.
Wilber, K. (1997). An integral theory of consciousness. *Journal of Consciousness Studies, 4*(1), 71–92.

Stefan Brunnhuber, is a medical director, endowed chair for sustainability, trustee at the World Academy of Arts and Science, full international member of the Club of Rome and member of the European Academy of Arts and Science.

The Present of an Ancient Notion: Eudaimonic Well-Being as a Concept for Sustainable Entrepreneurs?

Salvatore Lavecchia

Abstract This contribution intends to show that a sustainable eudaimonic well-being is inseparable from an appropriate representation of the human self. This assumption is based on the original meaning of the Greek substantive eudaimonía, presupposing that human beings manifest their true self in their encounter with the world. Relying on some suggestions stemming from Plato's work, particularly regarding the notion of good and the Socratic concept of midwifery, the aforementioned true self may be characterized as an eminently dialogical self, whose modality of consciousness, transcending any first, second, and third person perspectives, is qualified by unegoistic freedom, that is, by intrinsic openness and self-givingness toward the harmonious manifestation of other beings. This representation of the true self is in fact consonant with an unprejudiced observation of our perceiving/knowing self, through the observation of which we become immediately capable of abandoning the current, both solipsistic and dualistic image of the self as a localizable point shrunk in itself and separated from the world. A representation of the self can be thus attained, in which the very self-phenomenal consciousness may be perceived as a possible source of inexhaustible generativity concerning sustainable eudaimonic well-being, for both individuals and for the world.

1 Focusing on the True Self: What Does *Eudaimonía* Originally Mean?

In continuity with some directions of past (Norton, 1977) and present research (Vittersø, 2016; Las Heras et al., 2022), this contribution proposes to emphasize the original Greek notion of *eudaimonía* as a possible threshold to a path on which

S. Lavecchia (✉)
Dipartimento di Studi Umanistici e del Patrimonio Culturale, Università degli Studi di Udine, Udine, Italy
e-mail: salvatore.lavecchia@uniud.it

M. Mühlböck (ed.), *Sustainable Transformation and Well-being*, CSR, Sustainability, Ethics & Governance,
https://doi.org/10.1007/978-3-031-75566-8_2

eudaimonic well-being can lead to both individual fulfillment and the realization of common good, thus revealing itself as a fruitful impulse for sustainable entrepreneurship as well as, in general, for any sustainable individual or collective activity.

Instead of giving a—necessarily too selective and approximate—summary regarding the history of the Greek notion of *eudaimonía*,[1] in consonance with some paths of recent research on well-being (Ryff & Singer, 2008; Schlegel et al., 2009), the following considerations will concentrate on the close connection of this notion with the true self of the human being. This choice provides, namely—hopefully this will become clear through the following considerations—, the best guarantee against any misunderstanding of eudaimonic well-being in the direction of narcissistic self-affirmation.

The Greek term *eudaimonía* means in fact, literally, "having a good (*eu*) *dáimon*." It therefore indicates a condition in which persons are capable of manifesting harmonious consonance with their own *dáimon*, whereas *dáimon*[2] is meant on the one hand as the good spirit who hypostasizes the destiny revealing the best potentialities of a person and on the other hand as the true self of a person. In both cases, the good *dáimon* is perceived as a being connecting the human with the divine world. *Eudaimonía* consists, namely, in the flourishing of the aforementioned potentialities, which makes a human being similar to divine beings. However, human flourishing attains its most accomplished modality whenever a human being is capable of consciously manifesting in every action, through authentic intelligence and rationality, the nature of the true human self, which nature is indeed divine, and thus makes the human being capable of a posthuman transformation—of *daimonification* (David Litwa, 2020) or *divinization* (Miller, 2011).

2 From Aristotle Back to Plato

When searching for a continuity with ancient thought, present research on eudaimonic well-being presently focuses on Aristotle (Ryff & Singer, 2008; Badhwar, 2016; Fowers, 2016; Moore, 2019), according to whom the best kind of life, and hence the highest form of *eudaimonía*, consists in the conscious activity of intellect/mind (*nous*), that is, of the divine component, of the true self of human being (Aristotle, *Nicomachean Ethics* X. 7-8, 1177a12-1179a32 and Long, 2011). The here considered activity can manifest itself as both theoretical and practical wisdom, whereas, due to the limits of human condition, theoretical wisdom (philosophy) as the ultimate aim of life cannot involve the elimination of care and exercise regarding practical wisdom (Kraut, 2022). Nonetheless, Aristotle was not the first

[1] *Eudaimonie* (2023) provides a thorough, quickly accessible introduction to the notion of *eudaimonía* as well as to its history, also containing a generously informative list of references regarding both primary sources and related bibliography in German and English.

[2] Andres (1918) contains the most comprehensive collection of literary and philosophical sources concerning the Greek notion of *dáimon*.

philosopher to point to the intrinsic relationship between *eudaimonía* and intellect/mind as the true self of human being. In fact, Plato provides the first explicit identification of intellect/mind (*nous*), meant as the true self of human being, with the *dáimon* who is presupposed in the notion of *eudaimonía* (Plato, *Timaeus* 90c4-6: cf. Lavecchia, 2006, 231–236; David Litwa, 2020, 45–56). Furthermore, Plato's work, much more than Aristotle's, contains important suggestions concerning the characteristics of the self connected with *eudaimonía*, which may be significantly helpful for delineating a concept of the *eudaimonic self* being consonant with sustainability. However, before concentrating on those suggestions, a general consideration regarding our current image of the self is important, in order to understand whether this image is consonant with an unprejudiced observation regarding the self, as well as, in the case of a dissonance, to reflect on the possibility of substituting our current image of the self with a more congruous one. As we will see, the proposed consideration will indicate the most productive way of integrating the suggestions coming from Plato into a reflection concerning eudaimonic well-being in the present time.

3 Solipsistic or Dialogical Self?

Our current image of the self is consonant with the image of the I which Ludwig Wittgenstein in his *Tractatus Logico-Philosophicus* connects with solipsism: "The I in solipsism shrinks to an extensionless point and there remains the reality co-ordinated with it" (Wittgenstein, 1922, 5.64); this notion is in turn coincident with the representation of a subject constituting "the limit – not a part of the world" (Wittgenstein, 1922, 5.641).

In our encounter with the world, we habitually tend, in fact, to represent the perceiving, thinking, and understanding self as a point localizable within our body, whose interiority we define as separated from the exteriority of the world we are encountering. As a consequence, we are oriented to associate the modality of perception actualized by our self with the perspective of tunnel vision (Metzinger, 2009), which hopelessly splits the first person perspective from the second and third person perspectives, separating a postulated mere subjectivity of the perceiving consciousness from a postulated objective being. A contrastive dualism between the self and the world is thus generated, which presently determines and configurates every field of knowledge and action.

Compared to the just described notion of self, an unprejudiced observation concerning our phenomenal experience of perception discloses a radically different possibility of characterizing our perceiving self. Actually, the image of a localizable point does contradict even the most elementary consideration regarding perception: in every perception we are, in reality, unable to localize the perceiving self at any place in space or moment in time, neither within a supposed merely subjective interiority nor in a supposed objective exteriority (Scheurle, 2013). Moreover, both modalities of localization would involve absurd consequences: in the case of identifying the self with a merely subjective interiority—with a point shrunk in itself—perception would be a bare projection of our subjectivity, without any authentic encounter with the perceived being; in the case of reducing the perceiving self to external determining factors, the perceiving self would be an illusionary no one (Metzinger, 2003), and we would have no reason to be certain of any evidence manifested by this no one (Lavecchia, 2014d). With respect to an unprejudiced observation of perception, the most consistent image of the perceiving self is hence the image of a non localizable, nonphysical point—or, more precisely, *center of attention* (Ganeri, 2017)—which, transcending any first-, second-, and third-person perspective, both singular and plural, instantaneously generates an unlimited *sphericity* illuminating the perceived being, offering it the possibility of manifesting itself, whereas the point and the sphere constitute an instant unity, not a temporal succession (Lavecchia, 2014a, c, d). The aforementioned image preserves in fact both the indispensable openness of the self toward the perceived being and the—equally indispensable—remaining centered in itself which characterizes the perceiving self. In the absence of the first, the self would perceive and act as a solipsistic subject; without the second the self would dwindle down to the status of an inessential no one.

Which of the two considered representations/images of the perceiving self—the localizable point connected with tunnel vision and the non-localizable center/sphere—is more consonant with an unprejudiced observation of the self who encounters the world?

In effect, the image of the self connected with tunnel vision means a hopeless separation of the self from the world, that is, from any possibility of attaining knowledge which could be really significant with regard to both the life of the self in the world and to the world. In consonance with this image of the self, eudaimonic well-being—which obviously has to be understood as the well-being of the self—would never be capable of transcending the limits of an egoistic, narcissistic subjectivity: it would be characterized by an eminently dualistic gesture, which would negatively contract the consciousness of the perceiving, thinking, and acting self into an exclusively subjectivistic perspective, rendering it incapable of an unprejudiced encounter with phenomena in whatever dimension of life. On the contrary, the image of the perceiving self as non localizable center/sphere instantaneously manifests an intrinsic, unrestricted *dialogicity* in the encounter with the world, that is, in the encounter between interiority and exteriority as well as subjectivity and objectivity. Accordingly, the eudaimonic well-being of the self is in this case, like the self, intrinsically *transperspectival and dialogical* in its relation with the world, and therefore transcends any first-, second-, and third-person perspective, both singular and plural.

In consonance with the just formulated premises, if not conditioned through conceptual prejudices or preconceived representations—like the habitually internalized representation of the localizable point shrunk in itself—, the self, encountering the world, reveals itself as constitutionally capable of preserving in its consciousness both its own autonomous self-awareness/self-transparence and the possibility for the perceived world of autonomously manifesting its own transparence. This constitutionally, eminently *dialogical* self, not the solipsistic self, represented through a localizable point, can be assumed as a consonant fundament for any concept of eudaimonic well-being oriented toward a productive harmony between the self and the world. The delineation of a philosophical anthropology relying on this fundament—which, of course, cannot even elementarily be provided in this context—should be perceived, differently from the present situation, as an urgent prerequisite of any truly consistent research on eudaimonic well-being. This would, namely, give a decisive contribution to more and more deeply integrate an intrinsically dialogical gesture into the self-awareness of the person encountering the own interiority as well as acting in the world, thus making avoidable any merely egocentric interpretation of eudaimonic well-being.

4 The Platonic Notion of Good: Toward an Agathological Self

The above characterized notion of the *dialogical self* can help to eliminate any dichotomy not only between interiority and exteriority, subjectivity and objectivity, self and world, individuality and relationality/universality, but also between perception and knowledge, on the one hand, and ethics and virtue on the other hand. The here considered dialogical self can in fact be associated with a conception of the good capable of transcending the aforementioned dichotomy. The intended conception can be developed based on two key passages of Plato's work, *Timaeus* 29e-30a and *Republic* 506d6-509c (Lavecchia, 2012, 2014a). The great interest these two passages can awake is connected with the fact that they presuppose a not prioritarily contrastive or teleological notion of the good: here the good is understood neither as arising from the contrast with evil nor as a goal (*télos*) perceived as merely objective and transcendent in relation to the acting self—that is, to individual, personal consciousness; on the contrary, the good is characterized here through the gesture of an unconditional generosity, which disposes for the most unrestricted self-givingness, that is, for absolute openness toward the autonomous manifestation of another being.

Timaeus 29e-30a—the only passage in which Plato explicitly characterizes the notion of good—declares the reason for which the divine Artisan/Demiurge gives life and form to the cosmos, identifying this reason with his being good, whereas good means in this passage being unrestrictedly free from envy and therefore inclined to the highest form of generosity. The unconditionality, the absolute freedom characterizing the good, is here evidenced both by the fact that the consciousness and activity of the Demiurge is not determined—neither by any need or necessity, nor by an impulse toward self-affirmation, nor by any norm, imperative, commandment, nor by preceding experiences or by calculations regarding the future—and by the fact that, according to *Timaeus* 34b6-8, the cosmos generated by the Demiurge is characterized by complete autonomy; the cosmos is a self-sufficient being, whose life is independent from any external factor with regard to preserving itself, and is characterized by self-consciousness. In other words, the divine Demiurge is eminently good because he is capable, through unconditioned freedom, of opening a radically unrestricted space of manifestation for another free and autonomous being, that is, for a new world. *Republic* 506d6-509c paradigmatically illustrates the corresponding modality of consciousness through the analogy between the origin of every being and the sun. The origin of every being, transcending every form of being (*Republic* 509b8-10), is here identified with the supreme good because, in the same way that the sun is unconceivable without its manifestation through light, the supreme good is, beyond any dichotomy between transcendence and immanence, unconceivable as separated from its unconditional, unrestricted manifestation through being: from its absolutely free, generative openness to the transparent manifestation of *another* autonomous and free being. Therefore, in Plato's analogy, the light of the sun illustrates the unconfined transparence, unhiddenness of the supreme good, deriving from its unconditional self-givingness and producing the manifestation not only of merely objective being but also, with equal originality, of *intellect/mind* (*nous*, *Republic* 508b12-c2), that is, of a self-conscious self being capable of autonomously and generatively perceiving, knowing, and acting.

In the two just considered passages, the notion of good characterizes a particular modality of consciousness. This modality of consciousness can be described as paradigmatically, unrestrictedly *transperspectival and dialogical*, both transcending any determination of being and connoted by an unconditional, generative openness toward the manifestation of and the dialogical encounter with another autonomous consciousness or, in general, another being. The here characterized *agathological*[3] openness reveals itself as the fundament of both creative intelligence and true virtue, whereas virtue is here understood as the capability of generating relations which are harmonious and consistent for both the perceiving/knowing self and the beings and the world the self encounters or is going to encounter (Lavecchia, 2014a). Consequently, this agathological openness is on the one hand capable of generating eudaimonic well-being for both the acting self and the world, and is on the other hand completely consonant with the representation of the dialogical self described above (§ 3): the perceiving, knowing, and acting self would not be capable of veritably perceiving and knowing *another* being, nor would it be able to generate a harmonious, eudaimonic relationship with that being, if this dialogical, agathological openness were not intrinsically present in any act of unbiased perception and knowledge, that is, in any unprejudiced encounter with the world. In consonance with these premises, the here presupposed concept of good could become a key notion in any reflection concerning the perceiving and knowing self as transcending any tunnel vision, regarding both epistemic and ethical operativity.

[3] *Agathological* is here not exclusively related to the field of ethics or practical philosophy, but rather to a philosophical approach which, transcending any dichotomy between practical and theoretical lives, perceives a supreme good (*agathón*), qua modality of consciousness characterized by unconditionally dialogical generosity, as both ultimate origin and reason of a free and autonomous thinking (*lógos*). See Lavecchia (2014a).

Furthermore, this transperspectival, dialogical notion of good may be considered as a fundament of any authentic generation and manifestation of eudaimonic well-being, since it manifests most efficiently the dialogical gesture of the perceiving, knowing, and acting self.

5 Confidence, Wonder, and Eudaimonic Well-being: Encounter with the World and Experience of Birth

The above delineated *agathological* and thus dialogical self can be further illustrated through another passage of Plato's dialogues: *Theaetetus* 149a-151d3, the only source in which Socrates represents himself through the well-known analogy with midwives (Lavecchia, 2013, 2022).

Socrates helps intellectually and spiritually pregnant persons to give birth to true knowledge; however, interestingly enough, giving birth to one's own knowledge or wisdom is forbidden to him while acting as a midwife (*Theaetetus* 150c4-d2), although he has to be capable of distinguishing true from false knowledge, in order to really help to achieve a harmonious birth (150b2-3, c1-3, 151d2-3). In any case, the aforesaid prohibition does not imply that Socrates never gave birth to true knowledge; otherwise, the analogy with the midwives would be inconsistent, since according to Socrates midwives are required to have experience in giving birth (149b5-c2). In consonance with this analogy, Socrates *has* therefore given birth to true knowledge. The obvious reason why he is not allowed to do it while acting as a midwife is the fact that *other* persons—not Socrates—have to give birth to true knowledge autonomously instead of passively learning something from Socrates (150c6, d1-2, d6-7-e1). A deeper reason is the fact that helping another person to autonomously generate true knowledge presupposes a transperspectival, eminently dialogical modality of consciousness in the helping person. This modality is, namely, the only one able to transcend in every perception and situation the conditioning originated by inappropriately standardizing factors emerging from previous experiences: the perceiving self makes itself constantly and generatively *void* with respect to any already attained knowledge and thus opens an unlimited, not conditioning space of transparence for the inimitable individuality of the perceived being as well as for the way through which that individuality can manifest itself most transparently. Exactly to this modality of consciousness points the analogy with midwifery related to Socrates. In fact, it presupposes the same unrestricted, generatively generous *confidence* required from a midwife in the event of a birth, where, notwithstanding her expertise, if she wishes to avoid mistakes, that is, to manifest well-being, a midwife is never allowed to merely *deduce* from any past experience the singularity and uniqueness of the newborn, of the mother, and of the situation concomitant with the birth: an efficient midwife makes herself constantly *void* with regard to her previous experience, exclusively concentrating on opening an unlimited space for the manifestation of the individuality presently going to be born.

Based on those premises, Plato's representation of Socrates' midwifery can be considered perfectly consonant with the transperspectival notion of good deduced above from the analogy between the sun and the good. The unconditional confidence, to which the dialogical gesture of Socrates' midwifery points, appears, namely, as the ultimate ground of genuinely *good* choices, whereas *good* transcends here any dichotomy between epistemic and ethical domains, simply referring to a perceiving, knowing, and acting self able to generate an unconditioned space for the unrestricted harmonious manifestation of any individual being. In other words, the *agathological* modality of consciousness presupposed by Socrates' midwifery can be perceived as a paradigm for any person wishing to attain harmonious perception and generate harmonious relationships in any encounter with the world, thus producing well-being both for the self and for the encountered beings: harmonious perception and hence true knowledge and authentically good action cannot be generated through exclusive instantiation of exterior or interior norms, standards, and methods or through formalized procedures and disciplines, since, no matter how refined and how supposedly anticipatory they may be, they are always a result deduced from past experiences and therefore valid only in limited perspectives.

Socrates' dialogical, agathological voidness with regard to any preceding experience implicitly reveals that a monarchy of formalized procedures biases the individual encounter with the world. Socrates makes manifest that, on the contrary, both epistemic and ethical virtue involves the willingness of being unrestrictedly confident in the generative power of the *present* self and of the *present* individual situation the self is encountering: confident in the fact that authentic manifestation of the good, that is, genuine virtue and, consequently, well-being, no matter in which dimension of life, presupposes the willingness to generate an unlimited space for the manifestation of a *new birth* (Plato, *Symposium* 212a2-5). In other words, Socrates points to a modality of consciousness in which the above considered (§ 5), constitutionally dialogical gesture of our perceiving and knowing self becomes more and more a constant presence in every situation of life. This modality of consciousness can be additionally specified through a further passage of Plato's *Theaetetus*, in which Socrates considers the act of wondering as the origin of philosophy (155d1-5), whereas philosophy is here meant as striving for true knowledge and wisdom (*sophía*). Aristotle gives the best commentary to this passage in *Metaphysics* 982a30-983a20: wonder is the origin of philosophy—of striving for *sophía*, with the same meaning as in Plato's *Theaetetus*—because it is generated through the encounter with phenomena which are incommensurable with preceding experiences—that is, phenomena which are not deducible from any already acquired experience, discipline, method, procedure, and standard—, and thus stimulates to transcend the state of helplessness with which those phenomena are associated, encouraging the generation of authentically *new* perspectives of knowledge and life (Lavecchia, 2014b).

Aristotle's consideration regarding wonder can be perceived as an additional, implicit declaration of the fact that only an eminently dialogical modality of consciousness is able to attain an experience of authentic *newness*. Wonder can therefore be considered as another component indicating the unrestricted openness which characterizes the unprejudicial perceiving, thinking, acting self, not imprisoning it in limited perspectives, within the confines of tunnel vision, and thus revealing it as an eminently dialogical and agathological reality, through whose activity eudaimonic well-being can be inexhaustively generated, both for the acting individual and for the world.

6 Consciousness Determines Being! Freedom as Solution for Present Crises

In the present time, many approaches which aim at developing solutions for crises are focused on the presupposition that only through implementing worldwide, objectively evaluable control measures can human beings be oriented toward the good and, consequently, toward well-being. Such approaches presuppose, in other words, that transformations regarding consciousness and, consequently, action or entrepreneurship in the world can only emerge from objectively determining and determinable factors, since it is assumed that consciousness and the self are to be intended as merely emergent and therefore eventually illusory phenomena. As a consequence, if present crises are generated by, for example, lacking a sense of community, the aforementioned approaches presuppose that solely by implementing worldwide standardizing measures and programs, which aim at generating and fostering the aforesaid quality, crises might be resolved and more and more well-being generated.

The aforesaid tendency, which reveals an exclusive credence in centralized planning strategies, as well as an equally radical distrust in the positive power and creativity of concrete individual selves and of communities, leads to the emergence of some paradoxes. A most conspicuous conceptual paradox emerges, for instance, from the presupposition that creating a permanently traceable and trackable digital self for every human being in the world would foster a sense of community. In fact, this assumption pretends to stimulate communitarian values based exactly on the representation of the self as a localizable point shrunk in itself, which, as we saw, Wittgenstein—rightly—characterized as consonant with solipsism, that is, with the most unproductive form of subjectivism! Therefore, is the assumption really unquestionable that identifying human beings with digital selves will generate a sense of community and not the contrary?

Interestingly, predominant approaches to present crises do not take seriously the productive implications a consistent philosophical anthropology could generate in coping with crises. According to the preceding considerations, consistency would be manifest in an anthropology able to develop an image of the—both inwardly and outwardly—perceiving self which would be consonant with the intrinsic transperspectival and dialogical nature of the self. This anthropology—which would necessarily understand the human being as triarticulated in body, soul, and spirit (Weger

& Wagemann, 2021; Weger et al., 2021)[4]—would not lead to a new form of negative, unsustainable anthropocentrism; on the contrary, it would eliminate from the self-perception of human beings those representations and images which have generated the existing negative anthropocentrisms, and would manifest the human being as constitutionally characterized by a positive orientation toward dialogical self-transcendence, that is, by unprejudicial, generative openness toward harmonious, and thus sustainable relations with the world. In sum, the here intended anthropology would reveal that the fundament for a sense of community—community not only with other human beings, but with every being in the world—is intrinsically present in *every* human being, since every human being *is* in principle a transperspectival, agathological, and thus eminently dialogical self, whose striving for authentic well-being will be never separated from a harmonious relationship with the world. According to these premises, every human being might be perceived as a potential sustainable entrepreneur if we are willing to separate the notion of enterprising from any automatic association with an activity aimed exclusively at individual advantages and profits. In other words, a consistent, unprejudicial philosophical anthropology would make manifest that the human modality of consciousness is intrinsically oriented toward dialogical community and hence toward sustainability. This would constitute a not negligible premise for genuine creativity in coping with present crises, which are in fact mainly associated with a missing sense of community and sustainability. It would, namely, encourage confidence in the fact that *every* human being is intrinsically able to autonomously generate a sense for community and sustainability, beyond any planning and standardizing intervention. This would furthermore implicate that crises would be encountered much more efficiently if individual awareness of the aforesaid premise as well as confidence in free individual initiative were fostered, instead of implementing more and more standardizing and collectivizing dynamics, more and more presupposing that individual well-being is tendentially, if not intrinsically, conflicting with the well-being of communities and of the world.

Of course, predominant approaches to crises are not based on the here presupposed confidence, since they have no confidence in the reality of the self. However, if taken seriously, this lack of confidence leads to manifest absurdities: why should we, namely, trust in no matter which conceptual elaboration, epistemic or ethical decision, and economic enterprise of a non-real self, that is, of a no one/nobody? Why should we, in general, trust in someone or something, since only a real and free self can decide to trust or not to trust? The absurdities implicated by the elimination

[4] With *spirit* I do not mean here a merely transcendent self but a dimension of consciousness concretely present in every human being, which makes the human being potentially capable, in *every* situation of life, of perceiving, thinking, knowing, and acting harmoniously with both their own self and the world, beyond any influence coming from sympathy and antipathy, pleasure and pain, joy and sorrow, as well as beyond any conditioning through past experiences or predetermined patterns. Through this dimension of consciousness, every human being is *free*, that is, potentially able to generate in every situation, beyond any computability, a truly *new and individual* approach and procedure for a harmonious encounter with the world, as well as to autonomously decide regarding agreement or disagreement with any result of experience.

of trust in the reality of the perceiving, thinking, understanding, deciding, acting, and enterprising self lead in truth to the irreducibility of freedom (see Esfeld, 2020): they lead to the irreducibility of phenomenal consciousness to any form of being and consequently to any form of standardization, algorithmization, and computation (Chalmers, 1996; Bayne, 2010; Dainton, 2011; Bitbol, 2014; Sparby et al., 2019). Is the radicality of present crises maybe associated with mistrust in the fact that in a world governed by peace and harmony, that is, in a world characterized by well-being, consciousness determines being and not the contrary? Would genuine well-being be possible if we do not perceive the intranscendible irreducibility of consciousness and freedom? Can a truthful concept of sustainable entrepreneurship be separated from the aforesaid irreducibility?

7 Freedom, Eudaimonic Well-Being, and Sustainable Entrepreneurship

If entrepreneurship is not to be identified with any more or less complex automatism, it cannot be realizable without an intrinsically free, confident, and wondering self, who, in the encounter with the world, is constantly open for the birth of authentic newness. This openness is impossible without freedom, since it involves no predetermination and prejudice before phenomena and events transcending preceding experiences, procedures, theories, and methods. However, the here intended freedom does not mean narcissistic self-affirmation. Freedom means here, in truth, the real presence of a person who is unrestrictedly confident in the possibility of autonomously manifesting a *good* modality of consciousness, that is, an unrestricted generosity and self-transcendence. In fact, the self of the aforesaid person perceives itself as free not only from every other being but also from any own *selfness* (cf. Ganeri, 2017), and thus capable of transforming every perception in the experience of a wondering encounter with a new birth. Therefore, the self of this person is not a solipsistic but an *agathological* self: a good *dáimon*, a spiritual being leading to decisions and choices which are good for both the acting individual and the world, or, in other words, an *eudaimonic self.*

In this perspective, *eudaimonía* and well-being cannot be perceived as something confining the individual in the limits of an egoistic subjectivity—the not very appropriate translation of both *eudaimonía* and well-being with "happiness" (Badhwar, 2016; Musikanski et al., 2020) may actually orient in this direction—, since the good *dáimon* is here a reality which intrinsically effects the harmonious integration of the individual in the world. Accordingly, *eudaimonía* and eudaimonic well-being appear as notions which are per se productively consonant with sustainability. They presuppose in fact not a self encapsulated in its own subjectivity, and consequently striving toward a narcissistic self-affirmation, but an eminently *dialogical and agathological* self, who, beyond any spatial or temporal localizability, is constitutionally capable of an unrestricted openness toward "meeting the needs of the present without compromising the ability of future generations to meet their own needs" (United Nations General Assembly, 1987, Chap. 2, Paragraph 1).

References

Andres, Fr. (1918). Daimon. *Paulys Realencyclopädie der classischen Altertumswissenschaft* (Band S III, pp. 267–322). Retrieved August 20, 2023 from https://de.wikisource.org/wiki/RE:Daimon

Badhwar, N. K. (2016). *Well-being. Happiness in a worthwile life*. Oxford University Press.

Bayne, T. (2010). *The Unity of consciousness*. Oxford University Press.

Bitbol, M. (2014). *La conscience a-t-elle une origine? Des neurosciences à la pleine conscience: une nouvelle approche de l'esprit*. Flammarion.

Chalmers, D. J. (1996). *The conscious mind. In search of a fundamental theory*. Oxford University Press.

Dainton, B. (2011). *The phenomenal self*. Oxford University Press.

David Litwa, M. (2020). *Posthuman transformation in ancient Mediterranean thought. Becoming angels and demons*. Cambridge University Press.

Esfeld, M. (2020). *Science and human freedom*. Palgrave Macmillan/Springer.

Eudaimonie. (2023, August 17). Wikipedia. Die freie Enzyklopädie. Retrieved August 19, 2023 from https://de.wikipedia.org/wiki/Eudaimonie

Fowers, B. J. (2016). Aristotle on Eudaimonia: On the virtue of returning to the source. In *Vittersø 2016* (pp. 67–83).

Ganeri, J. (2017). *Attention, not self*. Oxford University Press.

Kraut, R. (2022). *Aristotle's ethics. The Stanford encyclopedia of philosophy*. Retrieved August 21, 2023 from https://plato.stanford.edu/archives/fall2022/entries/aristotle-ethics/

Las Heras, M., Grau-Grau, M., & Rofcanin, Y. (Eds.). (2022). *Human flourishing. A multidisciplinary perspective on neuroscience, health, organizations and arts*. Springer.

Lavecchia, S. (2006). *Una via che conduce al divino. La homoiôsis theô nella filosofia di Platone*. Vita & Pensiero.

Lavecchia, S. (2012). Agathologie. Denken als Wahrnehmung des Guten oder: Auf der Suche nach dem offenbarsten Geheimnis. *Perspektiven der Philosophie, 38*, 9–45.

Lavecchia, S. (2013). Vertrauen in die Ich-Geburt. Führung in Sokrates' Horizont. In G. Werner & P. Dellbrügger (Eds.), *Wozu Führung? Dimensionen einer Kunst* (pp. 73–81). Karlsruher Institut für Technologie Scientific Publishing.

Lavecchia, S. (2014a). Agathological realism: Searching for the good beyond subjectivity and objectivity or on the importance of being platonic. *Etica & Politica, 16*, 533–549. https://www.openstarts.units.it/bitstream/10077/10742/1/LAVECCHIA.pdf. Also in G. De Anna, R. Martinelli (Eds.), *Moral realism and political decisions* (pp. 29–50). Bamberg: University of Bamberg Press 2015

Lavecchia, S. (2014b). Chi invidia non conosce meraviglia. Note su *thaumazein* ed essere *aphthonos* in Platone e Aristotele. *Thaumàzein, 2*, 169–194. https://rivista.thaumazein.it/index.php/thaum/article/view/23/24

Lavecchia, S. (2014c). Das Ich und das Gute. Ansätze einer Licht-Philosophie in Anknüpfung an Novalis und Platon. *Perspektiven der Philosophie, 40*, 9–46.

Lavecchia, S. (2014d). Trasparenza o illusione dell'autocoscienza? Metafisica della luce e neurofilosofia. In F. Fabbro (Ed.), *Neuroscienze e spiritualità. Mente e coscienza nelle tradizioni religiose* (pp. 74–91). Astrolabio-Ubaldini Editore.

Lavecchia, S. (2022). Erkenntnis und Wissenschaft als Geburt der Freiheit: Betrachtungen zu Sokrates' Hebammenkunst. *Perspektiven der Philosophie, 48*, 129–140.

Long, A. A. (2011). Aristotle on *eudaimonia, nous*, and divinity. In J. Miller (Ed.), *Aristotle's Nichomachean ethics. A critical guide* (pp. 92–114). Cambridge University Press.

Metzinger, T. (2003). *Being no one. The self-model theory of subjectivity*. MIT Press.

Metzinger, T. (2009). *The ego tunnel: The science of mind and the myth of the self*. Basic Books.

Miller, P. L. (2011). *Becoming god. Pure reason in early Greek philosophy*. Continuum.

Moore, C. (2019, April 8). What is Eudaimonia? Aristotle and Eudaimonic wellbeing. *PositivePsychology.com*. Retrieved August 19, 2023 from https://positivepsychology.com/eudaimonia/

Musikanski, L., Philips, R., Bradbury, J., de Graaf, J., & Bliss, C. (2020). *Happiness, well-being and sustainability. A course in system change*. Routledge.

Norton, D. L. (1977). *Personal destinies. A philosophy of ethical individualism*. Princeton University Press.

Ryff, C. D., & Singer, B. H. (2008). Know thyself and become what you are: A Eudaimonic approach to psychological well-being. *Journal of Happiness Studies, 9*(1), 13–39.

Scheurle, H. J. (2013). Das Gehirn ist nicht einsam. In *Resonanzen zwischen Gehirn, Leib und Umwelt*. Kohlhammer.

Schlegel, R. J., Hicks, J. A., Arndt, J., & King, L. A. (2009). Thine own self: True self-concept accessibility and meaning in life. *Journal of Personality and Social Psychology, 96*(2), 473–490. https://doi.org/10.1037/a0014060

Sparby, T., Edelhäuser, Fr., & Weger, U.W. (2019, October 22). *Frontiers in Psychology, 10*. https://doi.org/10.3389/fpsyg.2019.02250. Retrieved August 21, 2023 from https://www.frontiersin.org/articles/10.3389/fpsyg.2019.02250/full

United Nations General Assembly. (1987, March 20). *Report of the world commission on environment and development: Our common future*. Retrieved September 1, 2023 from http://www.un-documents.net/our-common-future.pdf

Vittersø, J. (Ed.). (2016). *Handbook of eudaimonic well-being*. Springer.

Weger, U., & Wagemann, J. (2021). Soul, Spirit, and consciousness in psychology and philosophy: Traditions, current views, perspectives. (editorial). *European Psychologist, 26*(2), 71–72.

Weger, U., Sparby, T., & Edelhäuser, F. (2021). Dualistic and trichotomic approaches in psychological enquiry. *European Psychologist, 26*(2), 85–95.

Wittgenstein, L. (1922). *Tractatus logico-philosophicus* (Fr. P. Ramsey, Trans.). Kegan Paul, Trench, & Co.

Salvatore Lavecchia is a professor for ancient philosophy at the University of Udine as well as teacher in the master's program "Meditazione e Neuroscienze" (Udine) and "Contemplative Studies" (Padova). Main research interests are as follows: the Socratic dialogue, also in its relationship to any field of counseling; ancient aesthetics and notions of art; the notion of self and mind in Plato and Platonism; the metaphysics of light in Plato and Platonism; and the development of a philosophy of the self being capable of generating a productive dialogue with both neurosciences and with Eastern as well as Western spirituality.

How Purpose Drives Sustainable Business Transformation and Well-Being: Practical Wisdom of Taoism

Liangrong Zu

Abstract In the ever-changing world of modern business, the ancient philosophy of Taoism provides a valuable guide to achieving sustainable business transformation. This chapter explores how business can move beyond the confines of traditional, profit-focused models by adopting a purpose-centric approach rooted in Taoist principles. The shift toward this purpose-driven strategy incorporates Taoist teachings such as balance, harmony, and sustainability. By following the Middle Path of Tao, organizations can successfully merge ethical decision-making, employee welfare, and long-term prosperity that respect both the individual and the group, as well as the corporation and the planet. The Taoist-influenced, purpose-driven model thus emerges not just as an ethical duty but also as a practical wisdom that fosters lasting customer loyalty and sustainable profitability. The chapter also discusses the paradigm shift in business from profit-focused to purpose-oriented, sustainable strategies. It underlines the increasing need for businesses to focus beyond profits and explores how integrating Taoist wisdom can promote sustainable growth and transformation in contemporary businesses. The chapter also discusses how purpose and the wisdom of Taoism can contribute to the improved well-being of business leaders.

Keywords Taoist wisdom · Sustainable business · Transformation · Purpose-driven · Leadership · Effectiveness and efficiency

---

L. Zu (✉)
Vinovo, Italy
e-mail: lrzu@ungyla.org

M. Mühlböck (ed.), *Sustainable Transformation and Well-being*, CSR, Sustainability, Ethics & Governance,
https://doi.org/10.1007/978-3-031-75566-8_3

1 Shifting Paradigm: From Profit-Driven to Purpose-Driven Sustainable Business

In the constantly evolving world of business, the traditional approach that focuses on a one-size-fits-all, profit-centric model is progressively becoming outdated and less effective. As we continue to progress deeper into the twenty-first century, there is an increasing demand for businesses to expand their focus beyond just generating profits. The call for businesses to incorporate a wider range of considerations and to serve a broader purpose is growing louder and more insistent. Companies are being strongly encouraged, and in some cases even required, to transition from an exclusive focus on profit generation to a more comprehensive model that includes a clear purpose and a commitment to sustainability. This shift in expectations and demands is not just a passing trend or a temporary response to the latest business fad. Instead, it represents a necessary and significant evolution in the way that businesses operate and define their success.

This evolution is driven by the demands and needs of the modern world, which increasingly values sustainability, social responsibility, and a clear sense of purpose alongside financial success. It reflects a broader change in societal values and priorities, and businesses that want to remain competitive and successful will need to adapt and evolve in response.

In the present global climate, an increasing cadre of enterprises are awakening to the realization that they possess a noteworthy role in surmounting the world's grand challenges. This insight is not merely a passing observation, but, rather, it is becoming a cornerstone principle—shaping their strategic direction and defining their very existence. Their survival and continued relevance are no longer solely pegged to their capacity to churn out profits but are increasingly hinged on their ability to make material positive contributions to society and the environment.

The birth of this novel perspective has given rise to the purpose-driven, sustainable business model, a paradigm shift in the world of commerce. At the heart of a purpose-driven business lies a lucid, overarching purpose and meaning—a beacon that guides its operations and decisions—transcending the traditional confines of mere profit-making.

This innovative model lays equal emphasis on three fundamental aspects—the people involved, the planet we inhabit, and the profits generated, a concept encapsulated by the triple bottom line (Elkington, 1998). Businesses subscribing to this ethos are committed to generating beneficial ripple effects on a broad spectrum of stakeholders. This includes their employees, customers, the communities they operate in, and the environment at large.

Simultaneously, these purpose-driven entities are also acutely aware of the need to maintain financial viability. Thus, they strike for a delicate balance—creating a positive societal and environmental impact while ensuring the sustainable generation of profits. This emerging purpose-driven model is rapidly becoming a central tenet in the world of business, reshaping conventional norms and expectations, and presenting a promising blueprint for the future.

The journey from being a profit-driven business to becoming a purpose-driven, sustainable business is not a straightforward one. It requires a fundamental shift in mindset and operations. However, embracing this transformation can offer numerous benefits, such as improved brand image, increased customer loyalty, better employee engagement, and potential for long-term profitability. Businesses like Patagonia and Unilever are excellent examples of successful purpose-driven companies. They have demonstrated that businesses can indeed thrive while making a positive impact on the world. The shift from profit-driven to purpose-driven sustainable business is more than a mere change—it's an evolution. In the face of our world's mounting challenges, businesses must rise to the occasion and take the lead in driving positive change. By doing so, they can ensure their own survival while contributing to a more sustainable and equitable world.

2 The Profound Connection Between Purpose, Well-Being, and Taoism

In the modern world, where stress, anxiety, and a lack of fulfillment are prevalent, many individuals seek a deeper understanding of their purpose and a path to well-being. One philosophy that offers profound insights into these aspects of human existence is Taoism. Originating in ancient China, Taoism provides a holistic approach to life, emphasizing harmony, balance, and the cultivation of inner peace. I will explore the intricate relationship between purpose, well-being, and Taoism, highlighting the wisdom and practices that can guide individuals toward a more fulfilling and balanced life.

2.1 The Essence of Taoism

Before delving into the connection between purpose, well-being, and Taoism, it is essential to grasp the fundamental principles of this ancient philosophy.

Taoism, an ancient Chinese philosophical and religious belief system, is one of the most significant ideologies globally. It perceives the universe as an intricately interconnected entity where no element is independent of another. The universe operates under a set of unchangeable laws, culminating in a constant flux of transformation. At its core, Taoism revolves around the concept of the Tao, which can be translated as "the way" or "the path." The Tao represents the underlying principle of the universe, the source of all existence, and the natural flow of life. It is believed that by aligning oneself with the Tao, individuals can find harmony, peace, and a sense of purpose.

2.2 The Pursuit of Purpose

One of the central themes in Taoism is the pursuit of purpose. Taoist teachings emphasize that every individual has a unique role to play in the grand tapestry of existence. This role is often referred to as one's "true nature" or "authentic self." According to Taoist philosophy, discovering and embracing one's purpose is the key to living a fulfilling and meaningful life.

Unlike many modern approaches to purpose, which often focus on external achievements or societal expectations, Taoism encourages individuals to look inward. It suggests that true purpose arises from connecting with one's inner self, understanding one's desires, passions, and talents, and aligning them with the natural flow of life. By doing so, individuals can tap into a deep sense of fulfillment, contentment, and joy.

2.3 The Path of Well-Being

Taoism recognizes that purpose and well-being are intertwined. The pursuit of purpose is not seen as a separate endeavor but rather as an integral part of cultivating overall well-being. In Taoist philosophy, well-being encompasses physical, mental, emotional, and spiritual aspects of life. It is the harmonious balance of these elements that leads to a state of optimal health and contentment.

To achieve well-being, Taoism offers various practices that promote balance and harmony. These practices include meditation and the cultivation of virtues such as compassion, humility, and gratitude. By engaging in these practices, individuals can enhance their physical vitality, calm their minds, cultivate emotional resilience, and deepen their connection with the Tao.

2.4 The Wisdom of Wu Wei or Nonaction

Wu Wei is a central concept in Taoism that has significant implications for both purpose and well-being. Wu Wei can be translated as "nonaction" or "effortless action." It refers to a way of being in which individuals align themselves with the natural flow of life, rather than resisting or forcing outcomes. It is not about inaction or laziness but rather about acting in harmony with the Tao, without attachment or excessive striving.

In the context of purpose, Wu Wei encourages individuals to let go of rigid plans, expectations, and the need for external validation. Instead, it invites them to trust their intuition and embrace the spontaneous unfolding of life. By doing so, individuals can discover unexpected opportunities, make meaningful connections, and fulfill their purpose in ways they may not have anticipated. In terms of well-being, Wu

Wei offers a powerful antidote to stress, anxiety, and burnout. It teaches individuals to surrender to the present moment, releasing the need to control or manipulate outcomes. By cultivating a state of flow and acceptance, individuals can reduce the mental and emotional burdens that often lead to suffering. This state of Wu Wei allows for greater peace, joy, and overall well-being.

3 Taoist Principles for Purpose-Driven Leadership in Transforming Sustainable Businesses and Well-Being of Leaders

In the previous sections, I delved into the ideas and connections between Taoism, purpose, and well-being, as well as the combination's influential role in boosting sustainable business transformation. This section will provide various principles for leaders keen on nurturing and leveraging their purpose and improving their well-being.

3.1 The Power of Knowing Thyself: Self-Awareness and Effective Leadership

Modern society places a high emphasis on self-awareness: "knowing ourselves" and "understanding our true identity." This self-discovery and acceptance are considered by numerous traditions, including some Western psychology schools, as a crucial element to personal well-being and a significant step toward individual transformation. Certain social theorists and activists also believe that without a profound level of self-awareness and accountability, it would be challenging to address the various social and environmental issues that humanity presently faces. What advice does Taoism provide in this aspect? How can we express our virtue and comprehend our true selves in a way that links us with the rest of our world? Lao Tzu instructs Taoist leaders not to just know themselves but also to transcend themselves, as he states in TTC, Chap. 33:

> Those who know others are smart,
> Those who know themselves are wise.
> Those who can overcome others are strong,
> Those who can overcome themselves are powerful.[1]

Lao Tzu counsels a prudent leader to invest time in self-discovery and identifying their passion. He encourages introspection to unearth one's potential and

[1] The English version of the Tao Te Ching in this paper was quoted from the book *The Wisdom of the Tao: A Novel Translation of Tao Te Ching with Illustrations of Chinese Calligraphy* by Liangrong Zu, Amazon, April 2023.

purpose. If you identify and align with this inner current, you can impact the world significantly while deriving pleasure from the journey. Lao Tzu's insights on self-awareness resonate with the ancient Greek maxim "know thyself," which emphasizes understanding one's internal dynamics, both positive and negative aspects, as well as individual strengths and weaknesses. Wisdom-laden sayings like these, regardless of their temporal or geographical origins, share a common thread: they inspire one to undertake challenging and extraordinary deeds that promise substantial rewards. Lao Tzu's teachings perfectly exemplify this notion.

Why did the ancient sages attach a great importance to "know thyself"? "Know thyself" is always an immensely difficult task. There are the four impediments to knowing thyself, which include everyday distractions that encourage a superficial and nonanalytical existence; psychological obstructions that shade our eyes from the glare of troublesome truths; hedonism, the lack of courage to face and address painful and inconvenient realities; and the distorted imagery of mass media presentation (Soupios & Mourdoukoutas, 2014). Those who are courageous to know themselves must be able to:

- *Explore the dark side of their inner being, the frauds, and deceits that nurture a counterfeit reality*
- *Reveal psychological deficiencies and troublesome truths*
- *Cast a harsh and unforgiving light at the self—exonerating falsehoods*
- *Commit to an agenda of spirited self-indictment—an honest self-discovery process designed to dispel self-induced frauds*

Possessing the right practical skills is one thing, but genuine leadership sophistication resides in the more demanding realm of cultivating self-awareness and leading yourself first. This involves steering toward a fixed point of personal and organizational purpose, journeying with those who choose to follow you, and charting a course beyond everyone else's visible horizon. To be an effective leader, you must constantly evaluate yourself, neither overvaluing your strengths nor undervaluing your weaknesses. It requires humble self-reflection. As self-awareness and honesty are inseparable, let others help you identify your blind spots as you strive to improve both yourself and the organization (Burnison, 2016).

As per Lao Tzu, an exceptional leader not only comprehends himself and his internal world but is also aware of others and understands the external world. This is because self-awareness and awareness of others are the keys to happiness, personal satisfaction, and successful relationships. If we aspire to be successful and content, we must take accountability for our emotions, thoughts, and actions rather than blaming others for our decisions. Interestingly, as we become more self-aware, we become less self-conscious, worrying less about our appearances or others' perceptions of us. This enables us to be more selfless and better equipped to collaborate with, support, and contribute to others. Self-awareness also enriches our inner life and gives us the ability to identify and concentrate on what truly matters to us—the people we cherish. Only those who understand their inner and outer worlds have the power and strength to surpass themselves, fulfill themselves, and evolve into a mindful leader.

With a firm foundation in mindfulness of self and others, as well as self-leadership, leading others becomes significantly easier. If you combine these skills with strategic and tactical management tools and techniques such as defining vision and strategy, and applying situational leadership, you will have a solid framework for success. The more these skills are used in leading others, the greater your capacity for compassion and wisdom will be. Collectively, these elements form a dynamic, continuous, and ever-evolving development roadmap that defines mindful leadership.

3.2 *Authenticity and Influence: Self-Cultivation*

The Chinese sage, Lao Tzu, consistently underscores the importance of authenticity and integrity in life and leadership. He advises wise leaders to uphold integrity, avoid hypocrisy, embrace authenticity, and reject vanity. Authentic leaders, or purpose-driven leaders, are individuals of high integrity who are dedicated to transforming organizations for a sustainable business. They are genuine in their desire to serve others through their leadership, focusing more on empowering their teams than on personal gain. Authentic leaders devote themselves to personal growth, understanding that leadership requires a lifetime of development. The path of self-cultivation is crucial to developing authenticity and achieving life's ultimate purpose, a central concept in Taoist philosophy. Self-cultivation allows leaders to reflect their core values and contribute to the common good. Lao Tzu emphasizes the importance of self-cultivation and its impact on teams, communities, organizations, and beyond. He says in TTC, Chap. 54:

> He who cultivates himself with virtue becomes authentic.
> The family that is influenced by his virtue becomes harmonious.
> The organization that is influenced by his virtue becomes sustainable.
> The country that is influenced by his virtue becomes prosperous.
> The world that is influenced by his virtue becomes peaceful.

Those leaders who hold onto the great Tao cultivate themselves and are authentic. Those who are authentic have the power to influence others, and those who are influential have the capacity of leading others. Leadership is more about influence and enabling the right things to happen rather than lining up a shot and making it happen. Real power of leaders is derived from influence through authentic leadership rather than the perception of authority. Lao Tzu's wisdom tells us that we are all connected to and interdependent on each other in the universe. These connections and interdependence stretch like an incredibly interwoven and complicated tapestry. Each of us exists within this tapestry. Our thoughts and actions are like a stone dropped in a pond and can create ripples that travel outward. Everything we do and think affects the people in our lives and their reactions in turn affect others. The choices we make have far-reaching consequences. Each of us carries within us the capacity to change the world in small ways for better or worse. We can use the principles of the ripple effect to magnify our actions and their effects.

Lao Tzu's wisdom also implies that if you want to be a positive and influential leader in the world, first you must get your own life in order. Ground yourself in the universal principle so that your behavior is wholesome and effective. If you do that, you will earn trust and respect and be a powerful influence. Your behavior influences others through a ripple effect. A ripple effect works because everyone influences everyone else. Powerful people are powerful influences. If your life works, you influence your organization. If your organization works, your organization influences the community. If your community works, your community influences the nation. If your nation works, your nation influences the world. If your world works, the ripple effect spreads through the cosmos. It is important to remember that your influence begins with you and ripples outward. So be sure that your influence is both potent and wholesome.

3.3 Leading from the Behind: Transcending Ego

In the following section, I will examine the three types of leaders as conceptualized by Lao Tzu, each representing a different mindset. One such type is the leader with a fixed mindset. These leaders are characterized by their egocentricity and perpetual pursuit of self-enhancement. They operate under the assumption that some individuals are inherently superior, and they constantly feel the need to demonstrate and exhibit this superiority. Instead of nurturing the growth of their team, they exploit them to satisfy their ego. This pattern ultimately leads to the detriment of their organizations. The fixed mindset provides insights into the origins of colossal egos, their functioning, and their self-destructive tendencies (Dweck, 2006).

Of course, not all leaders fall into this category. There are many leaders with a growth mindset who are driven by a purpose greater than their personal careers and ambitions. However, aligning with this greater cause and serving others necessitates leaders to overcome their ego. According to Lao Tzu, humility in service to others is a fundamental virtue that leaders should cultivate. In Chaps. 7 and 66 of the Tao Te Ching (TTC), Lao Tzu uses the metaphors of natural phenomena—heaven and earth, river and sea—to illustrate the qualities of leaders who transcend their ego for the sake of serving others:

> Heaven is eternal and Earth everlasting.
> The reason that Heaven and Earth exist eternally is.
> Because they do not live for themselves (unselfish nature),
> They can sustain life in the universe (sustainability).
> Therefore, great leaders (learning from Heaven and Earth)
> Place themselves behind, and yet, find themselves ahead.
> Lay their life down, and yet, find their life goes on.

Give no thought to themselves, and yet, end up fulfilling themselves.[2]

The reason why the rivers and oceans, being the source of water,
Can become the king of the hundreds of smaller streams
Is because they lie low.
They can become the leaders of all streams.
Wise leaders who wish to guide people from above,
Must, in their words, humble themselves before them.
And if they want to lead people in the front,
They must, in their actions, willingly follow them from behind.
When wise leaders are above people, people do not feel oppressed.
And when they stand in front of people, people do not feel obstructed.
Therefore, the world gladly supports and follows such leaders
Without growing weary or resentful.
Because wise leaders do not contend, no one can contend against them.[3]

Lao Tzu characterizes two types of effective leadership driven by purpose or ideals. In TTC Chap. 7, he outlines the concept of altruistic leadership. This type of leader aims to enhance the well-being of their followers. According to Lao Tzu, altruistic leaders are inspired by heaven and earth; they exist not for their own sake but for others. They prioritize the interests of their people over their own, avoiding selfishness and focusing on the rights and interests of others. As a result of their altruism, they earn the respect and loyalty of their followers, thereby achieving their own fulfillment and purpose.

In TTC Chap. 66, Lao Tzu discusses the importance of humility in leadership. He points out that servant-leaders, who have the humility, bravery, and wisdom to acknowledge that they can learn from those with less power, actively seek the ideas and unique contributions of those they serve. In doing so, servant-leaders foster a culture of learning and encourage followers to reach their maximum potential.

Lao Tzu also advocates for a leadership approach where leaders guide from behind, not from the front. However, this doesn't mean he opposes leaders who lead from the front or above. It may seem contradictory, but it's the essence of leadership. When leading from the front, leaders must be humble in their actions and words while also prioritizing the interests of their people. Leaders who embody humility and selflessness do not compete with or take credit from others. Instead, they are prepared to assume risks and responsibility in times of crisis. Consequently, when they lead from the front or above, their followers do not feel oppressed or harmed, leading them to willingly and enthusiastically follow their leaders.

Servant-leaders are those who adhere to the wisdom of leading from behind, as per the thoughts of Lao Tzu. As stated in TTC, Chap. 49, servant-leaders should embody the qualities he describes:

Great leaders have no self-interest,
but take people's interests as their own.
They do good to those who are good,
They also do good to those who are not good,

[2] TTC, Chap. 7.

[3] TTC Chap. 66.

Thus, virtue of goodness radiates, reflects, and spreads.
They trust those who are trustworthy,
They also trust those who are untrustworthy,
Then, virtue of honesty radiates, reflects, and spreads.
When great leaders lead people and transcend egotism,
People can become good and honest.

Lao Tzu emphasizes the necessity of two attributes in servant-leaders: compassion and integrity or trustworthiness. Compassion is equated with power, specifically the power to do good, while trust is viewed as a principle underlying power. Servant-leaders must possess the ability to manifest goodness, foster a trusting organizational culture, and effect change in the world through acts of kindness. Those who embody compassion and trust can set aside self-centeredness and ego, focusing instead on the thoughts and needs of others. They diligently nurture these essential virtues in their lives. Servant-leaders distinguish themselves from average individuals by treating everyone with equal respect, irrespective of their moral standing or reliability. They exhibit magnanimity and a deep-seated sense of honor. When making life decisions, servant-leaders opt for the noblest path, striving for personal excellence and the betterment of others. Their pursuit of greatness is internal, a virtue that compels them to seek honor and greatness in their lives, even in the face of adversity.

3.4 Humility, Service, and the Pursuit of Excellence: Level 5 Leadership

Having purpose with humility is the finest approach for us humans to ardently pursue a noble cause without jeopardizing the potential harm our efforts might cause. This is a critical attribute for leaders driven by purpose. When our passion for a worthy cause transforms into self-righteousness, we lose our ability to learn from our errors or even to acknowledge when we are wrong. We must continually ensure that the rightful satisfaction gained from a purposeful life does not turn into pride, which has been acknowledged, rightfully so, as a deadly sin (some would argue the deadliest). Pride is harmful to us, to those we aim to aid, and ultimately to the purpose we strive to achieve. The only way to dodge this ongoing threat is to maintain a balanced view of ourselves (Damon, 2003).

In Taoism, humility is a fundamental virtue that is nurtured and developed, and it is critical for sustainable leadership. Humility, according to Taoist teachings, is the rejection of asserting authority or striving to be first in any endeavor. It views boldness as a dismissal of wisdom and a hasty attempt to engage in situations prematurely. Humility, along with compassion and moderation, is one of the three treasures or virtues held by those who adhere to the Tao. In Chap. 39 of the Tao Te Ching, Lao Tzu outlines the characteristics a prudent leader should embody:

Humility forms the basis of nobility.
Lowliness provides the foundation for greatness.

Therefore, ancient leaders referred to themselves humbly
As orphans, solitary, and unworthy.
But were they truly lowly? No, they were not.
Thus, the highest honor is no honor at all.
Do not seek to shine like jade but remain steadfast like a stone.

Lao Tzu presents guidance to a prudent leader, suggesting that they should not strive to be glossy like jade but rather be unadorned like a stone. Meaning, one should lead a simple and modest life, similar to a basic, solid stone. Lao Tzu perceived humility as being in sync with nature and the universe's wisdom. Taoist wisdom considers simplicity and humility as the foundations of greatness. Evidently, the characteristics of a stone align with modern "level 5 leadership," a term introduced by Collins (2005). This leadership style is regarded as the pinnacle of managerial development and thought.

Collins argues that the key ingredient that allows a company to become great is having a "level 5 leader": an executive in whom genuine personal humility blends with intense professional will. He describes "level 5 leadership" as the starting point in the process to becoming great. "Level 5 leaders" channel their ego needs away from themselves and into the larger goal of building a great company. It's not that "level 5 leaders" have no ego or self-interest. Indeed, they are incredibly ambitious—but their ambition is first and foremost for the institution, not themselves. "Level 5 leaders" are fanatically driven, infected with an incurable need to produce sustained results. They are resolved to do whatever it takes to make a company great, no matter how big or hard the decisions.

Collins bolsters his points by referencing "level 5 leadership characteristics," using the Taoism yin and yang theory. The yin aspect is portrayed through personal humility, showcasing qualities of modesty and avoiding public praise. It represents quiet determination, relying more on inspired standards than charisma to motivate. It focuses ambition on the company, not oneself, and sets successors up for future success. It takes responsibility for poor outcomes rather than blaming others or circumstances. The yang aspect, on the other hand, is about professional will. It plays a key role in progressing from good to great. It shows unwavering determination to achieve the best long-term results, regardless of the difficulties. It aims to build a lasting great company and won't settle for anything less. It gives credit for the company's success to others, external factors, and good fortune.

Besides the stone metaphor, Lao Tzu also used water as a symbol to represent humility and willpower of great leaders. The characteristics of water are also synonymous with "level 5 leadership." Lao Tzu detailed the qualities of humility and selflessness water embodies in TTC, Chap. 8:

The supreme goodness is like water.
Water benefits all creatures, and yet it does not contend.
It stays in the low place that others dislike,
Thus, water comes nearest to the Tao.
Great leaders, like water,
Stay in the right location.
Keep their mind empty and inclusive.
Serve with compassion.

Communicate with integrity.
Lead with justice.
Manage with competence.
Take the ride on the flood.
Great leaders do not contend, therefore,
They end up being free from apprehension.

Lao Tzu also illustrated the quality of an unwavering resolve water embraces:

Nothing in the world is softer or weaker than water,
Yet nothing is better at overcoming the hard and strong.
This is because nothing can surpass water.
The soft overcomes the hard, and
the weak overcomes the strong.[4]

The water metaphor illustrates that it's not "genius" that fuels success but rather a distinct blend of passion and sustained determination. Individuals who foster and adopt the attributes of stone and water show courage in taking responsibility during an organizational crisis, thereby transforming into excellent leaders. These are the leaders that Lao Tzu urges to embody the traits of water:

Those who bear the country's disgrace can rule the country.
Those who bear the country's misfortune can lead the world.[5]

Leaders who embody the characteristics of water—adaptable yet tenacious, modest yet brave—are known as "level 5 leaders." These individuals rise to the occasion during times of adversity, safeguarding their people and nation. Crises are the ultimate test of leadership, often making or breaking a leader. Many leaders and their organizations have crumbled under such pressure, while others have thrived, demonstrating their true potential. Navigating an organization through tough times is exceptionally challenging. Similar to being in a battle, crises push us to our limits due to the unpredictability of the outcomes. Not only do we need to employ all our wisdom to steer our organization, but we must also summon our inner strength and determination to persevere (George, 2009).

3.5 Doing the Right Thing Right: Sustainable Leadership

From the analysis above, we can deduce that purpose refers to the reason for something's existence or why a certain action is taken or created. To determine whether an entity has a purpose, it is beneficial to be aware of the relationship between the exchange of resources and the expected or perceived value it brings. Effectiveness is achieved when resources are utilized purposefully. For instance, time is a resource that everyone possesses, but it is finite. Hence, it is our most precious resource. This may not seem apparent because most of us spend a significant amount of time on

[4] In TTC, Chap. 78.

[5] ibid.

tasks that don't serve a meaningful purpose. Becoming conscious of our values and what is important in life can guide us toward living with a greater sense of purpose. Understanding the "why" behind a business or organization's existence, their purpose, is crucial for effectiveness. Thousands of years ago, Lao Tzu recognized the connection between effectiveness (purpose) and efficiency and the positive influence of purpose on organizational longevity. From TTC's Chaps. 27 and 54, I've extracted three key factors for a sustainable enterprise: recognizing your purpose (*why*), maximizing your human and material resources (*what*), and leveraging your strengths (*how*):

> He who excels in traveling leaves no traces.
> He who excels in speaking makes no slips.
> He who excels in reckoning uses no devices (and yet no mistakes are made).
> He who excels in locking doors uses no bolts,
> And yet the doors cannot be opened.
> He who excels in knotting uses no cord,
> And yet the knots cannot be unraveled.[6]

> He who excels in building an enterprise never fails.
> He who excels in sustaining the enterprise never loses it.
> He who succeeds the enterprise can carry it on for generations.[7]

The invisible wisdom of Taoism teaches us that whatever actions we take in our career, organization, or leadership can be driven by a purpose. Having a clear purpose allows us to function effectively and efficiently. Lao Tzu illustrates the influence of purpose on performance using the metaphor of individuals in various occupations. For instance, a mindful traveler's purpose is to leave no trace of his journeys; a proficient speaker's purpose is to avoid mistakes in his speech; a competent accountant's purpose is to avoid errors in calculations, even without the use of a calculator; a skilled locksmith's purpose is to secure doors without the use of bolts; and a good knot maker's purpose is to maintain the strength of knots, even without the use of ropes.

In the passages above, leaders with a purpose need to grasp the connection between effectiveness and efficiency. Effectiveness refers to the ability to achieve a goal or produce the desired or expected outcome. On the other hand, efficiency is about performing tasks in the most optimal way with minimal waste of time and effort. From a Taoist point of view, effectiveness is associated with the efficacy of the Tao, while efficiency relates to strategy. The Tao has its unique efficacy, which is indicated in the title of Lao Tzu's later work TTC or "The Classical Book of Tao (Way) and Te (Virtue)." Here, Te (virtue) can also be interpreted as "efficacy," embodying the Tao's power or "force." From a societal viewpoint, Te symbolizes morality or doing what is right, while from an individual perspective, it denotes inner strength, efficacy, and integrity.

[6] TTC, Chap. 27.

[7] TTC, Chap. 54.

How does this relate to leadership? Misguided efficiency—prioritizing efficiency over effectiveness or being efficient at doing the wrong thing—is a prevalent error among leaders. Leaders focused on efficiency often neglect the external environment, and the firm's operation becomes their primary concern. In contrast, leaders who prioritize effectiveness perceive their organization as a part of a larger entity, as a single piece of a puzzle. Understanding the influence of effectiveness (purpose) and efficiency on leadership requires leaders to adopt the second key principle of Taoism—maximizing resource potential. As Lao Tzu advises:

> An excellent leader always
> Makes the most of everyone, no one is left behind.
> Makes the most of everything, nothing is left useless.
> This is called invisible wisdom.[8]

An effective leader who has concern for all beings must devise strategies and tools that are efficient in order to accomplish their objectives. Efficiency is directly related to reducing waste. If something is perfectly efficient, it will produce no waste. This is the ideal that Lao Tzu advocates for a leader—one who concentrates on maximizing human potential and resource efficiency. The creation and preservation of effective and productive human and natural resources are the initial steps toward a responsible and sustainable business. Human resources are indispensable for the success of a business.

4 Conclusion: The Path to Transformation and Well-Being: Purpose and the Taoist Way

As we discussed above, in today's fast-paced and often chaotic world, finding a sense of well-being has become increasingly important. Many people seek fulfillment, happiness, and a sense of purpose in their lives. While there are countless ways to approach this quest, two perspectives that offer valuable insights are purpose and Taoism. Purpose gives our lives direction and meaning, while Taoism provides a philosophy grounded in harmony and balance. By exploring the intersection of purpose and Taoism, we can uncover a path toward transformation and improved well-being.

Taoism, originating from ancient China, is a philosophy that emphasizes living in harmony with the natural order of the universe. The fundamental concept of Taoism is the Tao, often translated as "the way" or "the path." It is the underlying principle that governs all things, representing the natural flow and balance of life. Taoism encourages individuals to align themselves with the Tao, embracing simplicity, spontaneity, and a deep connection with nature. While purpose and Taoism may seem distinct, they share a harmonious connection when it comes to enhancing well-being. Both perspectives focus on finding balance, authenticity, and a deeper

[8] TTC, Chap. 27.

understanding of oneself. By integrating purpose and Taoism, individuals can embark on a transformative journey toward greater harmony and fulfillment.

Taoism teaches us to simplify our lives and let go of unnecessary complexities. By stripping away distractions and external pressures, we can create space to connect with our inner selves. This simplicity allows us to reflect on what truly matters and uncover our core values and passions. It is through this process of simplification that purpose begins to reveal itself. Taoism encourages individuals to cultivate mindfulness and awareness in their daily lives. By practicing present-moment awareness, we can observe our thoughts, emotions, and actions with clarity. This heightened awareness enables us to discern what brings us joy, fulfillment, and a sense of meaning. It allows us to make conscious choices aligned with our purpose, rather than being driven by external influences.

Taoism teaches us to observe and learn from the natural world. By immersing ourselves in nature's rhythms and patterns, we gain a deeper understanding of the interconnectedness of all things. This connection with nature can inspire us to align our purpose with the greater harmony of the universe. Just as nature follows its own course, we, too, can find our unique path and purpose by embracing the flow of life.

References

Burnison, G. K. B. (2016). *The leadership journey: How to master the four critical areas of being a great leader*. Wiley.

Collins, J. (2005). Level 5 leadership: The triumph of humility and fierce resolve. Retrieved from https://hbr.org/2005/07/level-5-leadership-the-triumph-of-humility-and-fierce-resolve. Retrieved 20 April 2018, from Harvard Business Review https://hbr.org/2005/07/level-5-leadership-the-triumph-of-humility-and-fierce-resolve

Damon, W. (2003). *Noble purpose: The joy of living a meaningful life*. Templeton Foundation Press.

Dweck, C. S. (2006). *Mindset: The new psychology of success*. Random House.

Elkington, J. (1998). *Cannibals with forks: The triple bottom line of 21st century business* (2nd ed.). Capstone Publishing Ltd..

George, B. (2009). *Seven lessons for leading in crisis*. Jossey-Bass.

Soupios, M., & Mourdoukoutas, P. (2014). *The ten Golden rules of leadership: Classical wisdom for modern leaders*. AMACOM.

Zu, L. (2023). *The wisdom of the Tao: A novel translation of Tao Te Ching with illustrations of Chinese calligraphy*. Amazon.

Dr. Liangrong Zu is a specialist in CSR, business ethics, and Taoism leadership. He earned his PhD in business and management at Nottingham University Business School. Dr. Zu used to work in the central government and the United Nations for over 30 years. Dr. Zu is the author of many publications: *Responsible and Sustainable Business: Perspectives from Taoism's Wisdom and Responsible Management and Taoism, Vol. I: Managing Responsibly for Sustainable Business Development in the VUCA World and Vol. II: Transforming Management Education for Sustainable Development Goals*. In recent years, Dr. Zu has initiated an ambitious authoring project aspiring to bridge the gap between the Eastern and Western worlds by interpreting the enigmatic codes and profound mysteries of Eastern thought, including Confucianism and Taoism, through the lens of celebrated management and leadership.

Centennial Companies in Japan: Ethical Strategies for Sustainability and Well-Being

Maki Sato

Abstract Japanese business sectors have been known for their manufacturing efficiency, such as the *kanban* style invented and adopted by Toyota. What needs to be discovered is that there are around 43,000 companies that have operated and existed in Japan for over a 100 years. This chapter focuses on the business styles of the centennial enterprises in Japan, known as *shinise*, and investigates, based on literature reviews, why they were able to sustain their business for such a long time and what role their underlying management philosophies play in this context. Academics have long neglected their importance, since those philosophies are mainly regarded as a mixture of ideas borrowed from Shintoism, Buddhism, and Confucian teachings. However, this paper argues that such ideas have not only led to the inherent neutrality, diversity, and flexibility that have helped centennial enterprises to sustain themselves for over a 100 years but at the same time have nurtured inter-relationality and mutuality among the stakeholders which provide a solid basis for more collective wellness and individual well-being. By highlighting such management philosophies, this chapter intends to provide impulses for the business sector to be more sustainable and abide by ethics and morality that might lead to a healthier planet and a flourishing society.

1 Introduction

Businesses will remain essential for humans to thrive and flourish, even with some actors criticizing their role in the light of pressing social and environmental concerns, such as climate change and biodiversity losses. At the same time, their contribution to a more sustainable society is claimed. In other words, as we face common environmental and social challenges, every member of society must

M. Sato (✉)
Tokyo, Japan
e-mail: maki.sato@aya.yale.edu

M. Mühlböck (ed.), *Sustainable Transformation and Well-being*, CSR, Sustainability, Ethics & Governance,
https://doi.org/10.1007/978-3-031-75566-8_4

consider how to adapt to and mitigate the changes and challenges we face. As such, for the first time, the business sector is also required to transition to a paradigm where sustainability, understood as a notion that takes into consideration social and environmental concerns along with economic development, is paramount. Yet, some questions arise in this context: If the business itself is not resilient enough to survive under various unprecedented changes and challenges, how can it help build a more resilient society? Are there problems embedded in the capitalistic system and its mechanisms such as a focus on profit maximization that create an unhealthy economic inequality and ignore the well-being of the planet and the flourishing of the people?

What is clear is that a transition to a new paradigm will require that businesses intensify their presence and responsibility as members of the enlarged society, not only in the limited B-to-B or B-to-C relationship. At the same time, only highlighting the problems that derive from the competitive market simplifies the matter and sheds light on the negative aspects of capitalism. Looking at it from a historical perspective, the merits of the capitalistic systems also become apparent, and it becomes clear why modern societies have chosen capitalism over Marxist theories. However, we should not close our eyes to the deep-layered problems embedded in the capitalistic world. We should humbly allow those critical voices and take the opportunity to update and enhance the present capitalistic system. According to Uzawa (2009), a healthy economic system should preserve the merits of capitalism, which seeks opportunities for innovation, efficiency, free choices, and self-regulation, and at the same time pay more attention to how a culture of collective wellness and individualistic well-being could be intertwined and nurtured. But how could such a new paradigm be entered?

Japanese businesses have been known for their manufacturing efficiency and constant micro-improvement, such as the well-known *kanban* style and *kaizen*, which were invented and adopted by Toyota, a car manufacturing company. Less known is that Toyota initially started in 1926 as Toyota Industries Corporation, selling weaving machines to textile industries. Although it has yet to reach its centennial, the key to its long-term business success lies in the flexibility of the business strategy in changing its main product and its continuous effort to achieve efficiency. Turning the eyes from this well-known global company to less famous examples, one discovers around 43,000 companies in Japan that have existed for over a 100 years. The oldest company in the world, which has existed since 578 AD, is Kongō Gumi Co., Ltd. (金剛組), a construction company based in Osaka that has been building temples and shrines. In other words, modern business analysis has yet to reveal the secret of why Japan holds the world's oldest company and the most significant number of centennial companies.

In an unpredictable world with unprecedented piles of problems in the social as well as environmental sphere that are often described as VUCA—volatile, uncertain, complex, and ambiguous—companies need to find a flexible but simultaneously resilient path that keeps their business and their stakeholders flourishing (Mack et al., 2016). However, it is not only in the twenty-first century that we suddenly realized such VUCA conditions, as Totman (2005) argues. VUCA-like challenges have occurred throughout human history, and centennial companies have survived such challenging situations for centuries. Additionally, it is worth noting that centennial companies were thriving with their business models way before the concepts of capitalism, socialism, or Marxist communism appeared in the late 1800s. Thus, it is worth taking a closer look at factors that might be decisive for Japanese centennial companies' longevity and success.

Through literature reviews, this chapter sheds light on how Japanese centennial companies have thrived in challenging situations and circumstances throughout history. What is striking is that existing analysis on centennial companies often concludes that the key to longevity lies in managerial philosophy (management principle, *keieirinen*, 経営理念) or family precepts, *kakun*, and mission objective, *shaze* (家訓, 社是). By summarizing the managerial philosophy of the centennials, this chapter digs deeper into the underlying thoughts and philosophy, unique yet common to those centennial companies in Japan.

Furthermore, this chapter introduces good practices from selected centennial companies to showcase which managerial philosophy approaches have helped them not only survive for centuries but also to nurture a culture where collective wellness and individual well-being emerge. In this quest, two concepts become evident: The first one is known as sanpōyoshi (good for all three, 三方よし). It refers to good relations among the three actors in business practices: the company (seller), the customers (consumers, buyers), and the broader society. The second one is the concept of *intoku* (陰徳), which the author believes is the key by which businesses have successfully sustained themselves while also contributing to the collective wellness and individual well-being of all stakeholders.

2 *Shinise*, the Centennial Companies, and Their Present Status

Japan is home, globally, to most of the companies in business that have survived for over 100 years. As of September 2023, 43,631 companies in Japan have been around for over a century, accounting for about 58.9% of the world's total of 74,037 such companies (Teikoku Databank, 2023). The United States ranks second in being home to 21,822 centennial companies (29.5%), and Germany is third with 5290 companies (7.1%) (Nikkei, 2022). There are more than 4,000,000 companies in Japan in total as of 2022, which indicates that 2.65% of the companies in Japan are centennial companies (Teikoku Databank, 2023). These figures demonstrate the longevity of Japanese businesses and their unique status of being a nest for centennial companies despite various natural disasters and drastic social changes that Japan has faced in the past centuries. The oldest company in the world is Kongō Gumi Co., Ltd. (金剛組), based in Osaka, a construction company that was established in 578 AD. This company primarily designs, constructs, and repairs shrines, temples, castles, and cultural heritage buildings. When we consider companies outside of Japan, the oldest is St. Peter Stiftskulinarium in Austria, founded in 803 AD, which operates as a restaurant. Another notable old company is Staffelter Hof in Germany, a winery since 862 AD. In Japan, there is a saying that the essential three items in life are "clothing, food, and a place to live." The oldest companies may prove that the proverb well captures the essentials of human life. However, in Japan, with countless natural disasters, such as earthquakes, tsunamis, and typhoons throughout its long history, the construction and reconstruction of buildings that work as a hub to sustain society may seemingly remain the most important.

In Europe, there is an organization that calls itself the Henokiens Association (https://www.henokiens.com/), founded in 1981 by Marie Brizard, a French anisette (anise seed liquor) maker, that only allows companies to join with a minimum period of 200 years of family-owned business experience, with at least one member of the founding family still engaged with the company. The Henokiens Association is mainly composed of businesses from EU countries, and as of 2023, it has 56 members: 16 French, 14 Italian, ten Japanese, four German, four Swiss, two Dutch, two Belgian, and one Portuguese. The number of members is limited because of its original criteria of the founding members still being engaged in the present business. Surprisingly, the oldest company among the Henokiens is Hōshi (法師) in Japan, founded in 717, a family-owned hotel run for 46 generations. In contrast, the oldest European member company dates back to 1502, The Coatinc Company, which operates in Germany as a surface finishing industry. Overall, across 59 countries and regions worldwide, 8916 companies have been in business for more than 200 years, and nearly half of these (3913 companies) exist in Japan (Morishita & Tateoka, 2012), and among those, there are companies with a business history much longer than 200 years, as shown in Fig. 1.

Japanese scholars in business study extensively initiated research on those centennial companies in Japan in the early 2000s after facing an unprecedented bubble economy crash in the last decade of the twentieth century. Such centennial companies are defined as *shinise* (老舗), which means "long-established companies" in Japanese.[1] It is said that the name is derived from the Japanese word "to replicate (*niseru*)," with the attached meaning "to continue and replicate the family business of one's ancestors without ceasing" or "to build assets by gaining credibility through long business activities." Although they have an enterprising spirit and a spirit of "never changing," they are also "willing to change" without limiting their achievements. Such an attitude of flexibility is also significant to learning from the business continuity and sustaining perspective. Japanese business studies define *shinise* as "a company that has been in business for over a 100 years and has continued to prosper" (Maekawa, 2011, pp. 9–28). In this chapter, *shinise* is therefore used to indicate the centennial companies.

According to the research done by Teikoku Databank, in 2023, the most common industry among all *shinise* was the industry relating to brewing *sake* (Japanese rice wine), including the retail and wholesale of *sake*, which totals 2239 companies (it

[1] The Japanese language doesn't distinguish between single and plural. In this chapter, *shinise* can therefore either stand for a particular company or a group of companies that sustains its business model for more than a hundred years.

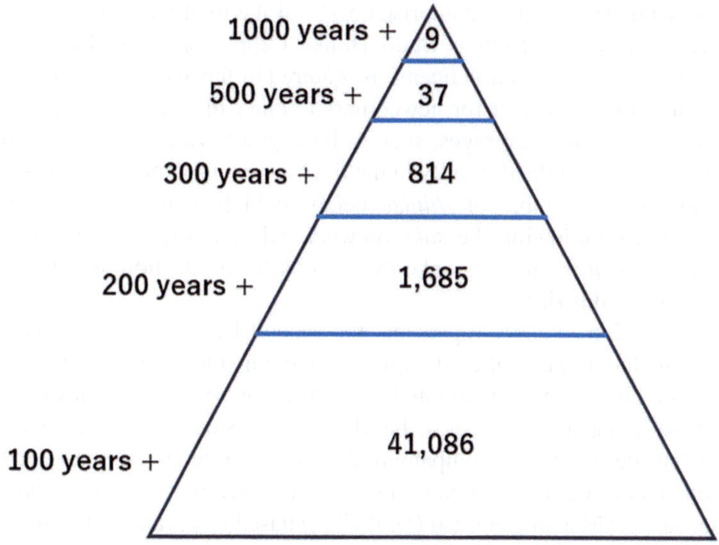

Fig. 1 Number of *shinise* based on their age. (Data source: Teikoku Databank, 2023)

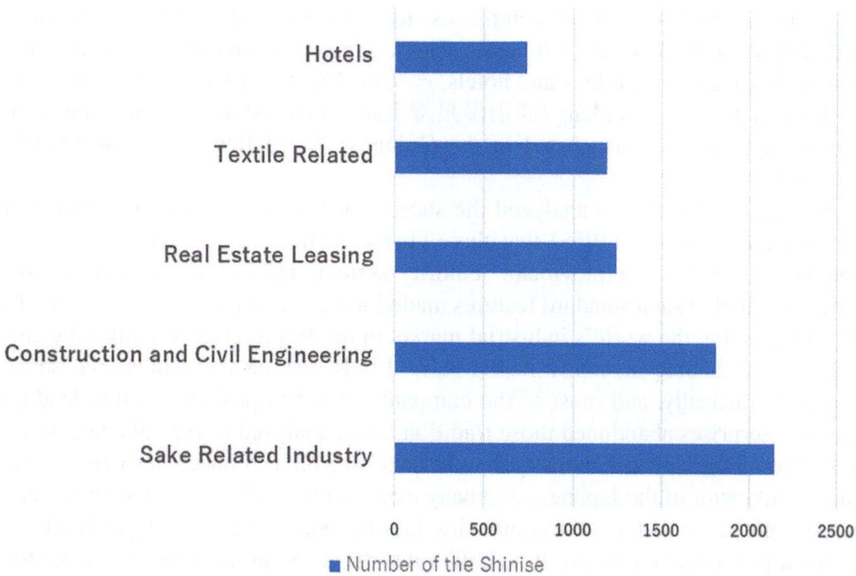

Fig. 2 Number of *shinise* by industries. (Data source: Teikoku Databank, 2023)

consists of 936 *sake* breweries, followed by 702 companies as *sake* retailers and 601 companies as *sake* wholesalers) (see Fig. 2). Figure 2 shows the top five industries among the 43,631 *shinise* as of 2023. Following those five significant industries are businesses in the confectionary industry, the lumber industry, and other miscellaneous industries that support the supply chain.

Sake-brewing-related firms comprised 5.1% of the total number of *shinise*. The oldest *sake*-brewing company is Sudō Honke (須藤本家), founded in 1141 in Ibaraki Prefecture. A production license is required in Japan to operate a *sake* brewery. Since it is more difficult for newcomers to enter the industry compared with other types of alcoholic beverages, such as beer, grape wine, and whisky, Teikoku Databank analyzes that the difficulty in the license-acquiring process may be one of the reasons for the number of *shinise* being so high in this industry (Teikoku Databank, 2022). Following the *sake* brewery industry, soy sauce breweries and fermented rice (*kōji* for making sake or miso paste) production are the other top brewing-related industries.

The construction industry represents the second largest number of centennial companies, including general civil engineering companies, with as many as 1962. As mentioned above, construction and reconstruction sectors have always been in high demand in Japan, where most buildings have been constructed with wood. Therefore, besides the oldest company in the world, Kongō Gumi, based in Osaka, various other construction company and civil engineering companies have been operating, such as Matsui Kensetsu (松井建設) based in Toyama. Matsui Kensetsu was founded in 1586 after Toyotomi Hideyoshi, the second *samurai* unifier of Japan during the Sengoku era, who took over Oda Nobunaga with damaged castles and temples elsewhere in Japan. The third largest industry is real estate leasing service companies, which total 1877 companies, followed by textile-related companies, including clothing retail and wholesale. The fifth most common industry among the *shinise* is Japanese-style inns and hotels, *ryokan*. The oldest *ryokan* in operation is Nishiyama Onsen Keiunkan (西山温泉慶雲館), founded in 705 in Yamanashi Prefecture. They are also listed in the Guinness World Records as the world's oldest inn.

Abegglen (1984), who analyzed the success of Japanese companies, argued in *The Japanese Factory* (1958) that three characteristics are common to Japanese companies—lifetime employment, seniority system, and in-house unions. It was long thought that such standard features made Japanese companies successful globally, influencing the world's industrial market in the 1980s. However, after the collapse of the bubble economy in the early 1990s, the Japanese industrial scene changed drastically, and most of the companies that had proliferated into leading global enterprises abandoned those traditions once analyzed as typical characteristics of booming Japanese industry. In other words, what was once interpreted as the successful factor of the Japanese company management style can no longer be seen in many global-scale Japanese companies thriving today. Since then, Japan has been facing a prolonged recession of so-called lost 30 years. In order to counteract the prolonged recession and to learn from the centennial companies, much of the research on *shinise* was conducted during the early 2000s to identify why Japan has the most significant number of centennial companies among other countries and the secret to their longevity. There are analyses done using quantitative data, but most of the research indicates that the qualitative analysis of the management philosophy of the centennial companies may hold the key to their longevity. The following section will summarize the outcomes of the existing research conducted by Japanese scholars based on literature reviews.

3 Conditions of *Shinise*

There are several reasons why Japanese companies have been able to continue for more than 100 years. Of course, one has to consider its historical and geographical background. Being a country of an archipelago with 14,125 islands scattered around the five main islands, it was not colonized or invaded by other countries, allowing companies to flourish with their relatively small local businesses. In addition, from a cultural perspective, the subtle influence of Confucius' ideology that emphasizes the importance of family rather than individuals' success was quite strong, with the custom of family succession being a common practice even among ordinary lay citizens. Those aspects might well work as a joint basis for the longevity of *shinise*. However, it does not explicitly explain why an outstanding number of around 43,000 centennial companies exist in Japan.

According to literature reviews, among others, there are three main points that are analyzed as key reasons for the longevity of *shinise* (Sone, 2019; Tamiya, 2022; Tanaka, 2015; Maekawa, 2011; Yokozawa, 2012). The combinations of these elements have enabled Japanese companies to continue for over 100 years.

3.1 Management Philosophy

Centennial companies' management philosophies (*keieirinen, kakun,* and *shaze*) are the prototypes of modern management principles. *Kakun* or *shaze* usually clarify and state the "purpose of existence" and the "ethical purpose," along with concrete *keieirinen* methods that clarify how to fulfill such purposes as a business, which is essential for *shinise*'s survival. Typical examples of such *keieirinen approaches* are "First do things right, and then the profit will come later (先義後利)" and "First let the customer use the product, and (if they like the product) the profit will come later (先用後利)," which derives from the famous motto among the pharmaceutical retailers from Toyama Prefecture. Daimaru (大丸) keeps its management philosophy of "first do things just and right" as a retail company founded in 1717 in Kyoto. During the Edo period, it expanded its business. However, they have provided food and clothing to the poorest of the poor since their establishment. Because of that, in 1837, when there was a massive riot by the citizens suffering from hunger, Daimaru was never attacked but was saved by the citizens (Tanaka, 2015, pp. 68–71).

Additionally, the principle of Ōmi merchants' philosophy of "benefits for all three sides (*sanpōyoshi*, 三方よし)" is also a famous principle adopted by countless centennial companies that emphasizes the importance of taking into consideration the benefits for the customer, vendor, and society as a whole at the same time (Tamiya, 2022, p. 139). To be precise, although the benefits may differ from what role the actors play in business transactions, the principle navigates the moral that the actors should share: the distribution of utility through business transactions. Precisely, sellers who sell the product or service should benefit from the transaction by receiving a fair price. Buyers who purchase the product or service should benefit by receiving a product or service that meets their needs and is worth the price they paid. The society or the wider community surrounding the transaction should also benefit from the business transaction between the merchants through the cascading effect from the overall economic activity, such as job creation, income gain, and other social contributions made by the business, like bringing happiness and satisfaction to the employees, customers, their family members, and other individual members of the society.

Some companies have interpreted and rewritten these lessons of their predecessors in a modern management principle. For example, the eighth-generation owner and president of the confectionary shop in Saitama Prefecture, Kameya (亀屋), founded in 1783 upon the 200th anniversary of the company's launching in 1983, rewrote their management principle of the *sanpōyoshi* in modern style as follows: "Our business keeps up together with the progress of the world; Be kind to all people." Yokozawa analyzes that the reinterpretation of old management philosophy is essential for the leadership team and employees to understand and grasp the importance and the role they play in society through their confectionary business, even in modern times (Yokozawa, 2012). President Yamazaki of Kameya takes every opportunity to explain the meaning of their management principle to employees: "*Shinise* has to be humble to be loved by the locals. If the locals do not love them, the shop will not survive. So be kind to everyone, not only to customers, business partners, but also to local people outside the small circle of our working environment." Such an attitude is crucial, since it is the basis of building trust between the business and society.

Trust building with clients and customers is believed to be one of the most critical factors for longevity. This is because with trust, clients and customers will gradually bring their problems to the manufacturers and vendors in the hope that they might have a better product to meet their needs or an idea to solve their problems. Such intimate communication between the customers and vendors/manufacturers/service providers only happens when the customers fully trust the people involved in their business transactions, and such conversation provides a massive impetus for the following product or service development that better meets the users' needs. Such relational understanding, as well as kindness, truthfulness, and trust building, is considered the most essential asset for *shinise* to be *shinise*.

3.2 Innovation Orientation Based on the Core Business

The business environment is always non-static, so both the core business and the products from the business change over time to meet the needs of the era and the needs of the customers. For a business to survive for a long time, it needs to have strong core competence, but at the same time it is essential to be creative and innovative in the core business. Fascinatingly, all the centennial businesses may also have been venturing companies in their early years when they started. Thus, the question arises: What is the difference between the venture and the centennial companies' mindset? Based on this overall question, a survey was conducted among 159 venture companies and 51 *shinise* in 1999 (Yokozawa, 2012). It was identified that both ventures and *shinise* are keen on cutting costs, improving existing products, and developing new production technology. However, *shinise* companies were more reluctant to expand and diversify their business.

Yokozawa analyzes that *shinise* companies put more of an emphasis on improving their existing business to better the products for the customers and maintaining reasonable profits to ensure the enterprise's survival rather than focusing on the expansion of their business. They are reluctant to invest in diversifying their business but prefer making investments to improve the existing business model. Diversifying and expanding the business takes many assets and is time-consuming until the new product or service starts to generate profits, thus making it costly and risky. Yokozawa argues that *shinise* avoid the risks of neglecting core business while the company's energy is invested in new activities. *Shinise*, in general, treasure the employees working for their business, and retaining their employees and sustaining their livelihoods is more important than expanding and risking their business. However, a few *shinise* have succeeded in diversifying their business.

Kao Corporation (花王), established in 1887, started as a soap and toiletry manufacturing company. Besides manufacturing soaps, detergents, cosmetics, and diapers, Kao also sold floppy disks from the 1980s to 1998. People might wonder why a soap company expanded its production line to floppy disks. Nonetheless, because floppy disks require chemical technology to develop surface-active agents, Kao's floppy disks became one of the most sold floppy disks in the global market. However, once the sale of floppy disks started declining, they withdrew from the business and are now once again focused on their original product line of various soap and toiletry goods.

Some companies succeed in continuing to expand and diversify but with different tactics. Noritake Company Limited (ノリタケ) was founded in 1876, starting as a retail company of Japanese products, and opened a shop in New York. Their business has developed in various businesses but has constantly expanded based on ceramics, ranging from a ceramic tableware company to high-end ceramic tableware, toilet bowls, insulators, and other special ceramic materials. However, upon its 100th anniversary, Noritake reorganized itself based on a "one business, one company" policy that made the line of command and responsibility much more apparent than before, enabling speedy decision-making. Now, their 30 domestic and 14 overseas group companies are branched out, and Noritake's headquarters is only responsible for the group company administration and development, overseeing the operation of the group and providing financial backup to ensure the further longevity of the brand, which Noritake believes is the company's greatest asset.

The willingness to change and the company's flexibility to respond to customers' needs are essential. Moreover, what is interesting about *shinise* is that they expand their business based on their core competence in the business field. For Kao, their chemical technology of the surface-active agent made them expand their business to a floppy disk manufacturer. However, as soon as the market started shrinking, Kao never hesitated to evacuate from the market itself. Noritake branched out its 44 companies to keep the business going so that one company would only be responsible for one field of a specific service line. Therefore, branched-out companies can focus on their competence and grow based on their core business.

3.3 Balance between Profit and Frugality

Shinise know how to balance tradition, what needs to be preserved, and innovation, what needs to be changed. Most *shinise* keep their management philosophy as tradition; although its wording might change in modern Japanese, the spirit of management philosophy stays. Such an approach promotes a culture of trust. With trust built between the customers and the local society, *shinise* companies' brand is stabilized, and the actual and potential customers link the core product or service to that brand that has established such a trustworthy reputation. Thus, there seems to be a flow of good information surrounding *shinise*. Although YKK, established in 1934, is not categorized as a *shinise* yet, their management philosophy of "circulation of goodwill (善之循環)" well articulates their intention that their business should play a part in a prosperous circulation of goods, including people's goodwill. For them, the role of a *shinise* is to be a node or base for such good circulation of goods to happen.

Their success has proven them right; even with some critics calling their business style stubborn, many *shinise* have abided by traditional management principles such as customer orientation, focus on core competencies, quality management, and risk management for such a long time. On top of that, *shinise* make various efforts to develop their human resources to sustain the business. After all, the driving energy for the company's survival depends on the employees' contribution to keep the business running. Without employees' everyday effort and love for the business and its products or services, the business model will not succeed. Thus, the shared philosophy that implies that "if you are not being kind to others or not working for the benefit of others, your prosperity will never happen" works equally for the employer and employees of *shinise*. Most *shinise* owners regard their employees as important as family members and prioritize maintaining their welfare. Because of this, only 1.6% of the companies among *shinise* (688 *shinise* as of 2022) are listed companies in the stock market, which allows them to work without stockholders' interference and the need to share their profit with them. Instead, *shinise* refrain from a profit-maximizing orientation and even share their profit that increases the employees' well-being or contributes to the local community through local festivals and revitalization projects. Through such intentional profit sharing with the extended communities to which *shinise* companies belong, they believe that their goodwill will return to them in a longer-term perspective. Thus, *shinise* cultivate an attitude of sufficiency that balances the quest for profits that is inherent to any business model.

4 Ethics and Morality as *Shinise's* Backbone for Sustainability

In Sects. 3.1, 3.2, and 3.3, we saw the three main components of the sustaining conditions of the centennial enterprises. At first glance, those philosophies just seem to be a mixture of ideas borrowed from Shintoism, Buddhism, and Confucianism that have culturally influenced the morality and philosophy of Japanese citizens over time. However, those teachings are intertwined and presented in a unique, harmonious style, making it difficult to tell which teaching ultimately dominates the *shinise* thoughts. Likewise, also the management philosophies of *shinise* are such integrated concepts, including ideas hybridized from those various teachings.

Such advanced integration of different teachings can be traced back to 604 AD, when Shōtoku Taishi (prince of saintly virtue) issued the Seventeen-Article Constitution (Kasulis, 2018, pp. 65–71). Article 1 of the Constitution starts: "Take harmony to be of the highest value and take cooperation to be what is most honored." Shōtoku and other intellectuals in that era were influenced by Confucius' and Mencius' teachings. He also wrote a commentary on three Buddhist sutras, although he himself came from the proto-Shinto Emperor family. The idea of harmony, which is originally Shinto's idea of coexistence, symbiosis, and co-prosperity (共存, 共生, 共栄) with other humans and other beings of the world, appears in the first article of this Constitution and still influences Japanese citizen until present days.

The teaching of "I only know I am sufficient (吾唯足知)" or "I am content with what I have" that is found in Ryuan Temple (竜安寺) in Kyoto is said to have been inspired by *Tao Te Ching* (道徳経) by Laozi (老子) on the thought of sufficiency (知足). Buddhist teaching also indicates in Shakyamuni's sutra on his last teaching (仏遺教経) that to "have small desire and know thy suffice (少慾知足)" is essential for true happiness and true fortune. Tanaka posits that the *shinise* exemplify this concept to such an extent that, despite operating within a capitalist market, the majority of *shinise* eschew the pursuit of business expansion, adhering to the belief in limitless possibilities and desires. Instead, they maintain a modest and sustainable business model within their local communities (Tanaka, 2015, pp. 164–170).

In Sect. 3.2, I pointed out that "First do things right, and then the profit will come later (先義後利)" is a typical management philosophy for *shinise*. The philosophy is taken from the teaching of Xunzi (荀子), who says, "The one who prioritizes justice and righteousness and who does not think about the profit in its first place is the one who prospers (先義而後利者栄)." Tanaka analyzes Confucius' teaching of virtue ethics (得), especially of its "five constant virtues (五常)" from the Song-Ming dynasty era that contributed to *shinise*'s virtue ethics (Tanaka, 2015, pp. 105–111). The *five constant virtues* are, namely, humaneness (仁, or benevolence, or goodness), righteousness (義), propriety (礼), wisdom (智), and faithfulness (信). These virtues are concerned with fostering and upholding good, meaningful, and mindful human relationships that shed light on regulating human activity and improving the person's moral quality. Until the Meiji period, *samurai* and lay citizens were equally educated on the importance of virtue ethics (道徳). Thus, Tanaka analyzes that the *five constant* virtue ethics must have been internalized in the founders and employees of the *shinise* and simultaneously clarified through their management philosophy.

The way of people (人の道), a mixture hybrid of Shinto, Buddhist, and Confucian teaching, was shared and popularized in the Edo period during the seventeenth and eighteenth centuries. Among many thinkers, Ishida Baigan (石田梅岩, 1685–1744) contributed the most to popularizing the virtue ethics for business people. His teaching was mainly about self-discipline, but concretely he highlighted the importance of honesty (正直), economy (倹約, or being thrifty), diligence (勤勉), and filial piety (孝行) in the business field. His school and teachings rapidly became popular and had expanded to 170 schools in 64 states by 1803. Furthermore, his students taught his teachings for more than 100 years to business people throughout Japan. Baigan's teachings survived and were crystallized by Shibusawa Eiichi (渋沢栄一, 1840–1931), who is known as the "father of Japanese capitalism." Shibusawa introduced Western capitalism to Japan after the Meiji Restoration and is well known for his book *Analects and Abacus* (論語と算盤). Stating the importance of Confucius' virtue ethics in modern business models, he writes that an abacus is for profits, whereas analects are for morality and that he believes that we need both an abacus and analects in business and that those two should be intertwined and must be in pure conformity (Shibusawa & Moriya, 2012). Shibusawa's writings are still well read among today's business people, and it has been decided that his figure will be featured on the Japanese 10,000 yen banknote planned to be circulated in the summer of 2024.

The longevity of *shinise* is not only about capitalistic rationality that focuses on profit-making. However, their rationale is placed into relationality with employees, customers, vendors, and the society they belong to. Learning and adjusting based on market mechanisms also plays a crucial part, but what is unique about *shinise* is that their rationale is based on their effort to build a long-term relationship based on trust and kindness to others. Most of the time, they are not interested in expanding or diversifying their business. Instead, they are focused on a core competence, such as a product or service, to prolong and allow their business to survive for the benefit of employees and their everyday life. Profit-making is not always necessary, but doing justice and good to the local community and society is the goal they have in common, and that makes *shinise* the centennial company.

5 Case Study of Selected *Shinise* and the Shared Concept of *Intoku* for Collective Wellness and Individual Well-Being

In this section, three manufacturing *shinise* companies are selected to be further analyzed and to showcase the interconnection of *collective wellness* and *individual well-being*. Wellness addresses optimal health, whereas well-being describes a state of being healthy and happy and the overall well-balanced satisfaction of an individual who has a sense of purpose (Swarbrick, 2006). According to the WHO, well-being could also be described as a positive state of health and quality of life influenced by the individuals' social, economic, and environmental conditions (https://www.who.int/activities/promoting-well-being). In this section, drawing from those definitions, the author differentiates and defines the concepts of wellness and well-being in terms of the ideal state of and the achievable goals for the collective and individuals to clarify the role of *shinise* in society and to consider the individual people working with *shinise* as community members while achieving one's well-being.

While the concept of wellness generally refers to the physical health of an individual, the concept of well-being emphasizes much broader and deeper concepts like happiness and satisfaction that are achievable based on mental, emotional, and social health. In other words, *wellness* is the basis or process for one to attain a state of *well-being*. That said, well-being may vary significantly from person to person based on the sets of values and goals they have in their mind. Therefore, what contributes to one individual's well-being might not contribute to that of others. Furthermore, personal fulfillment, a sense of accomplishment, or social satisfaction varies based on an individual's state of life, set of values, and the cultural value that signals and unconsciously shapes what happiness is about. Thus, in this section, wellness is defined as an achievable collective goal, such as sustainable and prosperous management of a business, and well-being is defined as a set of goals and aims defined by individuals that varies from person to person based on their life stage.

As argued in the above sections, the longevity of *shinise* may certainly be linked to the interrelation of the conditions that sustain a business (Sect. 2) and the moral or ethical attitudes that ground its managerial philosophy (Sect. 3). In other words, managerial philosophy could be translated as the shared concept of wellness crystallized in words (*logos*). Philosopher Miki Kiyoshi argues that *pathos* (emotions and motivations) requires *logos* to clarify the aim and a body that brings to life the combination of *pathos* and *logos* (Miki, 2023). Considering Miki's philosophy, managerial philosophy can be regarded as the combination of pathos and logos, or in other words the passion and ideal achievements expressed in words, of the founders and descendants of the *shinise*. In this regard, *shinise* is a body that implements their *pathos* and *logos*, or their managerial philosophy.

Three manufacturing companies have been selected to investigate this further: Nishikawa (西川), Toraya (虎屋), and Suzuhiro (鈴廣). All three are centennial companies still managed by the founding family. Each company is strongly committed to keeping the quality of their product and tradition in their respective business fields for over a 100 years. In addition, each company has respected and kept its traditions and business. Simultaneously, they have faced continuing challenges to be innovative in response to changes over time. In this section, to exemplify the realization of collective wellness and individual well-being through *shinise*, the three companies are examined through literature reviews and the information available on their websites. Moreover, in summarizing the common attitudes of the centennials, this section touches upon the concept of *intoku* (陰徳) as the grounding concept for achieving collective wellness and individual well-being which are both connected to inter-relationality and mutuality in the endeavor toward a sustainable society.

5.1 Nishikawa

Nishikawa[2] is a company founded in 1566 that initially started its business as peddlers (Gyosho, 行商) based in Omi (近江). Their signature product has long been a mosquito net dyed in light yellow-green that the founder's son invented. Mosquito

[2] https://www.nishikawa1566.com

nets already existed in the Edo period (1603–1867). However, the idea of dying them in a light yellow-green added a new value to mosquito nets and interior designs for the summer season in Japan. By the late Edo period, Nishikawa's mosquito nets even appeared in the writings of the famous novel writer Ihara Saikaku. Today, the company specializes in providing products designed to support comfortable sleep and offers a wide range of bedding-related products, including mattresses, pillows, blankets, covers, and bedsheets. Along with a commitment to "good life, good sleep" and as a provider of comfortable and good-quality sleep, which has evolved over the centuries up until today, they have been recognized as one of the most well-known bedding-related suppliers.

The company philosophy or motto is "sincerity, kindness, and mutual prosperity (誠実, 親切, 共栄)." Nishikawa states that the motto is reflected in the attitude toward all people surrounding Nishikawa, including customers, retailers, and suppliers, and states that "mutual prosperity" can only be achieved through "sincerity" and "kindness" in human relationships based on the respect for humanity (https://www.nishikawa1566.com/company/philosophy/). The company philosophy derives from the principle of Ōmi merchants' philosophy of "benefits for all three sides" (Tamiya, 2022) (see *sanpōyoshi in* Sect. 3.1). Nishikawa has gone beyond the concept of *sanpōyoshi* and has been conscious of employee satisfaction and sharing values of the company since the seventh generation from the founder in the mid-Edo period. The owner of that time made a "rule of one-third" and distributed 30% of its net profit to its employees (三ツ割銀制度) (Tamiya, 2022, pp. 144–146).

The company strongly commits to its employees and offers them an internal education and certification system that has been in place for about 20 years. Thus, employees become experts on sleep and can earn certifications such as pillow advisor, asleep advisor, and sleep master. Nishikawa also has a certified internal examination, which measures practical skills in bedding knowledge and sales techniques.[3] Through this internal education and certification system, Nishikawa aims to improve employees' skills and motivate them. That also benefits the customers by having qualified salespersons share their knowledge in every Nishikawa brand store.

Due to concerns about the limited resources of down feathers, Nishikawa has also been taking the initiative in the bedding industry to promote their recycling. The company collects used down feathers and cleans and recycles them accordingly. They also accept used down feathers in the bedding of other brands for recycling. Virgin down feathers that do not qualify for Nishikawa's bedding material are upcycled to downcoats through a newly created relationship with the apparel industry. Furthermore, Nishikawa supports the Inner Mongolia tree replantation project, since the region is well known for cashmere production used for high-quality blankets. A good life is based on good sleep. Nishikawa's managerial philosophy exemplifies that the company's well-being includes the well-being of the individuals. They do this through their internal education and certification system with the sustainability of the extended product chain in their mind, which includes future generations.

[3] https://www.nishikawa1566.com/company/certify/

5.2 *Toraya*

Toraya is a long-established confectionery of *wagashi*, a traditional Japanese confectionery. The company was founded in the early sixteenth century in the late Muromachi period by the Kurokawa family.[4] It is particularly known for its sweet bean jelly, *yokan*. The oldest documented record of Toraya appears in 1600 from the temple record of Myoshinji, a large temple complex in Kyoto. It was a purveyor to the imperial court during the reign of Emperor Go-Yozei (1586–1611). By the late seventeenth century, Toraya was already a well-known confectionery to feudal lords in many provinces outside Kyoto. However, in 1788, when the ninth-generation owner was in charge, it faced difficulty after a devastating fire in Kyoto. Along with transferring the capital from Kyoto to Tokyo in 1869, Toraya moved its primary operations to Tokyo, finally settling in Akasaka in 1964. It opened its shop in Paris in 1980, and today Toraya has three factories and approximately 80 shops throughout Japan.

After the 1788 fire, to overcome the business difficulties, the ninth-generation owner compiled a set of rules (Okitegaki, 掟書) for employees aiming to reform its business by not only encouraging frugality but also by putting effort into internal suggestion systems and employee education (https://assets.toraya-group.co.jp/toraya/corporate/chronology/pdf/04_1-1.pdf, p. 21). The ending chapter of their company history, titled "Five Centuries of Toraya," concludes that Toraya has survived for the last five centuries with the management philosophy of challenge, involvement, and commitment (*idomu*, *kakawaru*, *kodawaru*) and argues on the importance of changes without losing trust; coordinating interest and creating mutual trust among various stakeholders, such as customers, employees, and business partners; and changing what needs to be changed but not changing what does not need to be changed.[5]

Toraya's business philosophy is to "delight customers by serving delicious *wagashi*," taking into account the Earth, producers of the agricultural products, makers of the ingredients, customers, and Toraya as an interconnecting node of those actors, continuing efforts to sustain their business into the future. Although the technology evolves over time, Toraya does not rest on its laurels but continually experiments and strives to make better confectionery using the best ingredients available today and considers its mission to deliver delicious sweets to customers around the world. One of its initiatives that exemplifies the interconnectedness created around Toraya is the close relationship with producers, which Toraya places great importance on.

In order to make and provide quality sweets, it is a crucial task for Toraya to secure the best ingredients stably. Thus, Toraya has established a good relationship through contract farming with the producers (farmers) of a unique variety of white beans called *Fuku tora shiro* since 1927, along with *Erimo shozu* red bean farmers

[4] https://www.toraya-group.co.jp

[5] https://assets.toraya-group.co.jp/toraya/corporate/chronology/pdf/16_4-3.pdf, p. 381.

in Hokkaido. It also directly contracts with *kanten* (agar) makers in Nagano and Gifu prefectures and farmers in Okinawa for *kokutou* (raw sugar), which is essential for *wagashi* making. Toraya also sends their employees to let them experience and be educated on how farmers and producers make the ingredients. It states that these interconnected relationships are based on a sense of gratitude for being nurtured by the nature and climate of Japan, farmers, and producers established throughout regional connections.[6]

Besides their main business, Toraya has Toraya Bunko (Toraya Archives), which was established in 1973.[7] Toraya, which has served the imperial court for many years, has a wealth of historical documents, sample books of sweets, and antiquities. Therefore, Toraya Bunko serves as a repository for a wide range of materials related to *wagashi*, including ancient records, documents, literature, and articles. These materials are catalogued and preserved to contribute to the dissemination, transmission, and creation of *wagashi* culture. In summary, Toraya works not only as a maker of *wagashi* but also serves as a hub for preserving *wagashi* culture for future generations through contract farming, educating the employees, and creating delicious confectionery for customers to taste *wagashi* as their product.

5.3 Suzuhiro

Suzuhiro[8] is a manufacturing company of *kamaboko* (processed seafood product commonly known as fish cake), founded by the Suzuki family in 1865 in Odawara. They produce various seafood products, including traditional *kamaboko*, *datemaki*, dried fish, and local beer. Suzuhiro also owns several restaurants to show existing and potential customers how to cook and eat their products. Suzuhiro believes all life is interconnected and vibrant and states that food (shoku, 食) improves human life. Thus, it sees its role as an intermediator to transfer the life of the ingredients (fish) to the life of its customers through its business as a *kamaboko* manufacturer. Their attitude makes them committed to contributing to the natural resources (the Earth), the people, and the future.

Suzuhiro's management philosophy is encapsulated in the phrase, "Being a *Shinise,* yet not being a *Shinise*."[9] The contradicting statement represents Suzuhiro's commitment to upholding its centennial tradition while continuously embracing innovation. Suzuhiro believes in the importance of maintaining a steadfast attitude toward their business, such as treating their customers with honesty and sincerity. However, simultaneously, they are determined to courageously make changes where necessary, such as in their work methods: keeping traditions as traditions but also

[6] https://www.toraya-group.co.jp/corporate/csr

[7] https://www.toraya-group.co.jp/corporate/bunko/

[8] https://www.kamaboko.com/

[9] https://www.kamaboko.com/corporate/philosophy/

incorporating new technologies and knowledge through innovations and research to better their products and manufacturing processes.

In addition to manufacturing and selling their product, Suzuhiro operates a *Kamaboko* museum where visitors can learn about the history and production process of *kamaboko* and experience making *kamaboko* themselves under the guidance of skilled craftsmen.[10] Besides educating the consumers, Suzuhiro offers an in-house education system called Suzuhiro Hitozukuri School for employees, based on initiatives from individual employees to learn new skills according to each employee's stage of career and career plans. Hitozukuri stands for "making humans more human," which shows Suzuhiro's mindset toward their employees as capital and treasure for their company.

Furthermore, with the sustainability of the natural environment and community in mind, Suzuhiro has endeavored to create a circular model for recycling natural resources. They create fertilizer from fish wastes produced during the manufacturing process of *kamaboko*, grow vegetables and fruits from the soil using the fish residue fertilizer, and serve those organic products in their restaurants. Suzuhiro aims to serve as a hub for the circular economy through *kamaboko* manufacturing, and Suzuhiro's positive attitude toward educating consumers and encouraging employees to learn new skills and knowledge leads to Suzuhiro's collective wellness and the individual well-being of the consumers and employees.

5.4 Grounding Concept of Intoku

The selected three centennial companies share a common commitment to maintaining their traditional business while adapting to the modern needs of their business partners and customers. They all strive to provide high-quality products and services that benefit their customers and contribute positively to society as a whole, including educating employees, caring for future generations, and committing to the sustainability of the natural environment. The concept of *sanpōyoshi*—good for customers, business partners, and society—emphasizes the importance of creating shared value through business, as social responsibility represents the basis for their managerial philosophy. It is notable that they have paid tribute to the modern context and have extended the concept of *sanpōyoshi* to the environment, which includes taking into account the wellness of future generations and leads to environmental sustainability in respective business practices.

Apart from the concept of *sanpōyoshi*, the shared common grounding notion could be summarized as *intoku*. The term *intoku* refers to the concept of "hidden virtue" or "unseen good deeds" and is derived from a traditional Asian philosophy that emphasizes performing good without seeking acknowledgement, recognition, reward, or reputation but to perform good because it is the right thing to do. It is

[10] https://www.kamaboko.com/museum/

often contrasted with *yōtoku*, or virtue or merit visible to others. *Toku* (徳) originally derives from Confucian ethics and is often translated as virtue or moral character. It is a fundamental concept that represents the moral quality of a person, including their virtues and ethical standards. However, Confucian ethics is not only about personal development through individual morality but involves social relationships. This is because a virtuous person is thought to naturally attract and inspire others through his/her *toku*, which leads to spreading moral influence. Confucian moral and virtue ethics is somewhat different from Western virtue ethics, since it is more focused on social and relational aspects of collectiveness, on how one's morality naturally impacts other members of society and eventually changes society into a virtuous society as a whole.

In Japan, the word *intoku* first appears in *Honchō Monzui* (本朝文粋), a Japanese book compiled by Fujiwara no Akihira (藤原明衡) in Chinese around the middle of the eleventh century in the Heian period. The book is a repository of model pieces in prose and poetry, featuring 39 genres that instruct an educated Heian man to master participation in court life and perform duties within the court bureaucracy. *Intoku* are often small, everyday good actions that may go unnoticed by others. Nonetheless, these unseen good deeds are thought to bring positive consequences, a natural outcome of a virtuous life. However small, they are thought to contribute to society's overall goodness and to create positive impact and harmony. The centennial companies seemingly incorporate *intoku* actions of subtle and unnoticed good into their company value system, creating a corporate culture that shares these *intoku* mindsets along with their managerial philosophy. As argued, managerial philosophy can be recognized as a combination of pathos and logos of the founder that directly translates what the company aims for and how it could contribute to society through its business. However, the notion of *intoku* is embedded in the centennial companies that are deeply internalized into what is written or visible through their business model.

Besides the *sanpōyoshi* concept, the centennials share the visible virtue that could be captured as a phenomenon. Those phenomena that are common among the centennial companies can be summarized as follows: (1) flexible strength that is adaptable to the changing society and is not bound by past successes, (2) solid in-house/internal employee education system that not only leads to standardized knowledge of the products but also to developing pride and attachment to the quality of the products they produce, and (3) to expand their business interest in more extended vision and broader space, taking in mind the environment and society beyond their business sphere, including future generations. Along with these common phenomena, there is a shared notion of *intoku*, which cares for the inter-relationality and mutuality with others (customers, business partners, etc.), the natural environment, and the future.

As argued, this chapter defines wellness as an achievable collective goal that can be shared and measured. In contrast, well-being is defined as a particular set of goals that varies among individuals. Centennial companies may have been successful over their hundreds of years of business models. However, the author does not think every company can be a centennial company by mindlessly following the case models and good practices of the centennial companies. For the company, employers, and employees to sustain their business practices, the importance lies in realizing how one is interlinked with other members of society, within and outside the company, as a small unit of society—which leads to the consciousness of the importance of everyday *intoku*.

Other members of society to include, besides the existing business partners and customers, are the silent voices of the natural environment and future generations. However, centennial companies seemingly internalize the views that their company is just one node of the complexity of the social web and that pursuing the company's benefits selfishly will not benefit them in the long run, since they are just one member of society as a whole. Thus, although it is not clearly stated, *shinise* has trust in *intoku*, small hidden good doings that may work accumulatively to change the world into a better place.

Centennial companies make us realize that collective wellness can only be achieved through collaboration with others and by maintaining the well-being of all the actors working to sustain the business. Moreover, the ability to fully internalize the concept of *sanpōyoshi* is a crucial key for *shinise* to sustain their business. In other words, not only the company itself but all the individuals in whatever relation with the company and the surrounding environment that sustains the company by creating raw materials—all these small contributions of the anonymous make the business happen while playing a part of the countless connections of life. Internalizing these notions through the concepts of *sanpōyoshi* and *intoku* leads *shinise* to build trust in society. Therefore, the grounding concept that treasures the interrelationships and mutuality, that sees every node of the network as a crucial member that sustains their business, and business attitude or managerial philosophy that acknowledges such relations may be the key to sustainability and creating collective wellness and individual well-being through business.

Summing up, through exploring *shinise's* management approaches and their underlying philosophical thoughts, concepts, and principles, valuable hints emerge that point to positive effects toward sustainable business transformation and well-being. What is essential in this matter is that s*hinise* manage to intertwine all elements in a unique and harmonious style (see Fig. 3).

6 Conclusion

Sustainable entrepreneurship emerges from "not harming the environment." However, in recent years, the notion has slightly changed into coexistence and co-prosperity, considering the symbiosis of business and sustainability. Thus, the corporate stance is changing from merely promoting the Environment Sustainable Governance (ESD) index to challenging the business to "take part in the solution that leads to better society and the planet." Several radical arguments that oppose capitalism and criminalize the infinite human greed and growth embedded in it are seen. Sadly, history has shown that neither socialists nor communists are good at building harmonious relationships with the environment and human welfare. The

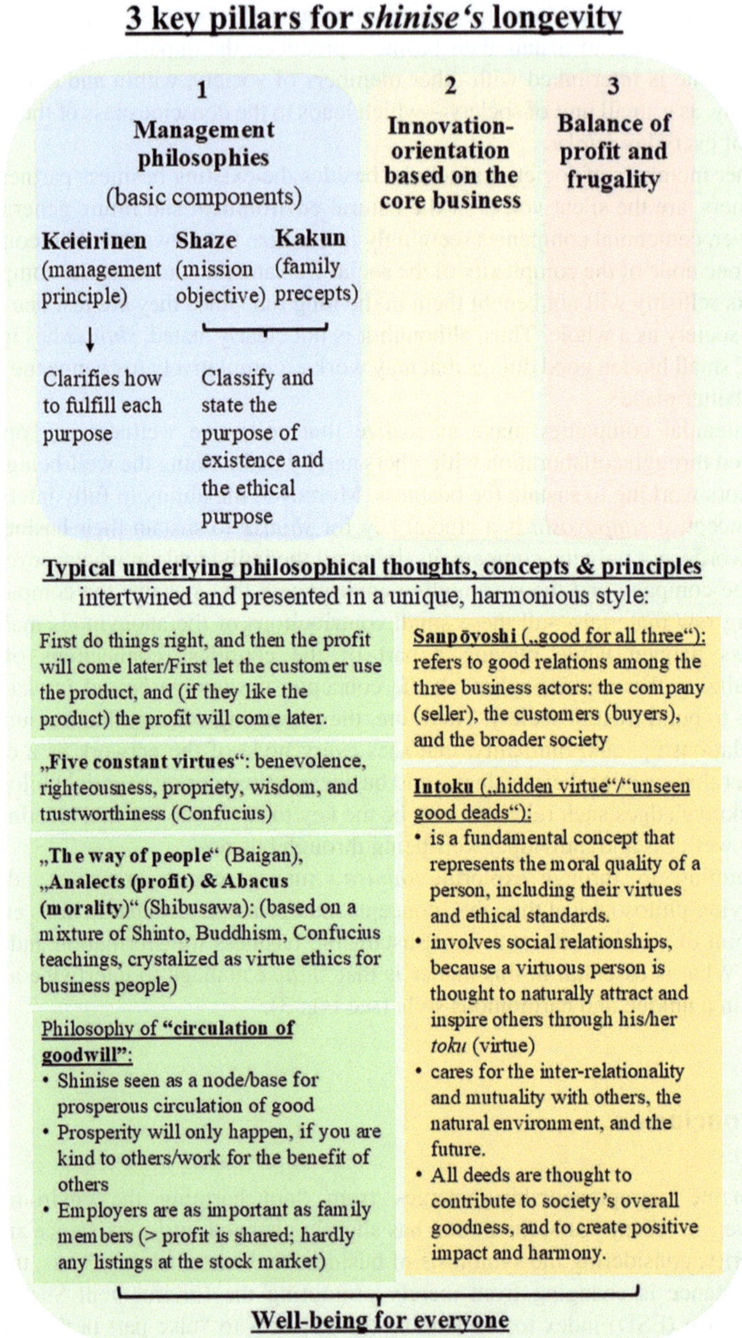

Fig. 3 Pillars of *shinise's* longevity and their implications for well-being

idea of sustainability has existed since the 1970s, when the Club of Rome presented the "Limits of Growth" (Meadows, 1972). The need for economic degrowth and decoupling economic growth and the use of natural resources was first argued for in the UNESCAP report published in 2012 (UNESCAP, 2012). However, not much has changed in conventional business practices. Thus, humanity and the business sector have been struggling to identify the concrete good and ideal action that can be taken to be sustainable while keeping society prosperous.

The long-term oriented practices of *shinise* have always linked business' requirements with societal needs and thus naturally lead to an optimal balance of community wellness and individual well-being. *Shinise* companies also teach us that sustainability in business is about morality and ethics throughout its business, which means (1) sharing the wealth with everyone in the community, (2) focusing on small areas and being competent in that business area, (3) being happy with being sufficient and not striving for the ultimate profit maximization (知足), (4) writing down the main purpose of the company as a management philosophy and sharing it among the employees, and (5) being just and being good to every actor involved in the business process and beyond. When humanity faces the unprecedented environmental crisis that is gradually affecting our global society, perhaps the hybrid integral philosophy of Shinto, Buddhist, and Confucius teachings that have helped support and sustain the Japanese centennial companies may provide us with a new perspective to the sustainably-oriented business. That hybrid and integrated management philosophy may lead to inherent neutrality, diversity, and flexibility in the concurrent business sector of the world. This chapter does not intend to argue that all businesses should follow the management philosophy of *shinise*. However, it may be vital to consider and incorporate new ideas and strategies to stay competitive and at the same time sustainable, leading to collective wellness and individual well-being in today's ever-changing business landscape.

Furthermore, this chapter has highlighted the two interrelated concepts of *sanpōyoshi* and *intoku*, which have contributed to centennial companies' longevity. Those underscore the value of social contribution and collaboration for achieving collective wellness and individual well-being rather than focusing solely on company profits or personal gains. The profound realization that business can only sustain based on inter-relationality and mutuality with a longer-term, more comprehensive, and broader view enables the individuals involved in the business to start caring for others. In this context, others are not limited to the human societal sphere, like customers, business partners, employees, and future generational customers, but are also inclusive of the silent voices of the natural environment, where raw materials are produced through priceless ecosystem service. Centennial companies in Japan exemplify how businesses can be morally virtuous through long-term relationships. They have prevailed over centuries because they have cared about the people involved and the surrounding environment from which they have benefited. Action that can be taken by one company might be small. However, when it accumulates, business can play a massive role in changing the world into a humane, livable, and sustainable society.

References

Abegglen, J. C., 1926-. (1984). The Japanese Factory /James C. Abegglen. Salem, N.H.: Ayer.

Kasulis, T. P., 1948–. (2018). *Engaging Japanese philosophy: A short history*. University of Hawai'i Press.

Mack, O., et al. (Eds.). (2016). *Managing in a VUCA world*. Springer.

Maekawa, H. (前川, 洋). (2011). *Textbook for Shinise studies* (老舗学の教科書). Doyukan (同友館).

Meadows, D. H., & Club of Rome. (1972). *The limits to growth: A report for the club of Rome's project on the predicament of mankind*. Universe Books.

Miki, K. (三木, 清). (2023). *Logic of creative imagination* (構想力の論理) (Vol. 1&2). Iwanami (岩波書店).

Morishita, A., & Tateoka, Y. (2012). Japanese style management and long-established businesses. 経営情報学会. *Japan Society for Management Information, JASMIN*, 181–184. https://doi.org/10.11497/jasmin.2012s.0.181.0

Nikkei BP. (2022). https://consult.nikkeibp.co.jp/shunenjigyo-labo/survey_data/I1-06/

Shibusawa, E., & Moriya, J. (2012). *Analect and abacus as guiding posts* (現代語訳渋沢栄一自伝:「論語と算盤」を道標として). Heibonsha (平凡社).

Sone, S. (曽根, 秀 一). (2019). *Mechanism of longevity of Shinise companies: Business system of the Temple and shrine construction companies* (老舗企業の存続メカニズム). Chuokeizai Publishing (中央経済社).

Swarbrick, M. (2006). A wellness approach. *Psychiatric Rehabilitation Journal, 29*, 311–314. https://doi.org/10.2975/29.2006.311.314

Tamiya, H. (田宮, 寛). (2022). *Companies that never closes: Rules of companies running business for a hundred years* (何があっても潰れない会社: 100年続く企業の法則). SB Creative (SB クリエイティブ).

Tanaka, M. (田中, 真 澄). (2015). *What is the difference between over 100 years continuing companies and others?* (百年以上続いている会社はどこが違うのか?). Chichi Publishing (致知出版社).

Teikoku Databank. (2022). https://www.tdb.co.jp/report/watching/press/p221003.html

Teikoku Databank. (2023). https://www.tdb.co.jp/report/watching/press/p231012.html

Totman, C. D., 1934–. (2005). *A history of Japan*. Blackwell.

UNESCAP. (2012). *Green growth, resources, and resilience environmental sustainability in Asia and Pacific*. https://www.unescap.org/sites/default/files/Full-report.pdf

Uzawa, H. (宇沢, 弘文). (2009). *Economic analysis of social common capital*. Cambridge University Press.

Yokozawa, T. (横澤, 利昌). (2012). *A research on Shinise companies* (老舗企業の研究). Seisansei Press (生産性出版).

Maki Sato received a PhD in global studies from the University of Tokyo. Maki is a fellow at The New Institute based in Hamburg, Germany. She also serves as a lecturer at Rikkyo University and a fellow at the University of Tokyo East Asian Academy for New Liberal Arts (EAA). Her research field is environmental philosophy. Based on her experience as a consultant for climate change and green growth for the UNESCAP, the Japanese ministries, and business sectors, her interest lies in achieving human flourishing in harmony with the living and nonliving existence of the world and how philosophy can help achieve such aims. Her present research interest expands to the philosophy of religion (Shinto and Buddhism) and cognitive philosophy through AI and robotics to question what it is to be human, not from a dichotomic perspective that segregates humans as superior beings, but from triangular aspects of inherent humanness: consciousness, embodiment, sound (music and language), and their interplays.

Part II
Transitional Chapter I

Paying Tribute to the Complexity of Human Beings: A Business Narrative Harmonizing the Well-Being of Individuals and Society

Marisa Mühlböck and René Schmidpeter

Abstract Sustainability might sound like an extensively used buzzword to some. In the business context, companies which report their sustainability activities are sometimes accused of greenwashing. Although research from various disciplines has shown that the image of *homo economicus* is a myth and a large number of enterprises have recognized the importance of the issue and developed their approaches further—from performing corporate social responsibility measures as an "add-on" to more integrated concepts that do not perceive purpose and profits as mutually exclusive objectives but an integrated "add-in"—the narrative of business as a profit-driven inhuman sector seems to prevail in society. This is of concern because narratives influence our beliefs, emotions, and behavior. In this article, we illustrate that adhering to the story of *homo economicus* is harmful in multiple ways. Drawing from insights from fields such as business ethics and interpersonal neurobiology, we take a particularly closer look at its negative effects for the well-being of individuals, by exploring different concepts of the self. At the same time, those concepts provide new perspectives such as the relational aspects of an individual's mental life. By integrating those insights, the paradigm of "all one" emerges and with it a new narrative of and about business which surpasses the mere alignment of individual and societal well-being but rather strives to harmonize the well-being of individuals in a manner that the well-being of society grows out of it.

Keywords Sustainability in business · Homo economicus myth · Narratives in business · Business paradigms · Corporate social responsibility (CSR) · Sustainable business transformation · Business ethics · Interpersonal neurobiology · Relational self concept · Dealing with complexity · Well-being of individuals · Collective well-being · Human flourishing

---

M. Mühlböck (✉)
Independent Researcher, Vienna, Austria
e-mail: research@marisamuehlboeck.com

R. Schmidpeter
Bern University of Applied Sciences, Bern, Switzerland
e-mail: rene.schmidpeter@bfh.ch

M. Mühlböck (ed.), *Sustainable Transformation and Well-being*, CSR, Sustainability, Ethics & Governance,
https://doi.org/10.1007/978-3-031-75566-8_5

1 Introduction

The business sector plays an important role in almost every sphere of our lives. It is firmly embedded in societal institutions and provides consumers with a wide variety of products and services. As a driver of innovation and by means of efficient use of resources, it has without doubt increased prosperity in many countries throughout the world (Pinker, 2018). At the same time, business activities have also caused negative effects, e.g., by consciously and unconsciously exploiting human and natural resources (Meadows et al., 1972). Thus, in its multifaceted role and due to its broad scope of action, business has a major impact in regard to our individual but also collective and planetary well-being.

Today, in many areas of the world, sustainability has been recognized as a key factor for creating well-being and ensuring that future generations will be able to enjoy it too. Along with development in society in many ways, business has undergone a sustainable business transformation (see Sect. 2), first mainly by adopting sustainability through various "add-on" measures, whereas today the focus is shifting more and more to integrating it into business models and strategies ("add-in"). Yet, the paradigm of the *homo economicus* as a guiding principle for our present narrative of and about business seems to prevail.

In this chapter, we shed light onto the key influence that narratives exert on both the behavior of business leaders and how society perceives business. We point to the manifold negative effects that wrong assumptions about economic agents' actions produce (Sect. 3) and present zeitgeisty images of the self, which help us unveil the true complexity of human beings. Drawing from that investigation, we sketch the concept of the *harmonic self,* which leads to creating the "all-one" paradigm as a basis for a new narrative that, we believe, better depicts reality and at the same time contributes to more individual and collective well-being (see Sect. 4).

In Sect. 5, we take a first step to applying the new narrative to business practice by advancing Auster and Freeman's (2013) concept of *poetic organizations* to the framework for *harmonic organizations.*

By recapitulating our investigations, the need to introduce a new inter- and transdisciplinary research field becomes apparent that, in line with the new narrative, further explores a new role for business in society. Section 6 closes this chapter with an outlook on such an endeavor.

2 The Long Shadow of *Homo Economicus*

For more than a hundred years, the paradigmatic status of *homo economicus* has been questioned (Kargol-Wasiluk et al., 2018), and over the course of the last 20 years, it has been branded a myth (Papadogiannis, 2014; Schultz, 2013), or dead (Gowdy & Iorgulescu Polimeni, 2005). Its image of acting rationally and solely for its self-benefit at all times has been critiqued as oversimplified (Fehr & Fischbacher,

2002) and even as a paradigm that "may be harmful to society in the real world" (Kargol-Wasiluk et al., 2018, p.33) or, in other words, may negatively impact the well-being of society and the planet. According to Schüz (2023), those who ignore that the principle of economic action is based on exchange, in which the interests of every exchange partner have to be integrated, serve only their own benefit and have the tendency to exploit the other.

Section 2.1 showcases new research fields which have been developed as a contrast to mainstream economics theories, which have focused on rationality, individualism, and the explanation of the economic world and on predictions of human behavior through equilibrium models, based on linear mathematical calculations (Laybourn-Langton & Jacobs, 2018; Kargol-Wasiluk et al., 2018). At the same time, management approaches have in reality undergone a transformation toward a stronger focus on integrating sustainability into business practices. Section 2.2 presents an overview of the key milestones in sustainable business transformation, first, by focusing on framework conditions (Sect. 2.2.1) and, second, by outlining the development at the operational business level (Sect. 2.2.2).

2.1 Science Contradicts the Old Paradigm

More than 20 years ago, Ben-Ner and Putterman (1998) advocated for a better understanding of economic behavior and organization theory with their book *Economics, Values, and Organization*. Supported by two Nobel laureates (Amartya Sen wrote the foreword and Douglass North an epilogue), they delivered impulses and ideas to go beyond rational choice theory and *homo economicus*. Anderson (2000) seconded that endeavor and encouraged scientists in economics to see new insights from other disciplines as a chance to better comprehend "the actual causes of social phenomena" (p. 200) rather than perceiving them as a threat to existing models.

Heterodox economics approaches such as behavioral (e.g., Fehr & Fischbacher, 2002; Kahneman, 2003; Tversky & Kahneman, 1974, 1981) or experimental (e.g., Smith, 1962) economics have proposed different perspectives acknowledging the fact that "economic behaviour is multifaceted and context-dependent" and that "the economic man does not always seem to be a self-interest-maximising being and, what's more, economic choices are often made on the basis of various motives, including emotions, social norms and values" (Kargol-Wasiluk et al., 2018, p. 34). Kaletsky (2009) goes as far as stating that "economics today is a discipline that must either die or undergo a paradigm shift—to make itself both more broadminded, and more modest" and that "[i]t must broaden its horizons to recognise the insights of other social sciences and historical studies" (para. 21). Such comprehensive academic critique points to the requirement of integrating new knowledge from other scientific disciplines in an open and transdisciplinary discourse about the future of economics and management science.

One such academic discipline which underlines that selfishness is not the central trait of human nature is psychology, in particular positive psychology (Seligman & Csikszentmihalyi, 2000; Harré, 2011). As a subfield of psychology, it has produced new conceptual frameworks and instruments to research and measure positive human functioning and flourishing.

Another scientific field which has been fueling the ongoing debate on the necessity of a paradigm shift in economics is complexity theory, which is generally based on the assumption of "the presence of a high number of interacting heterogeneous agents, the absence of any global controller, the presence of adaptation by learning and evolution, and the dominance of out-of-equilibrium dynamics" (Fontana, 2008, p. 4). Fontana (2008) underlined that "the theoretical and analytical tools of complex systems analysis have been seen by an increasing number of scholars as a promising route toward overcoming important weaknesses of the traditional approach to the representation and understanding of economic facts" (p. 2).

Consequently, the need for an interdisciplinary approach in economic behavior analysis becomes evident. Behavior economics already draws on psychology, sociology, and even neuroscience (Kahneman & Tversky, 1979; Simon, 1955) when showing that "deviations from self-interest have a fundamental impact on core issues in economics" (Fehr & Fischbacher, 2002, p. C1). Neuroeconomics is another interdisciplinary field which adds an additional layer by trying to explain choice behavior in the terms of brain processing (Glimcher & Fehr, 2014). It also seeks to shed light onto the complex relation to the emotional life of our brain in this context (Phelps, 2009; Lempert & Phelps, 2014). And ethics serves as a cross-functional umbrella which refers to all business activities in this matter (Schüz, 2019, 2021). In this article, we therefore take that need for a holistic and transdisciplinary perspective into account and build our investigation on this foundation.

2.2 *Sustainable Business Transformation in Practice*

Besides science, reality has shown that many entrepreneurs start businesses or make economic decisions which do not necessarily make strictly economic sense. This is why, for example, Sarasvathy (2001) developed the theory of effectuation, which explains the process that entrepreneurs follow to create new ventures, how they think about problems, and how they go about solving them. The logic of effectuation states that goals do not come first but are defined after assessing the available resources and the relationships they are enmeshed in. Based on the available resources, entrepreneurs, as "pilots in the plane," create their own opportunities by their actions.

One could also argue that entrepreneurs start their businesses because they see the world differently and want to remake or influence it through their actions. In general, they don't start companies and try to make as much money as possible, but they rather materialize their choice of life or set of values, attempt to create value, or shape their identity through their business activities (Spence & Schmidpeter,

2003). What certainly plays a role in influencing their behavior are external framework conditions such as rules or regulations but also societal expectations, which we will look at in the following section.

2.2.1 A Brief History of Sustainable Business Transformation

A sort of code of conduct which exceeds the corporate self-interest can already be traced back to the Middle Ages and was established mostly in Italy and the Hanseatic Cities through the model of the honorable merchant (Schwalbach & Klink, 2015). Klink (2008) describes its mechanism as a measure to secure the merchants' reputation. Compliance to the rules was judged by the citizens (in general each city's population was below 2000 inhabitants), and breaking the rules could, in the worst case, lead to social exclusion or appropriate punishments, respectively (Braudel, 1990). Kehnel (2022) showcases inspiring examples of sustainable entrepreneurship and management in the premodern era, and economist John M. Clark called for social responsibility in addition to economic responsibility as early as 1916 (Clark, 1916). Scientific discussion about a businessman's responsibility gained speed in the USA in the 1950s. Many refer to Howard R. Bowen's (1953) book *Social Responsibilities of the Businessman* as a key publication. In the late 1960s, the role of an entrepreneur as an active stakeholder in society aroused scientists' interest, and eventually empirical research on this subject began in the 1970s (Loew et al., 2004). This also marks the emergence of the term corporate social responsibility (CSR), even if its meaning and definition has been debated since then (Carroll, 1999). In 1972, the Club of Rome published its pathbreaking "The Limits to Growth" report, which is until today seen as a key driver for the global sustainability movement (Meadows et al., 1972).

In the 1980s and 1990s, the discussion on the subject intensified in the scientific community but also beyond, not least because of the growing environmental movement. A milestone in this time frame was the publication of the "Brundtland Report" in 1987 (Brundtland, 1987). It defines sustainable development as development which "meets the needs of the present without compromising the ability of future generations to meet their own needs" (p. 16). From that, Barbier (1987) derived his concept of "sustainable economic development," which aims at not only maximizing the goals of the economic but also those of the social as well as the biological and resource systems. Elkington (1998) later published his concept of the triple bottom line (TBL), which has been challenging companies to expand their reporting accordingly. Schüz referred to the equal importance of the economic, social, and ecological dimensions in his concept of "triple corporate responsibility" (1999), which he later expanded to the concept of "sustainable corporate responsibility" (SCR) (2023). The related model illustrates different degrees of sustainability and scopes of responsibility within its three dimensions identifiable with the aid of his SCR checklist.

Parallel to the development of responsible business, the term corporate citizenship (CC) was introduced in the 1980s and has since then been discussed in the public and scientific community (Crane et al., 2009). A business ethics boom started in the USA in 1991 with the "Federal Sentencing Guidelines," which allow judges to reduce punitive damages by up to 90% when the accused company has done everything to educate its employees in ethical awareness (Izraeli & Schwartz, 1998). In 1999, the former United Nations Secretary-General Kofi Annan initiated the "Global Compact," which is a call to companies worldwide "to align their strategies and operations with ten universal principles related to human rights, labour, environment and anti-corruption" (Kingo, n.d., para. 4).

However, serious debate on corporations' social responsibility in Europe has only started in the twenty-first century. This is mainly due to the more rigorous legal and institutional framework in many European countries when compared to the USA. As such, European companies are more frequently "forced" to comply (e.g., with the legal obligation to protect employees' rights), whereas their US counterparts take on such a responsibility voluntarily (Schmidpeter & Palz, 2008). Such country-specific differences are well mirrored and explained in Matten and Moon's implicit/explicit CSR concept, which offers a dual view on CSR by differentiating between "explicit CSR," "corporate policies that lead companies to assume responsibility for some interests of society," and "implicit CSR," "a country's formal and informal institutions through which the corporations' responsibility for society's interests are agreed and assigned to corporations" (Matten & Moon, 2005, p. 342).

At the EU level, the term CSR was officially mentioned in 2001 for the first time. A European framework for the social responsibility of enterprises was communicated in a green paper (European Commission, 2001a) followed by the mentioning of CSR as an important economic contribution for sustainability in the EU strategy for sustainable development (European Commission, 2001b) and the establishment of a multi-stakeholder dialog in 2002 (European Commission, 2003), as well as a European Alliance for CSR in 2006, a network of business community members which is still active and aims to promote CSR across Europe (BusinessEurope, n.d.). In 2011, the European Commission published a renewed strategy for CSR including a new definition referring to CSR as "the responsibility of enterprises for their impacts on society" (European Commission, 2011). Consequently, entrepreneurship has increasingly been facing expectations to transform from a mere business orientation to emphasizing to a much greater extent social impact creation and sustainable development (Smith & Sharicz, 2011). Thus, the sustainable development movement has been recognized as a major driver in the development of sustainable entrepreneurship as a new field in entrepreneurship (Shepherd & Patzelt, 2011).

In 2010, the *Journal of Business Venturing* published a special issue titled "Entrepreneurship for Sustainable Development" (Daneke et al., 2010). In 2015, the United Nations presented the 2030 Agenda for Sustainable Development, and all member states agreed to adopt them. The agenda includes 17 Sustainable Development Goals (SDGs), which focus on prosperity, people, and the planet, as well as economic, social, and environmental objectives. They were all defined to fulfill the global mission "to leave no one behind" (United Nations, 2024). Overall,

SDGs can be seen as a directive for all societal stakeholders and as such certainly also for businesses. The United Nations has also intensified their promotion to connect the SDGs with the ESG standards, a framework which refers to company behavior focusing on three pillars: environmental, social, and governance (ESG). First mentioned in the "Who Cares Wins" report (compiled by a group of banks and other investment firms upon invitation by the U.N. in 2004 (The Global Compact, 2004)) and the Freshfields report "A legal framework for impact: Sustainability impact in investor decision-making" in 2005 (Freshfields Bruckhaus Deringer, 2005), the set of standards has gained broad acknowledgement since then.

In 2020, the World Economic Forum and Big Four accounting firms released a white paper with standardized metrics for companies reporting on their ESG progress (World Economic Forum, 2020). At the end of 2022, a European Union directive specified that EU companies and non-EU businesses operating in the EU will be required to make corporate sustainability disclosures in relation to their alignment with an EU ESG-related taxonomy and audit sustainability data (European Parliament, 2022). In 2023, again at the EU level, a provisional agreement was reached on a directive on corporate sustainability due diligence. The aim of this directive is to foster sustainable and responsible corporate behavior and to incorporate human rights and environmental considerations in companies' operations and corporate governance, including their entire value chains (European Parliament, 2023). Figure 1 provides a historical overview of the above described development.

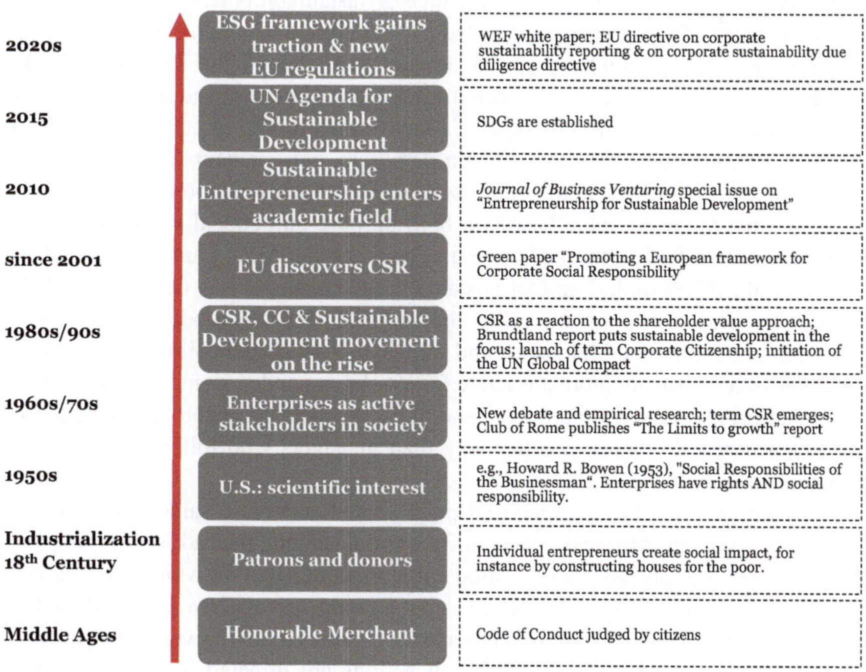

Fig. 1 The emergence of sustainable business transformation (based on Mühlböck, 2011)

2.2.2 From "Add-On" to "Add-In"

As early as the end of the 1970s, management professor Archie B. Carroll contributed to the debate on a company's social responsibility by identifying four types (economic, legal, ethical, and philanthropic responsibility), based on which he further developed his CSR pyramid, which became popular in the 1990s (Carroll, 1991), but was later subjected to a critical revision by himself (Carroll, 2016). Yet, for a long time, the concept of sustainability has rather been a mere "add-on" to business administration and management, in the sense of scattered CSR measures or the attempt to fulfill external rules, regulations, or social expectations by aligning social, environmental, and economic goals. One concept that corresponds to the idea of alignment and balancing those three pillars is the triple bottom line concept which has certainly made a valuable contribution to the advancement of sustainable business transformation. However, as long as a company just implements sustainability measures because it is externally forced on it, this might still lead to greenwashing and box checking (Elkington, 1998).

In recent years, sustainability has more and more become an integrated management paradigm which has affected management at different levels. This is reflected by the fact that all three dimensions of sustainability (economy, ecology, and social) have been embedded in innovative business concepts and management strategies as well as at the business administration level in various management disciplines such as finance, human resources, marketing, or supply chain management (Hahn, 2022; Schmidpeter, 2013–2024). As such, the idea of sustainability has been transformed from an "add-on" to an, as we call it, "add-in" approach in which entrepreneurship, sustainability, and profitability are no longer contradictory but become mutually conducive if the "sweet spot" between the interests of business and society is identified (which is in line with Schumpeter's ideas in his theory of economic development; see Schumpeter, 1934). In such a paradigm, entrepreneurs are successful if they align their interests and the interests of society. Such a comprehensive strategic process aims to create value based on both purpose and ethical values as well as economic interests. Recent studies (e.g., Danso et al., 2019; Durand et al., 2019; Hawn et al., 2018) underline that sustainable management has not weakened the competitiveness of business but rather the opposite. Business models and strategies which integrate the material issues and dimensions of sustainability have a higher risk-adjusted return, are more attractive for employees and customers, and are more resilient during crises. Entrepreneurs and managers who act according to their ethical perspectives do this in a better way than managers who follow a mere profit maximization doctrine.

That integration of entrepreneurial purpose and economic value has led to the development of new business concepts. To a certain extent, the "add-in" approach is covered by Porter and Kramer's shared value model (2011), in which they suggest that businesses move beyond CSR and gain a competitive advantage by including social and environmental considerations in their strategies. However, this concept is not yet a holistic approach in which all actions are directed toward a win-win situation for all stakeholders; rather, win-lose situations are still possible. In contrast, Freeman's stakeholder theory (1984) argues that in order to be successful a company must integrate the interests of all its stakeholders, rather than merely focus on those of the shareholders, into their strategies.

An integrative concept that focuses on a company's impact on society is the IOOI (input, output, outcome, and impact) method. It is a framework that was developed as a measurement tool by Bertelsmann Foundation (Riess, 2010) and is based on the "theory of change," which is widely acknowledged by the philanthropic and social entrepreneurship sector (Weiss, 1995). The impact, understood as the long-term effect on the environment, society, and business of any entrepreneurial measure, is perceived as vital to success. The outcome is defined as intended changes at the systemic level (e.g., implementation of a new mobility system through car-sharing platforms and new driving systems). Based on the intended impact and outcome, the necessary output can be defined and calculated using standard financial KPIs plus nonfinancial KPIs. The input describes what resources must be invested to reach the intended output. That is a process which fosters a way of thinking that encourages us to look at the present from the perspective of the future. It designs the overall strategic process from the long-term effects on broader society (impact), via its effects on systems and stakeholders to the output which is the directly measurable effect in terms of products and services. In the past, this model was mostly used at the project level. Yet, as a holistic concept, it could guide decisions in all business dimensions of a company (Schmidpeter & Funk, 2023; Schüz, 2021).

The underlying idea of all those concepts is a strategic innovative value creation process based on values and purpose. The narrative that comes with it is one of perceiving purpose and profit as two sides of the same coin. Instead of reducing everything to the one-dimensional perspective of money, the relevant questions for entrepreneurs are rather as follows: How to unfold the relationship with the stakeholders over time? What role do I play in the well-being of my stakeholders? And what sort of value do I produce for them? That narrative shift has already entered today's business sphere, where companies such as Toyota or industrial machinery manufacturer Barry-Wehmiller have clearly stated and integrated their purpose into their business strategies (Freeman et al., 2020). This is good news, since even though we do not postulate that the old ideas of economics are unimportant, the idea of stakeholders and purpose versus shareholders and profits is just a false narrative.

How does the "add-in" paradigm eventually manifest itself at the practical business level? As pointed out above, in order to be really integrative, such a thinking has to affect every business dimension and needs to be embedded in every entrepreneurial decision, e.g.:

- Human resources: To build a business environment that enables creative strategy development and implementation, the entrepreneur needs to foster an open and transparent business culture which is based on shared values and fair working conditions as well as a diverse workforce. From this perspective, diversity and inclusion play an important role in creating an inspiring workplace and promoting creativity and inclusive decision-making.

- Product development: Studies show (González-Viralta et al., 2023, Haba et al., 2023; Iqbal et al., 2023, Waqas et al., 2023) that green products and production processes are becoming important to the customers, which are sociologically seen as group of LOHAS (Lifestyle of Health and Sustainability) (Helmke et al., 2015). Thus, integrating the broad perspective of different stakeholders is key for the entrepreneur to design new products and develop innovative solutions for ever-faster changing markets. Addressing the societal and environmental challenges as well as developing new business models which create value for both society and owners/shareholders becomes key for sustainability with an "add-in" perspective.
- Finance and controlling: As financial institutions and investors are more and more interested in sustainability performance as well as the impact of the business on its environment (see, e.g., Sect. 2.2.1 regarding the growing importance of ESG in this context), the need for an integrative controlling perspective is increasing. Entrepreneurs need to make the true value creation of their business transparent to stakeholders. This triangle of strategy, innovation, and value creation drives entrepreneurship and business transformation based on a new and more holistic business narrative going far beyond the one-dimensional *homo economicus* paradigm.

Figure 2 provides an overview of the sustainable business transformation from an era of balance and alignment ("add-on") to integration ("add-in").

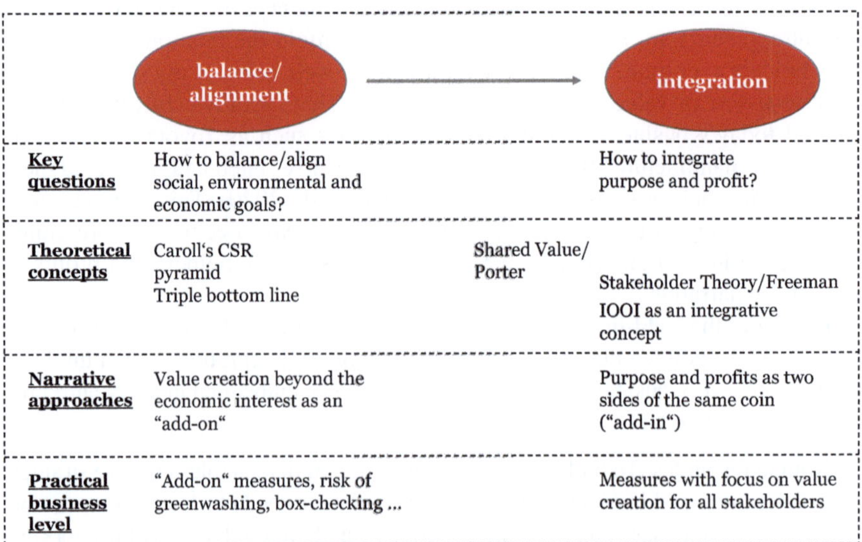

Fig. 2 From balance to integration

3 False Beliefs and Its Consequences

Summing up our previous investigations in theory and practice which underline the myth of *homo economicus*, the question remains why the image of business as a profit-driven inhuman sector still prevails: Deloitte's worldwide Gen Z and Millennial Survey 2023, which collected and analyzed data across 44 countries, shows that fewer than half of the respondents believe business has a positive impact on society (Deloitte, 2023). According to a 2023 Gallup poll, small businesses are the most trusted institutions in the USA (according to 65% of the respondents), but a majority of Americans view large businesses as self-serving and self-dealing with only 14% of respondents saying that they have a "great deal" or even "quite a lot" of confidence in big business (Saad, 2023). This shows that businesses, or at least large companies, are still perceived as purely profit-maximizing and self-interested, which is in line with the neoliberal paradigm and the narrative in many business textbooks.

Analyzing the previous paradigm shift to the neoliberal order, Laybourn-Langton and Jacobs (2018) explained its emergence and manifestation by disaggregating it into three levels:

> a. its intellectual and academic underpinning, particularly within economics; b. the discourses and narratives through which it was expressed in the wider public domain; c. the political policies and processes – notably the elections of governments – which enabled it to be implemented. (pp. 3–4)

While, as shown above, at least some schools of thought in the broad scope of economics have already dropped the illusionary paradigm of *homo economicus* and it can thus be claimed that the paradigm shift in economics is underway, it is the old narrative that seems to prevail in the broader society.

3.1 Why Narratives Matter

Why is this of concern? Narratives influence our beliefs or how we perceive ourselves, others, and the world in which we live (Collier & Tuckett, 2021; Breithaupt, 2022). They also affect us emotionally (Rutledge 2016; Dunlop et al., 2008). Those beliefs, perceptions, and emotions then influence the expectations and behavior of economic agents. Collier and Tuckett (2021) even claim that "narratives can be reset by policy communication" and "therefore have considerable potential to become the coordinating device to stimulate investment, analogous to the role that prices play in conventional analysis" (p. 4). In the context of sustainability and the future of our planetary health, according to Coulter et al. (2019), more shared future-oriented change narratives in society are required to realize proactive climate change adaptation measures. Roos and Reccius (2023) also stress the importance of narratives in relation to transformative and future thinking and thus for a sustainability transformation.

Which consequences are thus to be expected if we adhere to the narrative of mainstream economics? We can investigate this from two perspectives—from inside and outside of the microeconomic decisions-making sphere. First, even if the paradigmatic image of *homo economicus* at its finest is false per se, an entrepreneur or business leader who identifies with it might, as a result, feel drawn to focus on its narrow world in the short-term interest of the firm or even in their individual hedonic goal pursuit (i.e., remember Michael Douglas' award-winning role as Gordon Gekko in the famous 1980s movie *Wall Street,* who became an icon for many[1]). Schüz (2021) suggests the term "robber economy" for such a type of egoistic self-enrichment at the expense of all exchange partners involved. It counteracts an "economy of fair exchange" that has been sought after since antiquity and was further developed by Adam Smith (2012). Decisions those, let's call them "misguided," business leaders make will most likely not be organically compatible with society's collective long-term interest and planetary health. That reasoning corresponds in particular to what Bénabou et al. (2018) define as a "moral narrative," which is:

> any news, story, life experience or heuristic that has the potential to alter an agent's beliefs about the tradeoff between private benefits and social costs (or the reverse) faced by a decision maker, who could be himself, someone he observes, or someone he seeks to influence. (p. 2)

All that underlines that if we want business leaders to act ethically, the narratives which surround them should reflect that.

Let's now turn to a stakeholder's perspective outside of the microeconomic decision-making sphere. If such a person abides by the narrative of mainstream economics, they might have a rather negative image of business accompanied by a good portion of distrust. One might argue that society's reluctant stance against business could be based on negative personal experience, which is a legitimate argument. However, taking a closer look at the results of the 2024 Edelman Trust Barometer (Edelman Trust Institute, 2024), a gap is revealed when people are asked about their trust in business overall and the trust in their employer. Whereas trust in business overall reaches 63%, trust in employers increases by 16 percentage points to 79%. Thus, it seems that people have a more positive experience in their private relation to business, but when asked

[1] In a *Los Angeles Times* article screenwriter Stanley Weiser states: "As the years have gone by, it's heartening to see how popular the film has remained. But what I find strange and oddly disturbing is that Gordon Gekko has been mythologized and elevated from the role of villain to that of hero." The power of the narrative becomes even clearer when he reports "a typical example" of a business executive or a younger studio development person: "The movie changed my life. Once I saw it I knew that I wanted to get into such and such business. I wanted to be like Gordon Gekko" (Weiser, 2008).

the more abstract question about their overall trust in business to do what is right, fewer people have a positive image in their minds.[2] This appears to be a relevant gap of perception which needs to be investigated further to better understand the role of business in society. Research on the importance of trust as corporate capital has revealed the many effects at the microeconomic level such as higher profitability, growth, and sales on the one hand (Lins et al., 2016) and higher transaction, monitoring, and sanctioning costs caused by the absence of trust on the other hand (Kwon et al., 2012).

At the same time, the impact of distrust at the macroeconomic level has been declared every bit as undesirable for the economic performance of nations (Knack, 2001). Scientists investigating the field of social capital have underlined the importance of trust as a major factor for economic development and a nation's prosperity (Fukuyama, 1996, 2001; Covey & Merrill, 2011; Covey, 2022). Thus, it seems that old theoretical descriptions of business as self-interested and profit-maximizing undermine trust in business in general.

As such, most of the present business theories seem counterproductive in a sense that they influence reality in a negative way. This influence of theory on reality is often neglected in positivistic sciences. In particular, in social sciences, the ontology and epistemology chosen influence the results on empirical research (see arguments from critical realism, e.g., Archer et al., 1998; Bhaskar, 2014). Thus, the narrative and the normative assumptions on which the research is based have far-reaching consequences on the research results as well as on the behavior of humans in reality (see, e.g., reflexivity theory; Soros, 2013).

Summing up, not letting go of the old paradigm will be harmful in multiple ways, by (a) leading to wrong economic predictions at the microeconomic as well as the macroeconomic level, (b) steering human behavior in a misleading direction to exploitation of humankind and nature, and thus (c) precluding or at least decelerating a sustainability transformation. A deep dive into those aspects should be the subject of separate research endeavors. In the scope of this article, we will argue that the old paradigm further produces (d) negative effects for the well-being of individuals, since our sense of self is tightly connected to the narratives we adhere to. We will explore different concepts of the self that could serve as the basis for a new narrative of and about business that at the same time supports a sustainability transformation and pays tribute to the complexity of human beings.

---

[2] Even if we take into account that it is not solely business institutions who employ people, we believe that the percentual gap with this item is not due to employers in other sectors, since the percentages measuring trust in NGOs (59% of the respondents), government (50%), and media (50%) are below the value reached by business overall (62%).

3.2 *Narratives and Concepts of the Self*

The self and its related "phenomena" have been extensively researched in social science, in particular in psychology (Thagard & Wood, 2015). In this context, narratives have been found to serve the shaping of one's identity or view of the self. Exploring the different functions of narratives, Schiff (2012) points out that:

> [t]hrough the enactment of stories, we learn about the basic facts of our existence: what a self is, the roles and desires of others, how the world works, the meaning and goals of life. These stories are resources for understanding who we are and the meaning of our existence. (p. 41)

In philosophy, the development of concepts of the self can be traced back to the Greek philosophers, in particular Plato and Aristotle (1987), as well as to Eastern philosophy such as Buddhist philosophy (Dimkov, 2020; Oh, 2021). It would go beyond the scope of this chapter to list further concepts of relevance in this context. Yet, we would like to refer to Salvatore Lavecchia's chapter in this book, who, based on Plato's work and the Socratic concept of midwifery, draws an image of the "true self" which "may be characterized as an eminently *dialogical self*, whose modality of consciousness, transcending any first, second and third person perspectives, is qualified by unegoistic freedom, that is by intrinsic openness and self-givingness towards the harmonious manifestation of other beings" (Lavecchia, 2025, p. 15). To him, such a notion of the self aligns with a sustainable eudaimonic well-being for both the individual and the collective. As a contrast, he describes the consequences of what he identifies as the current image of the self as follows: "A contrastive dualism between the self and the world is thus generated, which currently determines and configurates every field of knowledge and action" (p. 17).

4 Creating a New Narrative

Our previous investigations have underlined the urgency of creating a new narrative that better depicts reality. The narrative which emerged in the course of sustainable business transformation from "add-on" to "add-in" has left its impact within the business sphere through the strategic integration of purpose and values in all their business dimensions. Yet, we postulate that truly paying tribute to the complexity of human beings requires digging deeper. Therefore, we will next introduce two concepts of the self which we believe qualify as a basis for a new and even better narrative for the sake of the individual and the collective. We will first look at Freeman and Auster's (2011) concept of the *poetic self*, which is built on the image of an *authentic self*. Authenticity is depicted as a creative process where self-creation and community creation are intertwined. Then, we will draw on the field of interpersonal neurobiology (IPNB), which reveals insights about the relational aspects of our mind with major implications for how we perceive ourselves and others in the world and how we act on this.

4.1 Problems of Authenticity and the Poetic Self

A starting point for the development of the concept of the *poetic self* can be found in what Auster and Freeman (2013) call "problems of authenticity." We often perceive authenticity with a positive connotation, in the sense of authentic people being seen as trustworthy. Thus, one could argue that striving to implement a culture of authenticity in an organization might be a relevant factor in preventing greenwashing. Yet, this seemingly obvious correlation has to be handled with great caution. George (2003), for instance, shares practical experience with dilemmas and conflicts that come with ethical deliberations in the context of authentic leadership from practical experience. From a business ethics' perspective, Freeman and Auster (2011) identify several problematic aspects in the context of values as key factors for the idea of responsible leadership.

They see one problem of authenticity in the fact that creating a value fit through aligning individual and organizational values does not automatically benefit the well-being of society. Although a business concept that successfully integrates purpose and profit might be likely to correspond to the creation of positive impact for society, we cannot be sure about this. Still, the company might act authentically if a value fit is reached (see, e.g., the armaments industry or the mafia).

Furthermore, Freeman and Auster's (2011) main investigation of the problems of authenticity refers to the ethical theories of internalism and externalism. Both point to difficulties in being authentic, since "'[k]now thyself' is easy to say and hard to accomplish" (p. 19). However, their lines of reasoning offer two different perspectives, which Freeman and Auster demonstrate by reflecting the assumption "that acting authentically is simply a matter of knowing (Plato) one's values and then acting on them (Aristotle)" (p. 17). Along with internalists, they refrain from assuming that values are clearly discernible by individuals or organizations or that it is mainly values that define the self. At the same time, they recognize the challenges which externalists point out, viz., that human beings are influenced by psychological forces which might superpose values. Similarly, Schüz (2019) postulates "depth ethics" that analyzes one's own shadow side in business activities in order to dissolve such challenges through mindfulness for the well-being of the whole.

Suggesting a "middle ground" between the theories of internalism and externalism, Freeman and Auster (2011) propose a view of the authentic self that, on the one hand, acknowledges the fact that human beings are complex and, on the other hand, "takes into account the mutuality and paradox of recognition, and the permanent tension between self and other, autonomy and connection, which comes with mutuality" (p. 21). They argue that the definition and declaration of authentic values is a lifetime endeavor and as such see authenticity as a long-term, ever-changing process, "as a creative project, one where we strive to create a life that is imbued with the process of trying to live in an authentic way" (p. 16). For them, values can serve as a starting point for exploring an authentic self, but they further encourage engaging in that creative project and as such in a dialog with our past, our relationships with others, and our aspirations for the future.

To pay tribute to this more nuanced view, they introduce the notion of the "poetic self," which is based on Harold Bloom's (1997) idea of the "strong poet":

> who literally sees the world in a way that is different from others, but is also embedded in a number of communities. As it represents the intersection of our values, our past, our set of connections to others, and our aspirations, the idea not only refers to individual aspirations but includes also community aspirations. Thus, the poetic self embraces the idea of simultaneously creating self and community. [...] Authenticity becomes the project of finding this unique expression of our own humanity that takes account of both individual (and intrapsychic) and community (and inter-subjective) aspirations. (Freeman & Auster, 2011, p. 21)

Consequently, they see self-creation and community creation "as two sides of the same coin," where self is created "in part by creating connection, and as we create connection, we create self" (p. 22)—a consequence of mutual recognition and the permanent tension between self and other.

We will further explore the implications of such a "more nuanced view of authenticity and values" for the idea of responsible leadership in Sects. 4.3 and 5.

4.2 The Self from the Perspective of Interpersonal Neurobiology and Its Implications for Well-Being

Lavecchia's (2025) image of the *dialogical self* as well as Freeman and Auster's (2011) concept of the *poetic self* already take into account various relational aspects of the self and as such differentiate themselves from the image of *homo economicus*. That corresponds to some key insights from neurobiology, a discipline that has revealed the biological mechanisms by which the nervous system mediates behavior. In the 1990s, US psychiatrist Daniel J. Siegel gathered a group of 40 experts from different scientific fields such as neuroscience, psychology, systems theory, and relationship studies who together developed a framework called interpersonal neurobiology (IPNB). Over the course of this multidisciplinary debate, they demonstrated how the mind, the brain, and relationships integrate, what this means for our sense of self, and how that affects individual and collective well-being (Siegel & Drulis, 2023).

Consensus of what we define as the mind has long been missing. Siegel and Drulis (2023) point out that "[a]t a minimum, one reason we cannot replace the term 'mind' with the terms 'brain' or 'brain activity' is that our subjective sense of being alive is simply not the same as the electrochemical energy flow of the nervous system" (p. 10). Thus, even if mind processes are dependent on brain processes for their emergence in our mental lives, they are not the same. The IPNB expert group eventually came up with a definition in which they characterize the mind as "an emergent, self-organizing, embodied and relational process that regulates the flow of energy and information" (p. 12). With that they take into account that the emergence of the mind includes our relationships with people and the planet. In reference to the sense of self which is constructed by our mind, Siegel and Drulis (2023) thus suggest a systems view of mind in which:

> the individual sense of 'me' is not the whole story of who we are – who we are is also a relational 'we'. In spatial terms, the me is inner; and the we is inter. Inner and inter are each an aspect of the system of energy flow we are saying is the system of mind. (p. 10)

What implications does such a view of the self hold? First, there are major consequences at an individual level. Siegel and Drulis (2023) refer to the relevance of the relational aspects of our mind for our mental well-being, since "when the self is fractured, the psyche aches" (p. 5). At the same time, they locate further harmful implications at the collective level:

> A consilient perspective in science is that the disconnection inherent with the solo-self construction of modern culture may be the heart of the hidden symbol/concept/category of separation at the root of many of the major challenges to well-being for humanity, and for the natural world, at this time of the Anthropocene Era. (p. 22)

Siegel (2022) further points to the mathematical principles of complex systems when exploring the mind and takes into account that:

> a complex system is composed of interacting parts, often called nodes, and each node interacts with other nodes via their linkages. […] If a node functions as if it is the totality, that it belongs only to its nodal part in the system, it will behave in a disconnected way, interdependence will shut down, and the whole complex system will lose its ability to adapt and learn. Its self-organization toward harmony will be compromised, and instead the system will move toward chaos or rigidity. (p. 52)

Siegel and Drulis (2023) conclude by pointing to the importance of honoring differences and cultivating emotional linkages, through which further neural integration in a person is created. According to them, this process of differentiation and linkage can be called integration. They identified nine dimensions in which such integration could take place.[3]

4.2.1 Harmony as a High-Level State of Integration

The insights emerging from the field of IPNB strongly point to the benefits of an integrated brain, since the level of integration corresponds to the experience of well-being. In this regard, Siegel and Drulis (2023) underline that "neural and relational integration mutually support one another" (p. 28) and that "integration within

---

[3] Integration of consciousness (e.g., the various objects of awareness), bilateral integration (the two sides of the brain), vertical integration ("interoception"/the interior sensations of muscles, bones, and organs), memory integration (past, present, future), narrative integration (e.g., how we make sense of something and find meaning), state integration (layers of self-experience which involve our individual bodies, our connection with other people, and our being a fundamental part of nature), temporal integration (e.g., our longings for certainty, permanence, and immortality and the reality of life's uncertainty, transience, and mortality), and identity integration (e.g., to learn to adjust an identity lens as we realize the reality of our inner me, our inter we, and our intraconnected identity as "MWe"); more details on these nine dimensions can be found in Siegel and Drulis (2023) and Siegel (2012).

emerges from integration between" (p. 26). The path toward harmony can thus be walked by striving to more and more integrate at different domains.

The introduction of the notion of harmony in this context is interesting. In the sense of IPNB, it can be understood as "a high-level state of integration." If we transferred this concept to the business sphere, the path toward harmony might find a basis, for example, in integrating purpose and profit ("add-in" perspective) but would offer many more levels for further integration. From such a perspective, harmonization in the sense of sustainable business transformation could be described as the constant attempt to identify and honor the uniqueness of each part—whereby the definition of what qualifies as "a part" becomes a key task of business leaders—and to harmoniously link them, in a manner that well-being for people and the planet is created. This idea should be at the heart of the creation of not only more holistic but more harmonious business models in the future. A consequence of such an endeavor might be a sort of mindset shift, since through differentiation business leaders recognize the abundance of available opportunities instead of an alleged scarcity. Such a new mindset might initiate a focus shift from competitiveness to collaboration and mutual effort. Rather than prolonging the limiting thoughts of the past, the new thinking thus reflects the plenitude of future opportunities. The individual is not a victim of scarcity but an enabler of abundance, which changes the perspective from a pessimistic to an optimistic view of the world and the self. It reconnects the individual well-being and the collective well-being and thus promotes a type of authenticity that is beneficial for society and the planet.

If we develop Freeman and Auster's (2011) concept of the *poetic self* as a creative project (see Sect. 4.1) further and connect it with the idea of harmonization, we could describe authenticity as a creative project of integration by which the well-being of the individuals is harmonized so that the well-being of society and the planet grows out of it. We therefore postulate that such an idea will be helpful not only to develop a narrative which encourages leaders to more purposeful and impactful thoughts and actions but to strengthen trust-building toward business in society.

4.3 The Harmonic Self *and Key Ideas for a New Narrative*

Now, closing the circle and coming back to our investigation about business's image in society, we are ready to underline that the prevailing narratives are not suitable enough to integrate the many revelations on human behavior that different scientific fields as well as insights in the business reality have provided us with. Furthermore, acknowledging the insight that the autonomous and isolated self is an illusion is probably tantamount to the most radical attempt to reject the paradigm of *homo economicus*.

It comes with a fundamental shift of perspective from an individual to a relational or, as Lavecchia (2025) called it, a *dialogical self*. Taking all this into account, we will now discuss its implications for sustainable business transformation or, in other words, how business's view of its role in society and society's view of business could develop, so that a higher level of well-being is reached at the individual and the collective levels, and which elements could support the creation of a more suitable narrative for and about business in this matter.

Drawing from the previously presented concepts of the self, we propose the concept of the *harmonic self* as an anchor point for a new narrative of and about business. As a first conceptual foundation for the *harmonic self*, we suggest the following key elements.

The *harmonic self* acknowledges that:

- **Our inner world is complex. The mind is not only embodied but also relational:** As pointed out in Sect. 4.2, the mind goes beyond its embodied brain activity. Taking into account the relational context and, as such, how a person can harmoniously build, manage, and nurture their connections with other individuals is an essential preventive factor against "psyche ache" or in other words a key well-being enhancing factor. A new narrative based on a *harmonic self* must therefore abandon a limiting view of individuals or individual organizations as separate entities.
- **Complexity requires our capacity to differentiate and link:** The outer world is complex too. Drawing from different dimensions of integration, we attempt to propose corresponding steps of integration with regard to the business sector.
  - **Integrating how we relate to and connect with others:** Such a process of differentiation and linkage could happen, for instance, between various stakeholders and their perspectives, e.g., between a decision-maker and their team members, between an entrepreneur and their shareholders, or at the same time with environmental activists. This is how relational integration emerges "as individuals are honored for their differences and connections are established with respectful, compassionate communication" (Siegel & Drulis, 2023).
  - **Integrating the past, present, and future:** Drawing from Siegel and Drulis' (2023) memory dimension, we could then look at the entrepreneurial experiences of the "long-ago past," actual ones in the present ("near past"), and business aspirations for the future. This aligns with Auster and Freeman's (2013) inquiry framework for *poetic organizations,* on the one hand, to explore a company's historical values or in other words to study the past and its impact on where a company is today and, on the other hand, to examine its aspirational values, which means an organization's hopes and dreams and how they see themselves making a contribution to their stakeholders and society.
  - **Integrating different narratives:** Interestingly, Siegel and Drulis (2023) have identified the narrative dimension itself as an opportunity toward higher integration and thus higher well-being. Consequently, this should also be possible to realize by differentiating and then linking different business narratives (e.g., business is about profits only vs. business is about creating value and impact at all levels).

In alignment with those examples of integration, we propose that the overarching question for a *harmonic leader* in business is as follows: What or at which levels do we need to differentiate and link elements in order to harmonize our business relations and processes so that the well-being of society and the planet as well as everyone's individual well-being become a natural output of this inquiry?

- **Authenticity emerges only through a lifelong creative process and ongoing dialog about key values:** With the concept of *poetic organizations*, Auster and Freeman (2013) took the notion of the *poetic self* to the organizational level of a business and proposed an anchor point for organizations that strive to nurture conditions in which authenticity is likely to emerge. They do this by implementing a "joint process of ongoing dialogue and conversation" which they call values through conversation (VTC). VTC thus offers a more nuanced view of and debate about four interconnected types of values that take into account the complexity of organizational values beyond mere value fit. These types of values are (1) introspective values (values which might be revealed by exploring how individuals in an organization examine, contemplate, and reflect on their collective self and the world around them), (2) historical values (values which show how individuals in an organization study the organization's past and its impact on where they are today), (3) connectedness values (values which describe how individuals in an organization connect with each other, how they lead and follow, and their beliefs about effective processes and how they should work), and (4) aspirational values (values which reveal the organization's hopes and dreams and how they see themselves as making a contribution to their stakeholders and the greater good over time).

4.4 The "All-One" Paradigm as a Basis for a New Business Narrative

In combining the ideas which we have provided above and establishing an integrated picture of reality, eventually, a new story about the very nature of business emerges in which the notions of sustainability and responsibility are tied to the creation of harmony through differentiation and linkage, or integration.

Recognizing the fact that our mind is much more than mere brain activity and taking into account the relational aspects of our mental life leads us to a new narrative beyond the notion of our isolated selves. It's a story of connection, of us all not only sitting in the same boat when it comes to dealing with the consequences of our actions, but a permanent energy flow of interactions with others and with the planet that has major impact on our thoughts, feelings, and behavior. In a nutshell, it is neither an "add-on" nor an "add-in" but becomes a story of "all one."

That is not a tale of individuals blending into each other but rather an acknowledgement of the uniqueness of each human being. Everyone has their unique function but is at the same time connected to everything else, like a complex organism whose survival is based on the harmonious interplay of all its parts. Organizations, and as such businesses, consist of individuals and are consequently the integrated version of their uniqueness that can be expressed by the purpose of the company. The legal form represents only the vessel for it, but purpose materializes through human action. Those actions mirror how they really see their role in society and how they perceive themselves in the world and in relation to all other stakeholders.

To acknowledge that human beings are complex is consequently nothing to be damped down. It's rather something to be celebrated. And it is what shapes our whole life and thus also the business world. If we look, for instance, at the concept of passion, we do not find much alignment with the notion of *homo economicus*.[4] But how could Steve Jobs or any other famous entrepreneur have done what they did without passion? From a rational point of view and a mere profit-oriented perspective, he and his team should have given up a long time ago. In reality, practically nobody will start a business by trying to maximize shareholder value or by trying to do something that only makes economic sense. Most entrepreneurs do something despite the fact that it doesn't make any economic sense.

A business is a value-creating vehicle by nature. The challenge of such a creative process is to create enough value as well as the right kind of value in order to keep all relevant stakeholders on board. A value fit of the individual's values in a company and the company's own inherent values can only provide a basis for authenticity (see Sect. 4.1). What this process additionally requires is self-understanding and introspection, connecting with others, and the creation of aspirations. The above-introduced VTC process can only be performed with inspired, engaged, and energized people. As much as *poetic organizations* require a context that is nurturing and supportive of such a creative project, we believe the same to be true for *harmonic organizations*. We will propose a first framework for this concept in the next section.

Figure 3 outlines the path of sustainable business transformation from balance and integration to harmonization, or with a focus on the accompanying narrative—from "add-on" and "add-in" to "all one" where individual and collective well-being is harmonized so that value and well-being are created for everyone. We believe that in such a new paradigm the level of authenticity and with it trust will eventually increase organically.

What is now lacking are tools to materialize the paradigm of "all one" at the organizational level. We should keep in mind that the well-being of society is something we've got to create. It is something to be made, not found. Part of the problem is that we often mix that up. Someone, for instance, suggests that carbon credits will lead to a better society. And they might. Yet, the process of trying to force that onto individuals who don't see the point of it and are often in the grip of the old story might again lead to greenwashing and box checking.

[4] Interestingly, psychology has proposed two types of passion, obsessive and harmonious, "that can be distinguished in terms of how the passionate activity has been internalised into one's identity" (Vallerand, 2008, p. 2).

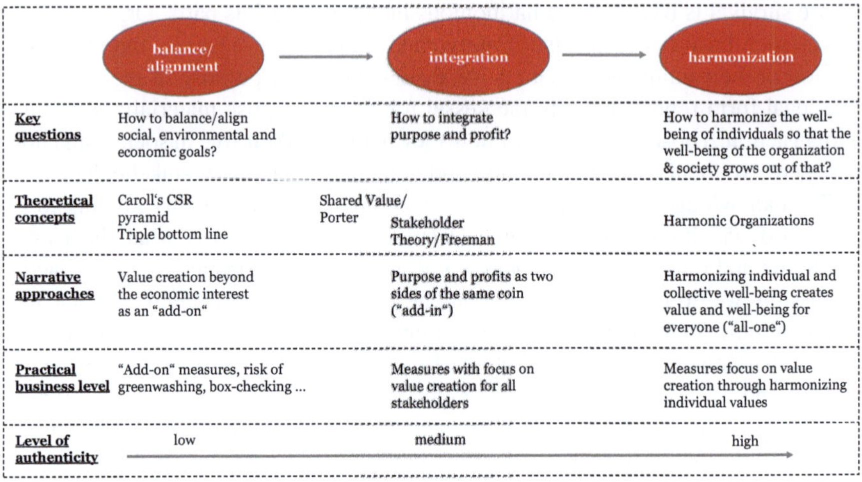

Fig. 3 From balance to harmonization and from "add-on" to "all one"

Consequently, a logical consequence of that new paradigm is the need to redefine the role of business leaders. With that comes a different set of responsibilities which in turn not only benefits the well-being of the organization and its stakeholders but also their individual well-being. It furthermore acknowledges the fact that leaders are human beings as everyone else, who struggle, who fail, who make wrong decisions, etc. At the same time, it underlines the fact that every unethical action, any activity that fosters separation, will in the end harm all of us. It thus encourages those who still try to "greenwash" to develop their sustainability activities further, and at the same time have those societal stakeholders who tend to blame and shame without thinking, reconsider before they strike out to provoke separation.

5 The New Business Narrative in Action

How could we eventually bring such a new narrative to life? And how could we build and successfully manage a *harmonic organization*? Building on Auster and Freeman's previously introduced VTC process, we attempt below to enrich their exploration of values by adding aspects and questions that align with the "all-one" narrative. It can be seen as a fine-tuning in the sense of what we call *harmonic organizations* (Table 1). While we believe that in organizations where different people work together to create value, a process such as VTC will be immensely beneficial to nourish authenticity, the focus of such an endeavor still lies on defining the organizations' value profile. What might still have gone unnoticed is the recognition of the individuals' states of values and well-being. It is therefore useful to acknowledge that human beings are quite complex and that "they are simultaneously rational, emotional, political, and social beings" (Auster & Freeman, 2013). All this

Table 1 Business leader's responsibilities in harmonic organizations—values and key questions (based on Auster & Freeman, 2013)

Business leader's responsibility	Value type	Key questions
Such a business leader would act as a space creator and space holder. They begin by creating such a space within themselves. As they acknowledge the well-being effects of the different levels of integration, they would make sure to improve their skills in this matter, e.g., through mindfulness practices. They could also dive into introspection with themselves about their state of differentiation and linkage for each level of integration.	Introspective values	What are my individual values? How do I live those values within the organization? How could I strengthen integration of the mind/further develop my skills in the nine domains of integration?
Then they create an integrative space for the individuals in the organization. This space allows for inner development, as described above. In such a process, factors such as the individuals' unique interests, innate drivers, and genuine talents are revealed. As those individuals recognize their unique expertise, they not only start to apply their newly acquired/acknowledged skills in their work context but as empowered experts become leaders themselves.	Connectedness values (introspective values)	How do we ensure that people working in/with our organization have a space for inner development? How do we encourage individuals to step into their own leadership role?
They create a space where a sense of belonging, an experience to be accepted for one's uniqueness, but at the same time being part of something bigger, is nurtured.	Connectedness values	How can we ensure that a sense of belonging is nurtured? How do we ensure that the uniqueness of the individual is not overlooked? How can we ensure that individuals feel safe to bring their uniqueness forward?

(continued)

Table 1 (continued)

Business leader's responsibility	Value type	Key questions
They study the past and the impact on where they are today and attempt to link their insights harmoniously.	Historical values	How do our interpretations of our organizational history converge and differ? How does our history shape the values, routines, structure, and processes we have today and what can we learn from it? How can we ensure an ongoing process of investigating and making sense of our history and connecting it to the present?
They finally strive to link the heterogeneous elements brought in by the individual stakeholders harmoniously and, out of that, develop a business model where the well-being of unique individuals (which is aligned with their subjective experience of life, their perspective on life and the world, and their agency) is harmonized so that well-being at the collective level grows out of it.	Aspirational values	How can we ensure a leadership context that provides suitable conditions for the harmonization of individual well-being? How do the unique hopes and dreams of individuals connect to the organization, our community, and the greater good? How can we use the uniqueness of the individuals for the benefit of the organization, our community, and the greater good? In this light, what is our purpose, and how do we create value for our stakeholders?
Overarching responsibility: They recognize that heterogeneity is not the problem but the solution to coping with today's high level of complexity and constantly strive to identify at which level they could cultivate the differentiation and linkage of different parts.		

must be taken into account even more when starting the process by differentiating individual parts or rather individual feelings, perspectives, talents, and dreams.

6 Conclusion and Outlook

The high degree of complexity that we face in today's world brings major challenges with it but at the same time opens up a huge window of opportunity. To us, the recognition of what science suggests, namely, that we are not rational and isolated but complex and related beings and that our individual and collective well-being is dependent on how we pay tribute to this, is a requirement for a positive outcome. That applies to all of humanity, yet it will be particularly relevant for the business sector, due to its broad scope of action. By means of their products and services but also as employers, companies interact and are intertwined with society, sometimes even worldwide. It will thus be crucial that business leaders take on the new responsibilities that come with those new insights. At the same time, society's willingness to pay them respect for such aspirations and actions will be vital too. A key step in this endeavor will be to adopt today's narratives accordingly. Our investigation has identified that it will be crucial to recognize that heterogeneity is not the problem but the solution to coping with today's high level of complexity if we succeed to integrate that plurality harmoniously. A prerequisite is thus that businesses constantly strive to identify at which level they could cultivate the differentiation and linkage of different parts such as thoughts and feelings, dreams and aspirations, ideas and narratives, and memories about the past and their impact on today. Connecting this overarching responsibility with the investigation of introspective, connectedness, historical, and aspirational values might provide a first framework on this path.

At the same time, we see the necessity to underline that we must not leave the development in such a direction with business leaders alone. A truly sustainable development will need to include large parts of the population in such an inner development movement, which is, for example, promoted by the Inner Development Goals initiative,[5] and will first and foremost require an integrative educational system from kindergarten to university. Still, what leaders of all kinds can do in that matter is to make sure that they provide spaces where their team members feel psychologically safe to develop at an individual level and are thus able to better cope with the complexity of our actual societal transformation. A good example in this matter is Lahey and Kegan's (2016) model of a "deliberately developmental organization (DDO)," which suggests that organizations create a workplace where the development and personal growth of their employees become part of their daily work environment (for more details, see the chapter by Von Leoprechting (2025) in this book as well as Welser (2025) as a practical example). Summing up, we hope that our first elements of a new narrative will inspire the academic debate and the practical progress of management and as such contribute to the well-being of entrepreneurs, society, and the planet.

What do the insights of our investigations now imply for further scientific work? To answer this question thoroughly, we should first clarify their implications for ontology. On the search for a new narrative for and about business, this chapter has looked at insights from different scientific disciplines. Our investigation about *homo economicus* has revealed economics' limited view of human behavior and its various negative effects. Yet, economics has produced valuable findings that have contributed to welfare in many ways. What we thus suggest in line with our findings about the power of differentiation and linkage is to apply that principle also when we outline what we believe to be a new inter- and transdisciplinary scientific agenda. From that, consequently, new questions emerge such as how to apply the principle of division of labor without alienation, how to realize international trade without exploitation, or how to establish markets as places of cooperation and fair exchange instead of destructive competition. A thorough effort of differentiation must thus include the identification of decisions and actions that contribute to the betterment of all stakeholders as well as those which still cause damage in whatever way. Only then, a correction can take place which aims at transforming the damaging behavior

[5] https://www.innerdevelopmentgoals.org/

in a way that nobody will be left behind. As business thus becomes a true enabler which also benefits the weakest, social progress gains speed.

The view of reality that comes with our investigations is therefore characterized by a strong heterogeneity in differentiation of the world. At the same time, taking account of the relational aspects that define our self and all parts of the world creates a reality of "all one," where only trying to understand disaggregated parts will not be sufficient.

As those assumptions draw from the logic of complex systems, we take a look at how complexity science is ontologically and epistemologically embedded. Yang (2024) gives an overview of present approaches and suggests critical realism as a suitable philosophy of science. He claims that:

> a satisfactory explanation of social complexity cannot be at the level of agential experience (the Empirical domain) or at the level of human and systematic events (the Actual domain) but needs to identify causal mechanisms (in the Real domain) of such events. (p. 1)

He argues that a critical realism approach "combines this depth ontology (that distinguishes the three reality domains) with epistemological relativism (that underscores the contingent character of knowledge claims)" (p. 1). Furthermore, Yang's critical realism approach suggests that at the epistemological level reducing complexity for human sense-making in the "Empirical" domain will not generate further complexity as it will not interfere with its ontological assumption of a stratified reality which consists of the three domains of reality.

Such an approach might provide the right ontological and epistemological background for further research that emerges in line with the new business narrative of "all one." It seems suitable to allow acknowledgement of the complexity in society, or in other words the outer world, but also the complexity of human beings, or the inner world of individuals. It supports a view of an open and ever-changing world. Thus, it refuses to describe or predict it but dedicates its research endeavor to generating an explanation of how reality works and how our actions might influence it. It takes the complexity into account but at the same time does not shy away from constructing scenarios that produce answers about expectations of the world. Yet, the definition of the most suitable research philosophy will need a more thorough investigation in the future.

We thus close this chapter with a first proposal of key questions which need to be answered to define a new research agenda that corresponds to the business narrative of "all one."

- Which new questions emerge?
- Which scientific disciplines could help us answer those and should consequently be integrated?
- Which stakeholder groups need to be taken into account as experts?
- Which research methodologies qualify to gain new knowledge?

Human beings are neither an "add-on" nor an "add-in," which as concepts both reflect a partial world. Rather, they "are" the world, and the world "is" them. Individual, society, and the planet are "all one." That understanding of interrelatedness and interdependency leads to a transdisciplinary fundament of economics and management science and as such to a "transdisciplinary responsibility" (Schüz, 2025) which, we believe, will foster the present sustainable transformation of mankind. It can be seen as another invitation to "exit from self-incurred immaturity," and, this time around, reason demands from us that we abandon the image of the purely rational *homo economicus*. By accepting that the world is one, we understand that we either win or lose together. Thus, we hope that the new business narrative which we have introduced in this chapter can lead to a better future for all. Welcome to a new enlightenment.

References

Anderson, E. (2000). Beyond homo economicus: New developments in theories of social norms. *Philosophy and Public Affairs, 29*(2), 171–200. https://doi.org/10.1111/j.1088-4963.2000.00170.x

Archer, M., Bhaskar, R., Collier, A., Lawson, T., & Norrie, A. (1998). *Critical realism. Essential readings*. Routledge.

Aristotle. (1987). *De Anima (on the soul)*. Penguin.

Auster, E. R., & Freeman, R. E. (2013). Values and poetic organizations: Beyond value fit toward value through conversation. *Journal of Business Ethics, 113*, 39–49. https://doi.org/10.1007/s10551-012-1279-5

Barbier, E. B. (1987). The concept of sustainable economic development. *Environmental Conservation, 14*(2), 101–110. https://doi.org/10.1017/S0376892900011449

Bénabou, R., Falk, A., & Tirole, J. (2018). *Narratives, imperatives, and moral reasoning* (Working paper 24798). National Bureau of Economic Research.

Ben-Ner, A., & Putterman, L. G. (1998). *Economics, values, and organization*. Cambridge University Press.

Bhaskar, R. (2014). *The possibility of naturalism: A philosophical critique of the contemporary human sciences (classical texts in critical realism)*. Routledge.

Bloom, H. (1997). *The anxiety of influence: A theory of poetry*. Oxford University Press.

Bowen, H. R. (1953). *Social responsibilities of the businessman*. Harper.

Braudel, F. (1990). *Sozialgeschichte des 15.-18. Jahrhunderts – Der Handel*. Kindler.

Breithaupt, F. (2022). *Das narrative Gehirn – Was unsere Neuronen erzählen*. Suhrkamp.

Brundtland, G. H. (1987). Report of the World Commission on Environment and Development: Our Common Future. United Nations General Assembly document A/42/427. http://www.un-documents.net/ocf-ov.htm

BusinessEurope. (n.d.). European Alliance for CSR – How to get involved. https://www.businesseurope.eu/european-alliance-csr-how-get-involved

Carroll, A. B. (1991). The pyramid of corporate social responsibility: Toward the moral management of organizational stakeholders. *Business Horizons, 34*(4), 39–48. https://doi.org/10.1016/0007-6813(91)90005-G

Carroll, A. B. (1999). Corporate social responsibility. *Business & Society, 38*, 268–293.

Carroll, A. B. (2016). Carroll's pyramid of CSR: Taking another look. *International Journal of Corporate Social Responsibility, 1*(1). https://doi.org/10.1186/s40991-016-0004-6

Clark, J. M. (1916). The changing basis of economic responsibility. *Journal of Political Economy, 24*(3), 209–229. https://doi.org/10.1086/252799

Collier, P., & Tuckett, D. (2021). Narratives as a coordinating device for reversing regional disequilibrium. *Oxford Review of Economic Policy, 37*(1), 97–112. https://doi.org/10.1093/oxrep/graa060

Coulter, L., Serrao-Neumann, S., & Coiacetto, E. (2019). Climate change adaptation narratives: Linking climate knowledge and future thinking. *Futures, 111*, 57–70. https://doi.org/10.1016/j.futures.2019.05.004

Covey, S. M. (2022). *Trust and inspire: How truly great leaders unleash greatness in others.* Simon & Schuster.

Covey, S. M., & Merrill, R. R. (2011). *The speed of trust. The one thing that changes everything.* http://ci.nii.ac.jp/ncid/BA85172096

Crane, A., Matten, D., & Moon, J. (2009). The emergence of corporate citizenship: Historical development and alternative perspectives. In G. Scherer & G. Palazzo (Eds.), *Handbook of research on global corporate citizenship* (pp. 64–91). https://doi.org/10.1007/978-3-531-91930-0_3

Danso, A., Adomako, S., Amankwah-Amoah, J., Owusu-Agyei, S., & Konadu, R. (2019). Environmental sustainability orientation, competitive strategy and financial performance. *Business Strategy and the Environment, 28*(5), 885–895. https://doi.org/10.1002/bse.2291

Daneke, G., Hall, J., & Lenox, M. (2010). Special edition: Entrepreneurship for sustainable development. *Journal of Business Venturing, 25*(5).

Deloitte. (2023). *2023 Gen Z and millennial survey. Waves of change: Acknowledging progress, confronting setbacks*. Deloitte.

Dimkov, P. (2020). The concept of self in Eastern and Western philosophy. In *Conference paper. 5th International e-Conference on Studies in Humanities and Social Sciences.*

Dunlop, S., Wakefield, M., & Kashima, Y. (2008). Can you feel it? Negative emotion, risk, and narrative in health communication. *Media Psychology, 11*(1), 52–75. https://doi.org/10.1080/15213260701853112

Durand, R., Paugam, L., & Stolowy, H. (2019). Do investors actually value sustainability indices? Replication, development, and new evidence on CSR visibility. *Strategic Management Journal, 40*(9), 1471–1490. https://doi.org/10.1002/smj.3035

Edelman Trust Institute. (2024). *Edelman Trust Barometer 2024*. Global Report.

Elkington, J. (1998). *Cannibals with forks: The triple bottom line of 21st century business*. New Society Publishers.

European Commission. (2001a). *Promoting a European framework for corporate social responsibility.* https://eur-lex.europa.eu/legal-content/EN/TXT/PDF/?uri=CELEX:52001DC0366&from=DE

European Commission. (2001b). *A sustainable Europe for a better world: A European Union strategy for sustainable development.* https://eur-lex.europa.eu/legal-content/EN/TXT/PDF/?uri=CELEX:52001DC0264&from=DE

European Commission. (2003). *EU multi-stakeholder forum on corporate social responsibility.* CSR EMS Forum, Publications Office. https://op.europa.eu/de/publication-detail/-/publication/72afe574-5cfe-11ea-8b81-01aa75ed71a1

European Commission. (2011). *Communication from the Commission to the European Parliament, the Council, the European Economic and Social Committee and the Committee of the Regions: A renewed EU strategy 2011–14 for Corporate Social Responsibility*. https://eur-lex.europa.eu/legal-content/EN/TXT/PDF/?uri=CELEX:52011DC0681&from=EN

European Parliament. (2022). *Directive (EU) 2022/2464 on corporate sustainability reporting.* https://eur-lex.europa.eu/legalcontent/EN/TXT/HTML/?uri=CELEX:32022L2464

European Parliament. (2023). *Corporate due diligence rules agreed to safeguard human rights and environment.* Press release. https://www.europarl.europa.eu/news/en/press-room/20231205IPR15689/corporate-due-diligence-rules-agreed-to-safeguardhuman-rights-and-environment

Fehr, E., & Fischbacher, U. (2002). Why social preferences matter – The impact of non-selfish motives on competition, cooperation and incentives. *The Economic Journal, 112*(March), C1–C33. https://doi.org/10.1111/1468-0297.00027

Fontana, M. (2008). *The complexity approach to economics: a paradigm shift.* Dipartimento di Economia "S. Cognetti de Martiis" Centro di Studi sulla Storia e i Metodi dell'Economia Politica "Claudio Napoleoni" (CESMEP). Working paper No. 01/2008.

Freeman, R. E. (1984). *Strategic management: A stakeholder approach.* Pitman Publishing Inc.
Freeman, R. E., & Auster, E. R. (2011). Values, authenticity, and responsible leadership. *Journal of Business Ethics, 98*(S1), 15–23. https://doi.org/10.1007/s10551-011-1022-7
Freeman, R. E., Parmar, B. L., & Martin, K. (2020). *The power of and: Responsible business without trade-offs*. Columbia University Press.
Freshfields Bruckhaus Deringer. (2005). *A legal framework for the integration of environmental, social and governance issues into institutional investment.* United Nations Environment Programme Finance Initiative.
Fukuyama, F. (1996). *Trust: Social virtues and creation of prosperity*. Free Press.
Fukuyama, F. (2001). Social capital, civil society and development. *Third World Quarterly, 22*(1), 7–20. https://doi.org/10.1080/01436590020022547
George, B. (2003). *Authentic leadership – Rediscovering the secrets to creating lasting value.* Jossey-Bass.
Glimcher, P. W., & Fehr, E. (2014). *Neuroeconomics. Decision making and the brain* (2nd ed.). Elsevier. https://doi.org/10.1016/C2011-0-05512-6
González-Viralta, D., Veas-González, I., Egaña-Bruna, F., Vidal-Silva, C., Delgado-Bello, C., & Pezoa-Fuentes, C. (2023). Positive effects of green practices on the consumers' satisfaction, loyalty, word-of-mouth, and willingness to pay. *Heliyon, 9*(10), e20353. https://doi.org/10.1016/j.heliyon.2023.e20353
Gowdy, J. M., & Iorgulescu Polimeni, R. (2005). The death of homo economicus: Is there life after welfare economics? *International Journal of Social Economics, 32*(11), 924–938. https://doi.org/10.1108/03068290510623771
Haba, H. F., Bredillet, C., & Dastane, O. (2023). Green consumer research: Trends and way forward based on bibliometric analysis. *Cleaner and Responsible Consumption, 8*, 100089. https://doi.org/10.1016/j.clrc.2022.100089
Hahn, R. (2022). *Sustainability management: Global perspectives on concepts, instruments, and stakeholders*. Rüdiger Hahn.
Hawn, O., Chatterji, A. K., & Mitchell, W. (2018). Do investors actually value sustainability? New evidence from investor reactions to the Dow Jones sustainability index (DJSI). *Strategic Management Journal, 39*(4), 949–976. https://doi.org/10.1002/smj.2752
Harré, N. (2011). *Psychology for a better world: Strategies to inspire sustainability*. Auckland, NZ: University of Auckland.
Helmke, S., Scherberich, J. U., & Uebel, M. (2015). *LOHAS-Studienübersicht* (pp. 91–97). Springer eBooks. https://doi.org/10.1007/978-3-8349-3719-3_4
Iqbal, A., Iqbal, M. S., Athar, A., & Khan, S. A. (2023). Impact of green marketing on consumer purchase intention: The moderating role of environmental knowledge. *Journal of Social & Organizational Matters, 2*(2), 43–58. https://doi.org/10.56976/jsom.v2i2.25
Izraeli, D., & Schwartz, M. S. (1998). What can we learn from the U.S. federal sentencing guidelines for organizational ethics? *Journal of Business Ethics, 17*(9–10), 1045–1055.
Kahneman, D. (2003). Maps of bounded rationality: Psychology for behavioral economics. *The American Economic Review, 93*, 1449–1475. https://doi.org/10.1257/000282803322655392
Kahneman, D., & Tversky, A. (1979). Prospect theory: An analysis of decision under risk. *Econometrica, 47*(2), 263–292. https://doi.org/10.1017/CBO9780511803475.003
Kaletsky, A. (2009). *Goodbye, homo economicus*. Prospect Magazine. April 2008 issue. https://www.prospectmagazine.co.uk/essays/52872/goodbye-homo-economicus
Kargol-Wasiluk, A., Wildowicz-Giegiel, A., & Zalesko, M. (2018). The evolution of the economic man. From homo oeconomicus to homo Moralis. Gospodarka Narodowa. *The Polish Journal of Economics, 293*(1), 33–57. https://doi.org/10.33119/gn/100566
Kehnel, A. (2022). *Wir konnten auch anders. Eine kurze Geschichte der Nachhaltigkeit.* Karl Blessing.
Kingo, L. (n.d.). *The UN global compact: Finding solutions to global challenges*. UN Chronicle. https://www.un.org/en/un-chronicle/un-global-compact-finding-solutions-global-challenges

Klink, D. (2008). Der Ehrbare Kaufmann – Das ursprüngliche Leitbild der Betriebswirtschaftslehre und individuelle Grundlage für die CSR-Forschung. *Zeitschrift für Betriebswirtschaft, 78*, 57–79.

Knack, S. (2001). *Trust, assocational life, and economic performance* (MPRA paper 27247). University Library of Munich.

Kwon, I. W. G., Hamilton, J. H., & Hong, S. J. (2012). Trust and transaction cost in supply chain cost optimization: An exploratory study. In K. Vaidya (Ed.), *Inter-organizational information systems and business management: Theories for researchers*. IGI Global. https://doi.org/10.4018/978-1-60960-768-5

Lahey, L., & Kegan, R. (2016). *An everyone culture: Becoming a deliberately developmental organization*. Harvard Business Review Press.

Lavecchia, S. (2025). The present of an ancient notion: Eudaimonic well-being as a concept for sustainable entrepreneurs? In M. Mühlböck (Ed.), *Sustainable transformation and Well-being. The role of business for individual and collective flourishing* (pp. 15–27). Springer.

Laybourn-Langton, L., & Jacobs, M. (2018). Paradigm shifts in economic theory and policy. *Intereconomics, 53*, 113–118. https://doi.org/10.1007/s10272-018-0737-4

Lempert, K. M., & Phelps, E. A. (2014). Neuroeconomics of emotion and decision making. *Neuroeconomics*, 219–236. https://doi.org/10.1016/b978-0-12-416008-8.00012-7

Lins, K. V., Servaes, H., & Tamayo, A. M. (2016). Social capital, trust, and firm performance: The value of corporate social responsibility during the financial crisis. *SSRN Electronic Journal*. https://doi.org/10.2139/ssrn.2555863

Loew, T., Ankele, K., Braun, S., & Clausen, J. (2004). *Bedeutung der internationalen CSR-Diskussion für Nachhaltigkeit und die sich daraus ergebenden Anforderungen an Unternehmen mit Fokus Berichterstattung* (Report no. GI2–46043/136). Bundesministerium für Umwelt, Naturschutz und Reaktorsicherheit. https://www.bmuv.de/fileadmin/Daten_BMU/Download_PDF/Wirtschaft_und_Umwelt/csr-studie_langfassung_bmu.pdf

Matten, D., & Moon, J. (2005). A conceptual framework for understanding CSR. In A. Habisch, J. Jonker, M. Wegner, & R. Schmidpeter (Eds.), *Corporate social responsibility across Europe* (pp. 335–356). Springer.

Meadows, D. H., Meadows, D. L., Randers, J., & Behrens, W. W. (1972). *The limits to growth. A report for the club of Rome's project on the predicament of mankind*. Potomac Associates – Universe Books.

Mühlböck, M. (2011). *Wirtschaftspolitik und Corporate Citizenship in Österreich. Das Potenzial von gesellschaftlichem Unternehmensengagement für mehr soziale Gerechtigkeit* [Dissertation]. Alpen-Adria-Universität Klagenfurt, Netlibrary. https://netlibrary.aau.at/obvuklhs/content/titleinfo/2409130/full.pdf

Oh, W. (2021). Understanding of self: Buddhism and psychoanalysis. *Journal of Religion and Health, 61*(6), 1–12.

Papadogiannis, Y. (2014). *The rise and fall of homo economicus: The myth of the rational human and the chaotic reality*. CreateSpace Independent Publishing Platform.

Phelps, E. A. (2009). The study of emotion in neuroeconomics. *Neuroeconomics*, 233–250. https://doi.org/10.1016/b978-0-12-374176-9.00016-6

Pinker, S. (2018). *Enlightenment now: The case for reason, science, humanism, and progress*. Viking.

Porter, M., & Kramer, M. (2011). Creating shared value. In *Harvard business review January – February 2011*. Harvard Business Publishing.

Riess, B. (2010). *Corporate Citizenship planen und messen mit der iooi-Methode. Ein Leitfaden für das gesellschaftliche Engagement von Unternehmen*. Bertelsmann Stiftung.

Roos, M., & Reccius, M. (2023). Narratives in economics. *Journal of Economic Surveys, 38*(2), 303–341. https://doi.org/10.1111/joes.12576

Rutledge, P. (2016). Everything is story: Telling stories and positive psychology. In E. M. Gregory & P. Rutledge (Eds.), *Exploring positive psychology: The science of happiness and well-being*. ABC-Clio Praeger.

Saad, L. (2023). *Historically low faith in U.S. institutions continues*. Gallup News. https://news.gallup.com/poll/508169/historically-low-faith-institutions-continues.aspx

Sarasvathy, S. D. (2001). Causation and effectuation: Toward a theoretical shift from economic inevitability to entrepreneurial contingency. *Academy of Management Review, 26*(2), 243–263. https://doi.org/10.5465/amr.2001.4378020

Schiff, B. (2012). *The function of narrative: Toward a narrative psychology of meaning*. Narrative Works.

Schmidpeter, R. (Ed.) (2013–2024). *Managementreihe CSR*. Springer Gabler. [book series].

Schmidpeter, R., & Funk, C. (2023). Sustainability as a management approach. In S. Idowu, R. Schmidpeter, N. Capaldi, L. Zu, M. Del Baldo, & R. Abreu (Eds.), *Encyclopedia of sustainable management* (pp. 3266–3269). Springer.

Schmidpeter, R., & Palz, D. (2008). Corporate social responsibility in Europa. In A. Habisch, R. Schmidpeter, & M. Neureiter (Eds.), *Handbuch corporate citizenship: Corporate social responsibility für manager* (pp. 493–500). Springer.

Schultz, D. (2013). Welfare Queens, calculative criminals, and the myth of homo economicus. In D. Schultz (Ed.), *American politics in the age of ignorance* (pp. 58–72). Springer Link. https://doi.org/10.1057/9781137308733_5

Schumpeter, J. A. (1934). *The theory of economic development*. Harvard University Press. (originally published in German in 1911).

Schüz, M. (1999). *Werte – Risiko – Verantwortung. Dimensionen des Value Managements*. Gerling Akademie Verlag.

Schüz, M. (2019). *Applied business ethics – Foundations for study and daily practice*. World Scientific Publishing.

Schüz, M. (2021). *Angewandte Unternehmensethik – Grundlagen für Studium und Praxis* (2nd ed.). Pearson.

Schüz, M. (2023). Sustainable corporate responsibility (SCR). In S. Idowu, R. Schmidpeter, N. Capaldi, L. Zu, M. Del Baldo, & R. Abreu (Eds.), *Encyclopedia of sustainable management* (pp. 3450–3461). Springer. https://doi.org/10.1007/978-3-031-25984-5

Schüz, M. (2025). Transdisciplinary Management of Polycrises – A philosophical approach. In M. Schüz (Ed.), *Transdisciplinary thinking and acting – Fundamentals and perspectives*. Springer-Nature.

Schwalbach, J., & Klink, D. (2015). Der Ehrbare Kaufmann als individuelle Verantwortungskategorie der CSR-Forschung. In A. Schneider & R. Schmidpeter (Eds.), *Corporate Social Responsibility. Verantwortungsvolle Unternehmensführung in Theorie und Praxis* (pp. 177–197). Springer Gabler.

Seligman, M. E. P., & Csikszentmihalyi, M. (2000). Positive psychology: An introduction. *American Psychologist, 55*(1), 5–14. https://doi.org/10.1037/0003-066X.55.1.5

Shepherd, D. A. & Patzelt, H. (2011). The new field of sustainable entrepreneurship: studying entrepreneurial action linking "What is to be Sustained" with "What is to be developed". *Entrepreneurship Theory And Practice, 35*(1), 137–163. https://doi.org/10.1111/j.1540-6520.2010.00426.x

Siegel, D. J. (2012). *The developing mind: How relationships and the brain interact to shape who we are* (2nd ed.). The Guilford Press.

Siegel, D. J. (2022). *IntraConnected: MWe (me + we) as the integration of self, identity, and belonging*. W. W. Norton.

Siegel, D. J., & Drulis, C. (2023). An interpersonal neurobiology perspective on the mind and mental health: Personal, public, and planetary well-being. *Annals of General Psychiatry, 22*(1). https://doi.org/10.1186/s12991-023-00434-5

Simon, H. A. (1955). A behavioral model of rational choice. *The Quarterly Journal of Economics, 69*(1), 99–118.

Smith, V. L. (1962). An experimental study of competitive market behavior. *Journal of Political Economy., 70*(2), 111–137. https://doi.org/10.1086/258609

Smith, A. (2012). *Wealth of nations*. Wordsworth Editions.

Smith, P. A. C., & Sharicz, C. (2011). The shift needed for sustainability. *The Learning Organization, 18*(1), 73–86.

Soros, G. (2013). Fallibility, reflexivity, and the human uncertainty principle. *Journal of Economic Methodology, 20*(4), 309–329. https://doi.org/10.1080/1350178X.2013.859415

Spence, L. J., & Schmidpeter, R. (2003). SMEs, social capital and the common good. *Journal of Business Ethics, 45*(1/2), 93–108.

Thagard, P., & Wood, J. V. (2015). Eighty phenomena about the self: Representation, evaluation, regulation, and change. *Frontiers in Psychology, 6*. https://doi.org/10.3389/fpsyg.2015.00334

The Global Compact. (2004). *Who cares wins. Connecting financial markets to a changing world.* https://www.unepfi.org/fileadmin/events/2004/stocks/who_cares_wins_global_compact_2004.pdf

Tversky, A., & Kahneman, D. (1974). Judgment under uncertainty: Heuristics and biases. *Science (New Series), 185*, 1124–1131. https://doi.org/10.1126/science.185.4157.1124

Tversky, A., & Kahneman, D. (1981). The framing of decisions and the psychology of choice. *Science, 211*(4481), 453–458. https://doi.org/10.1126/science.7455683

United Nations. (2024, December 2). Home - United Nations Sustainable Development. United Nations Sustainable Development. https://www.un.org/sustainabledevelopment/

Vallerand, R. J. (2008). On the psychology of passion: In search of what makes people's lives most worth living. *Canadian Psychology, 49*(1), 1–13. https://doi.org/10.1037/0708-5591.49.1.1

Von Leoprechting, R. (2025). The workplace as a challenge and a resource for adult development. In M. Mühlböck (Ed.), *Sustainable transformation and well-being. The role of business for individual and collective flourishing* (pp. 129–141). Springer.

Waqas, M., Qingfeng, M., Ahmad, N., & Iqbal, M. (2023). Green brands, customer satisfaction and sustainable performance in the Chinese manufacturing industry. *Management Decision, 61*(7). https://doi.org/10.1108/MD-09-2022-1251

Weiser, S. (2008). *Repeat after me: Greed is not good.* Los Angeles Times. https://www.latimes.com/archives/la-xpm-2008-oct-05-ca-wallstreet5-story.html

Weiss, C. (1995). Nothing as practical as good theory: Exploring theory-based evaluation for comprehensive community initiatives for children and families. In J. Connell, A. Kubisch, L. Schorr, & C. Weiss (Eds.), *New approaches to evaluating community initiatives*. Aspen Institute.

Welser, T. (2025). If I can really make an impact, then my life has been a success: A personal report on the transformation of a company and an individual. In M. Mühlböck (Ed.), *Sustainable transformation and well-being. The role of business for individual and collective flourishing* (pp. 303–310). Springer.

World Economic Forum. (2020). *Measuring stakeholder capitalism towards common metrics and consistent reporting of sustainable value creation.* White paper. https://www3.weforum.org/docs/WEF_IBC_Measuring_Stakeholder_Capitalism_Report_2020.pdf

Yang, Y. (2024). Complexity theory for complexity reduction? Revisiting the ontological and epistemological basis of complexity science with critical realism. *Journal of the Theory of Social Behavior*. https://doi.org/10.1111/jtsb.12412

Marisa Mühlböck holds a doctorate in social and economic science. Over the course of her career, she has gained extensive insight into different fields of work, which has fostered her understanding of diverse organizational structures, thinking patterns, and social context. As a former social entrepreneur, she also knows the entrepreneurial life from practice and incorporates this experience into her research. As a Research Country Co-Lead, she is involved in the Inner Development Goals initiative (IDGs), which is dedicated to researching and promoting inner development to achieve the U.N. Sustainable Development Goals. She also served as a founding board member of Verband für gemeinnütziges Stiften, Austria's first association representing and advocating for charitable foundations. Her present research interests are based on her theoretical background as well as on her practical experience and focus on sustainability in different areas, sustainable entrepreneurship, and well-being.

René Schmidpeter is a professor for sustainable management at the BFH in Bern and a scientific researcher at Parmenides Foundation in Munich. He also has been a guest professor/lecturer in Australia, England, China, India, and the USA. He is the series editor for Springer's CSR, Sustainability, Ethics and Governance books, editor of the Encyclopedia of Sustainable Management (ESM), and editor of the Dictionary of Corporate Social Responsibility (DCSR) as well as editor in chief of the International Journal of Corporate Social Responsibility (JCSR). His research and teaching activities focus on sustainable business transformation, international perspectives on sustainability, social innovation, and sustainable entrepreneurship with a special focus on the relationship between business and society. He has published widely and is the editor of the German management series on Corporate Social Responsibility at Springer Gabler and a scientific adviser to the FAZ-Institut and academic head of the annual Responsible Leadership Conference. Schmidpeter serves as an expert in many academic, economic, and political steering committees as well as on sustainability advisory boards. He has conducted research for and worked with the Federal Ministry of Education and Research, the European Union, the Bavarian Ministry of Social Affairs, the Anglo-German Foundation, the Bertelsmann Foundation, the Mercator Foundation, the Humboldt Foundation, the Federal State of Tyrol, and the Deutsche Forschungsgesellschaft (DFG). He was co-founder of M3TRIX in Cologne and Tetranomics in Munich.

Part III
Fields of Application for Sustainable Business Transformation and Well-Being

Philosophy as a Resource for Sustainable Business Transformation and Entrepreneurs' Well-Being

Yusuf M. Örnek

Abstract It is well known that businesspeople, investors, and managers tend not to occupy themselves with philosophy because they do not have the time to spend on such things. The Little School of Philosophy, founded in 2017 by the Antalya Businesspeople's Association (ANSIAD), proved that this prejudice is hollow. For more than 6 years, businesspeople have been meeting once a week in the evenings to read philosophical texts and discuss their relevance to the lives they live today, demonstrating that philosophy is not only the domain of academia. In the effort to combine the practical experience of business life with the theoretical side of philosophical thought, the participants found the opportunity to explore areas of experience that they were not previously aware of in relation to their own lives. The fundamental question of philosophical ethics "how should one live this life?" made the participants reconsider their relationships with people at all levels and realize that business life is not only based on the axis of profit and loss. It can easily be said that the lives of the participants, who were particularly influenced by the teachings of the philosophical schools of antiquity which taught that the union of *praxis* and *theoria* could be a kind of wisdom of life or the art of living, were enriched by the philosophical thinking. By enriching the philosophical texts they read in classes with philosophical excursions and public philosophy meetings and panels, these participants learned to examine words they hear or utter in daily life, such as justice, friendship, virtue, and freedom as concepts of philosophy. The experience gained at the ANSIAD Little School of Philosophy has demonstrated that the business world can benefit from philosophical thinking in many ways, both as individuals and as organizations, if the right communication approach is taken.

Keywords Businesspeople · Academic philosophy · Philosophy as wisdom of life · Well-being · Education

Y. M. Örnek (✉)
Antalya, Turkey
e-mail: yusuf.ornek@antalya.edu.tr

M. Mühlböck (ed.), *Sustainable Transformation and Well-being*, CSR, Sustainability, Ethics & Governance,
https://doi.org/10.1007/978-3-031-75566-8_6

1 Introduction

It is well known that the business world is far from philosophy and regards philosophical thinking and questioning as a kind of "sophistry." The widespread popular belief is that investors, entrepreneurs, and businesspeople are always busy with their work and have no time for philosophy. They are busy making money, and philosophy is ultimately an academic activity and belongs to the ivory tower of the university. How true is this common belief? Are businesspeople really far from philosophical thinking? And do they prefer to remain aloof? Is philosophy an activity that prevents them from working productively? Is that why they choose to stay away from philosophy?

In his *Lectures on the History of Philosophy*, Hegel gives the example of Thales of Miletus to illustrate how a philosopher can be a good merchant (Hegel, 1940): Thales, known as the first thinker and mathematician of Western philosophy, is rumored to have fallen into the pit in front of him while studying the movement of the stars on a magical Ionian evening. When he was told that he was incapable of seeing even in front of him and was subjected to ridicule, he showed his fellow countrymen what real foresight was like: Based on his meteorological research, he predicted that the olive harvest would be good the following year and bought all the olive trees around, hence becoming rich that year.

Thales is not the only example of a good entrepreneur among philosophers. As Plato said, just as a good philosopher can do any job, according to the example given in this article, successful people in business life can also be interested in philosophy and benefit from the world of philosophical thought.

2 The Birth of "ANSIAD Little School of Philosophy"

ANSIAD (Antalya Businesspeople's Association), one of the most respected non-governmental organizations in Antalya, was founded in 1990. The members of the association, who presently number 178, have an important place in the social life of Antalya, both with their commercial activities and their contributions to civil society. Of its members, 26% are engaged in construction and another 26% in trade, followed by industry with 15% and tourism with 12%, and the rest are employers and entrepreneurs in agriculture and the health sector. To be a member of ANSIAD, the company owner must have at least 25 employees and a certain level of turnover. Members are required to act in accordance with the rules of business ethics, use advanced technology, and be innovative. In addition, it is expected to contribute to the economy of Antalya with the turnover a member realizes, the tax paid, and the employment provided. As of today, 10% of ANSIAD members are women, and this is the highest rate among businesspeople associations in Antalya. The Education Foundation established by ANSIAD provides scholarships to university students in need. In 2023, the number of students granted scholarships was 186.

ANSIAD, which brings together the most active business people in the city, has contributed to many social responsibility projects; for instance, between 2006 and 2008, it identified the needs of 18 schools in the neighborhood where the most disadvantaged residents of the city live and met them. Between 2008 and 2010, it built nurseries in the gardens of 17 schools. Again in 2010, it built an apprenticeship school. In addition to Entrepreneurship Days, which have been organized for 20 years, it has paved the way for or supported many new investments through idea competitions among students from universities and high schools in the region. At the beginning of each academic year, many ANSIAD members visit high schools in the city and share their experiences.

I have been a member of ANSIAD since 1997. The company I was a partner in was a travel agency that had an important place in Antalya's tourism industry. As I was focused on tourism and business, I had switched from full-time teaching to a part-time position at the university. I was both lecturing at various universities and managing the company. Due to the consolidation of companies within the European tourism industry that began in the early 2000s, my company lost its main customer group upon the sale of the tour operator I used to serve. On top of that, when the financial crisis of 2008 came, the company went bankrupt, and I returned to my job at the university. The solidarity shown by ANSIAD members during this period, which was very difficult, both financially and psychologically, is worthy of every praise. In one of those meetings, the following occasion arose for the establishment of a philosophy school within the association: The president of the association asked me to present an evaluation of my experiences to the members, and I frankly explained all the mistakes I had made, large and small. At the end of the meeting, a member took the floor and asked me, how I had dealt with the difficult situation so well, saying that he had observed that some of the businesspeople who had experienced similar things had left the country and even committed suicide, whereas I had returned to my normal life. I replied emphatically that it was thanks to philosophy. Shortly afterward, the chairman of the board of directors of the association and a member of the committee responsible for education visited me and expressed their desire to learn philosophy.

To tell the truth, I was a bit hesitant at the beginning because I also had the common misconception about the general public's inclination toward philosophy. In addition, I had only taught young people at the university until that day, and I did not have much experience in adult education. After thinking about it for a few days, I prepared a training plan for the target group and told them that I had some conditions such as compulsory attendance, punctuality, discipline, reading the texts in advance, switching off smartphones during the lecture, and lectures in 90-minute blocks without a break. The ANSIAD Board of Directors accepted all my conditions and decided to establish a philosophy school within the association on March 1, 2017. I suggested avoiding ambitious names such as Philosophy Academy and instead using the more modest name "Little School of Philosophy." My proposal was accepted and the first class started on March 16, 2017.

The profile of the approximately 250 participants of the Little School of Philosophy and the profile of ANSIAD members do not mirror each other. While about half of ANSIAD members are engaged in construction and trade, only one-third of the participants of the Little School of Philosophy come from the construction and trade sectors. Participants from agriculture and the electricity/electronics sectors are represented by 8%, health sector by 12%, and insurance/accountants by 5%. In addition, 10% of the participants are lawyers and 6% are educators. The proportion of female participants is 65%; among them are also members' spouses who constitute 15% of all participants.

2.1 Courses and Contents

Each term of the Little School of Philosophy consists of a total of 20 lecture hours. One lesson lasts 45 min, and two lessons are held consecutively without a break. A semester consists of a total of ten teaching days but usually extends to 12–13 days. Classes are held once a week, on weekdays between 6:30 and 8:00 in the evening. During Covid, the classes, which were usually held in the meeting hall of ANSIAD, were first moved outdoors and then continued on the terrace of a relatively airy cafeteria with a sea view. Small snacks and drinks are allowed during the class, as participants usually come from work.

The courses consist of three modules: a beginner's class, advanced class, and text-reading seminar. The introductory class is a kind of introduction to philosophy. What is philosophy for? What are its main problem areas? In addition to a brief history of philosophy, systematic disciplines such as ethics, law, ontology, and metaphysics are introduced. The relevant chapters of the introductory philosophy book are read and reread in class. In philosophy lessons, students quickly realized that they were poor readers of texts in general. Reading was and still is often practiced in the lessons. The same applies to asking questions, making speeches, and participating in discussions. As a matter of fact, I can easily observe that the students who have participated in the philosophy lectures have made progress in this area by means of the questions they ask at the regular ANSIAD meetings.

After the first two semesters, we gradually started to read philosophical texts with the students. The first text was the Allegory of the Cave in Book 7 of Plato's *The Republic* (Plato, 1973). Since almost all of the participants had school-age children, the subject of education was of particular interest to them. "Education for what?" is a question that has been topical for thousands of years. The discussions on this question invariably brought the adult students in the classes to the topic of humanism, and it was argued that that is the main purpose of education.

Plato's extremely simple definition of justice in the same book remains valid today: "It is just to render every man his due" (Plato, 1973). This definition, which is easy to understand when it comes to the give-and-take between individuals, expands and deepens when confronted with the following question: How can justice be achieved in the relationship between the state and the citizen? The topicality and depth of this simple definition of Plato, who thought of philosophy in terms of concepts, was very effective for all students who were thinking about this subject.

The next book was *The Apology of Socrates*. Since there were usually a few lawyers in each class, it was appropriate to continue with this text. First, there was a brief introduction into Plato's general philosophy, and the basic concepts were then explained in more depth. Then the short text was read almost line by line, and it was explained how a philosophical text should be read and analyzed. Since people's reading habits are largely reduced to reading text messages on smartphones, reading and understanding philosophical texts was initially difficult for everyone. Over time, they became used to the use of technical terms, the weight of concepts, and the sequentiality of sentences. Where they had difficulties, these aspects were made comprehensible by explanations in the form of short questions and answers. Almost all these people, none of whom had studied philosophy before, said that they did not understand the texts when they read them alone but that they understood them when we read them together in class. I always reminded them of Nietzsche's quote: "My favorite reader should be calm and not read in a hurry" (Nietzsche, 1921).

The fact that Socrates had an ideal of humanity, that he did not give up honesty and justice, that he thought that only people with knowledge could be virtuous, that for this reason he put the god Apollo's commandment "Know thyself!" (Plato, 1976) at the center of his own philosophy, and that these thoughts, which came from thousands of years ago, are still very relevant today both astonished the participants and made them realize that philosophy is not just empty verbiage. I do not know to what extent the philosopher's statement "the life which is unexamined is not worth living" (Plato, 1976) led everyone to reflect on their own lives, but I am sure that many participants were influenced by this statement.

After Plato, we read certain chapters from Aristotle's *Nicomachean Ethics*. In Turkey, Aristotle is taught in high school logic classes and is generally known as a logician. When it was explained that he was the first teacher in Europe, not only a thinker but also a great scientist, the participants' desire to learn more about Aristotle increased. The participants were very interested in the philosopher's extraordinary analyses of the human psyche and life in general, as well as his evaluations that remain up to date. Aristotle's distinction between the virtues of character and the virtues of thought, his remarks on the importance of prudence and wisdom, and his quotation from Euripides about justice, "neither evening nor morning star is such a wonder" (Aristotle, 1976), surely had a greater impact on this specific audience than on university students. I can say that Aristotle has been the most popular philosopher in the lectures so far, especially for his writings on justice, love, and friendship.

After Aristotle's death, ancient philosophy continued in the form of various schools of ethics until the end of the fourth century AD. We have included three of these schools in our philosophy lessons and read their texts:

Firstly, we read and discussed Epicurus' letter to Menoikeus from Diogenes Laertius' book. Topics such as the importance of pleasure and pain in human life, the difficulty of finding the balance between these two, and the inappropriateness of fearing death were of great interest to the students. The following words of the thinker emphasized the indispensability of philosophy: "Let no one be slow to seek wisdom when he is young nor weary in the search thereof when he is grown old. For no age is too early or too late for the health of the soul. And to say that the season for studying philosophy has not yet come, or that it is past and gone, is like saying that the season for happiness is not yet or that it is now no more. Therefore, both old and young ought to seek wisdom" (Laertius, 1925).

Then we discussed the radical ideas of Diogenes, the greatest representative of Cynicism. Is it possible to live without needing anything, to lead the simplest life without fear of anyone and without needing anyone? It was the possibility of such a life of radical simplicity that interested these adult students, all of whom were wealthy people.

Stoic philosophy was the area we focused on the most. There were long discussions on the basic life principles of the Stoics, such as changing the perspective when looking at events, being prepared for the worst-case scenario, not getting caught up in the mistakes made in the past, separating what we have from what we don't have, and treating the people we encounter virtuously. The following questions, which is at the heart of philosophy lessons, was on the agenda almost every lesson: How should we live this life? How should life be lived? We read and discussed Emperor Marcus Aurelius' *Meditations* line by line. The philosopher's basic question was this: How do we achieve spiritual peace and happiness (eudaimonia) that nothing in the world can destroy? If even an emperor, whose life was spent on the battlefields, can retire to his tent in the evenings and question himself, that is, philosophize, then why can't I? This question came to life in everyone's minds. The reason why Stoic philosophy remains relevant today lies in the fact that it gives direct answers to the question of how life should be lived and advises making philosophy a way of life. When Marcus Aurelius writes, "how clearly is it borne in on thee that there is no other state of life so fitted to call for the exercise of philosophy as this in which thou now findest thyself" (Aurelius, 1975), he invites everyone to philosophize, to lead a philosophical life, to become an artist in life through philosophy: "The business of life is more akin to wrestling than dancing, for it requires of us to stand ready and unshakeable against every assault however unforeseen" (Aurelius, 1975).

After Marcus Aurelius, we read the texts *On Friendship* and *On Old Age* by Cicero, another Stoic philosopher, jurist, and politician. The idea that only good and virtuous people can be true friends was of great interest to everyone, and they began to ask themselves how many friends they had in this sense. This question also led adult students to question themselves about whether they were good and virtuous. Cato's words in *On Old Age* were also very interesting to the members who were over middle age. In this text, Cicero explains how unwarranted it is to be afraid of death and how futile it is to perceive old age as a disadvantage.

After so many texts on ancient philosophy, we turned to texts written by a contemporary Turkish philosopher. Ioanna Kuçuradi, the President of the Philosophical Society of Turkey, is the best known philosopher in Turkey. We discussed her recent texts on the philosophical foundations of human rights, and we saw how philosophical concepts such as democracy, secularism, tolerance, and justice are derived. Through these discussions, which took place within the framework of Turkey's present political problems and preparations for elections, the participants realized how important it is for words used in everyday language to become philosophical concepts.

Participants from the business world were most interested in ethics. In ethics courses, three basic understandings are emphasized: utilitarian ethics (J. Bentham, J.S. Mill), virtue ethics (Aristotle), and duty ethics (I. Kant). After discussing these conceptions, the following question is emphasized: What is a good deal? Is it maximizing the profit at once, or is it the sustainability of the trade? Furthermore, is it not an ethical requirement that all stakeholders, not just the buyer and seller, be taken into account in the transaction?

Why should I be fair? Why should I act honestly? What is justice? In the courses where questions such as these were addressed, students learned to question what they do and the decisions they make in their own practical work life through what they have learned theoretically in philosophy. I can give an example of this from Kant's ethics of duty: Kant's idea of the categorical imperative is based on doing the right action for the right reason. Our action may be right, but its motive may be wrong; then the action has no moral value. The important thing is to do the right action out of a sense of duty, i.e., to be able to say "my duty is to be honest." After we covered this topic, one of my students, who is the owner of a company that sells construction materials, said, referring to Kant's example of the prudent tradesman, "The reason we were honest with our customers was to protect the reputation of our company, but now I realize that this approach does not give moral value to our act of honesty; what is important is to say that being honest is our duty; for this reason we must be honest." After the lecture, this businessman gathered all his sales staff and told them that they had to change the motive behind being honest with customers. To be honest because of the assignment is to be honest for the right reason. This is an example of living life based on a principle, not on everyday desires, inclinations, or needs.

Some businesspeople invited me to their workplaces and had me give training to their staff, especially on this subject because they thought that lectures on ethics were very useful. In an industrial company employing 700 people and in an agricultural company employing 400 people, it was an extraordinary experience for both them and me to give lectures on the subject to all management staff.

2.2 *Philosophy Trips*

As an active association, ANSIAD has been organizing trips for its members and their families for a long time. Since the Little School of Philosophy was founded, almost all trips organized by the association have either had a philosophical content or philosophy has been the main subject of the trip. These are a few examples:

Our first big trip after the establishment of the Little School of Philosophy was to China. In the summer of 2017, we visited the temples of Tao, Confucius, and Buddha. In this way, I had the opportunity to inform the participants about Far Eastern philosophy.

In 2019, we visited Marx's house in Trier, Hegel's house in Stuttgart, Heidegger's cottage in the Black Forest Mountains, the theological school in Tübingen where Hegel, Schelling, and Hölderlin studied together, the house in Heidelberg where Jaspers and Heidegger had long conversations, and the place where Nietzsche lived in Basel.

In the same year, we travelled with a large group of participants to Ionia, the Aegean region of Anatolia, the birthplace of philosophy. We read the philosophical fragments of Heraclitus in Ephesus, Bias in the Priene theater, and Thales and Anaximandros in Miletus. Homer, who lived in the same region, was of course on our agenda throughout the trip, too.

In 2021, we resumed the trips we had interrupted due to the pandemic. First, we went to the region known as Phrygia in antiquity and visited Hierapolis, the birthplace of Epictetus, one of the greatest names of Stoic philosophy, and then Aphrodisias, where Alexander, one of the famous interpreters of Aristotle in the Roman period, lived.

Our next philosophical excursion was to Cappadocia. Here we discussed the teachings of the three great church fathers of Western theology, Basilius, Gregorius of Nazianz and Gregorius of Nyssa, as well as the philosophy of mysticism by visiting Hacıbektaş, the center of heterodox Islam.

Our last major overseas philosophical trip was the Athens-Delphi trip in the summer of 2023. It was fascinating to see the place where Socrates was imprisoned at the Athenian Acropolis, to see the sites of Plato's Academy and Aristotle's Lyceum in modern Athens, and to recall the texts read in our lectures there. But undoubtedly the most exciting was the Temple of Apollo at Delphi, the greatest oracle center of antiquity, located in an extraordinary landscape.

All these excursions enabled the participants to establish a connection between the texts they read in the lectures and the magical atmosphere of the ages in which the philosophers lived. Although it is difficult to get an impression of the past in today's vibrant environment, it was very useful to build a bridge between what was lived and what was put down on paper and conveyed to us. At the same time, the students who participated in the excursions were informed about how rich in cultural history the country where they were born and grew up is and that Anatolia is the birthplace of the science, philosophy, and theology of Western civilization. The horizon-opening characteristic of philosophy became more apparent in this way. I have no doubt that those who step into this horizon will look at universal civilization differently and will now question narrow-minded nationalism. I am sure that these adult students have gained a new awareness that it is a blessing to live in Anatolia, the birthplace of Western civilization.

2.3 Philosophy Meetings

With the start of the Little School of Philosophy, ANSIAD started to add and to gradually intensify a philosophical dimension to its regular meetings. For example, this feature was taken into consideration with regard to the conferences held at the universities in Antalya, Burdur, and Isparta within the scope of the annual Entrepreneurship Days: The theme of the conference in 2017 was *Philosophical Perspectives on Technology*, and that of 2018 was *Philosophical Foundations of Entrepreneurship*. In 2019, Mathias Schüz from Zürich, who is one of the contributors of this volume, gave lectures on entrepreneurship and responsible leadership in the three cities. Hundreds of students and faculty members attended the 2020 lectures on *Aesthetic Approaches and Cities* and 2021's lectures on *Degradation of the Natural Environment as an Ethical Problem.*

Also in 2021, ANSIAD organized the *Ethics in the Business World Panel* with the intense participation of businesspeople in Antalya, and at this meeting, speeches were made on media ethics, ethics in the health sector, ethics in the judiciary, and ethics in the neoliberal business world.

However, ANSIAD's biggest meeting on philosophy was the national congress titled *Philosophy in Turkey in the Centenary of the Republic*, organized by the Philosophical Society of Turkey between February 1 and 5, 2023. At this meeting, which was sponsored by ANSIAD members, hundreds of academic philosophers from many universities in Turkey had the opportunity to discuss the position of Turkish philosophy in 100 years. It is known that it is difficult to find financing for such large congresses. They are supported either by public institutions, universities, or research institutes. The last major philosophy meeting in Turkey was the World Philosophy Congress in 2003. In 2023, for the first time in 20 years, a major philosophy congress was organized with the financial contributions of ANSIAD members, i.e., an association belonging to the business world. It was also very pleasing that all the halls were constantly filled with eager audiences. Hundreds of citizens had the chance to attend public philosophy meetings for a week with the contribution of local administrations in Antalya.

3 Achievements and Results

At the National Philosophy Congress held in Antalya in February 2023, the ANSIAD Little School of Philosophy also held a session. The session chairman and speakers, all of whom were students of the school of philosophy, shared their experiences and achievements in philosophy lessons. A primary school teacher said the following:

> How did the courses at ANSIAD Little School of Philosophy affect my professional life? What was it like to study without taking an exam? I think the philosophy courses we take have an indirect effect on our professional, family and social life. In fact, the primary effect was to change 'I'. We make choices in every moment we live. Philosophy helps us to make

> the most appropriate choice by increasing our awareness. When we make the concern of being a virtuous person a source of motivation at a high level, when we try to strengthen our values, we experience more satisfying moments in our sphere of influence and with every person we communicate with.

A participant engaged in trade made the following contribution: "The impact on my profession has increased my awareness. I am engaged in the business of commerce, and I am in one-to-one contact with clients I saw that the already-existing virtues and values that I apply in my ongoing work are supported by the lessons here. I believe that the values I learned in philosophy classes are principles that should be applied."

A dentist expressed her connection with philosophy in the following words:

> I am a physician; I have been working as a dentist in Antalya for many years. Why did I choose to become a physician? When I think about it, I answer that I like to ask questions. I have been into asking questions since my childhood, that is why I chose medicine. If physicians cannot ask questions, they cannot diagnose and treat well. The beginning of medicine is asking questions. Who is medicine for? It is for human beings. It is for the well-being, happiness, health and long life of human beings. And who is philosophy for? Philosophy is also for human beings. Our questions are different, but our answers are the same. Medicine and philosophy should not be separated from each other. If physicians cannot see the whole and establish the necessary connections, that is, if they do not have a philosophical perspective, they are only a technicians. In order to be a good physician, I must be able to see the whole and establish connections. The School of Philosophy made a great contribution to me in this regard.

I have no doubt that there is a difference in the way the students' perspective on life has changed, after attending these courses. I think that there is a change in their lifestyles, in their private lives, in their family relationships, and in their approaches to business life. I can give another example from a meeting attended by the children of ANSIAD members: The children of many members who have graduated from university also work in family businesses. At a meeting attended by this second generation, one of the participants said that the most important change in the company and within the family in recent years was when his father started taking philosophy classes. This young man said that his father, who was a student at the Little School of Philosophy, had visibly changed the way he interacted with company employees and family members.

The ANSIAD Little School of Philosophy showed that philosophy has the possibility to go beyond the university and assume a pioneering role. "University philosophy" has been criticized since the second half of the nineteenth century, especially by Arthur Schopenhauer and Friedrich Nietzsche. At the beginning of the twentieth century, Martin Heidegger and Karl Jaspers formed a "unity of struggle" (Heidegger & Jaspers, 1990) against university philosophy. Philosophy in the academic world was a knowledge activity that spoke only the language that experts understood among each other, had nothing to do with practical life, and claimed to be a science. This frightened intellectuals and enthusiasts who were interested in philosophy but had not undergone an academic philosophical education. However, philosophy is an epistemic activity that can be realized by every individual with a healthy capacity for understanding. What needs to be taken into consideration when one wants to take this activity beyond the university is to speak in a language that people who have not studied philosophy can understand, to give examples of relationships between individuals, and to enable them to build connections between what they have learned in the classroom and what they experience in their daily lives. This is because those who are familiar with philosophy can see what others do not see and realize the connections they do not notice.

Philosophy can only be a wisdom of life or even an art of life if *theoria* and *praxis* are united. In the business world, there is only *praxis* and in philosophical thinking there is only *theoria*. A well-organized school of philosophy can contribute to uniting these two areas. Whereas *praxis* is superficial and belongs to daily life, *theoria* is deep and it is concerned with life's principles. This depth can be most useful for businesspeople. Living life according to principles means living a deep life, not a haphazard or superficial one. The business world can benefit more from philosophy understood in this way than a university student can. As with the participants of the school of philosophy, businesspeople are usually middle-aged and elder, i.e., they have a certain amount of life experience. They have experience with justice, virtue, honesty, etc. People who begin to question life and feel a kind of dissatisfaction during this questioning belong to this group. Their experiences in their private and professional lives bring the question of meaning to the forefront, and they do not know how to find an answer to it. This is the reason why the literature on personal development and the number of psychological counselors is increasing.

However, philosophy is not a therapy; it does not offer a ready-made medicine. Philosophy, with the questions it asks, allows those who ask the question to be shaken and to discover possibilities about themselves and life that they were not aware of until now. Academic philosophy as a field of specialization cannot help in this regard. The important thing is to be able to combine what the great philosophers in the history of philosophy, which is an inexhaustible spring, have to say with the problems of concrete life and to accompany them in asking philosophical questions, using a language that the participants can understand. It is necessary to be aware of the sensitivity of accompanying students on the path of philosophical thought.

In which direction will the ANSIAD Little School of Philosophy evolve? It seems difficult to say today, but a great distance has been travelled between the beginning of the project and today. I observe that the waves created by a stone thrown into a stagnant lake are growing. I receive requests from outside ANSIAD to offer lectures and I must turn them down. One way out could be for ANSIAD to periodically organize public discussions on present philosophical issues.

All this shows us that philosophy can be beneficial for businesspeople with the right language of communication. It is good for these people to be able to get rid of the pressure of daily life at least once a week for a short time, to "breathe fresh air" as they call it. In the coming semester, in two different classes of ANSIAD Little School of Philosophy, we will read various passages from Plato's *The Republic* and continue to discuss issues such as justice, virtue, and education. The students of ANSIAD, who have shown that philosophy has the possibility of extending beyond the academy, seem to have the will to be inspired by philosophy for a long time to come.

References

Aristotle. (1976). *Ethics* (J. A. K. Thomson, Trans.). Penguin Books.

Aurelius, M. (1975). *Meditations* (M. Staniforth, Trans.). Penguin Books.

Cicero. (1953). *Laelius on Friendship* (W. A. Falconer, Trans.). The Loeb Classical Library, Harvard University Press, pp. 103–211.

Cicero. (1953). *Cato the Elder on Old Age* (W. A. Falconer, Trans.). The Loeb Classical Library, Harvard University Press, pp. 1–99.

Hegel, G. W. F. (1940). *Vorlesungen über die Geschichte der Philosophie*, Erster Band (H. Glockner, Ed.). Fr. Frommanns Verlag.

Heidegger, M., & Jaspers, K. (1990). *Briefwechsel 1920–1963* (W. Biemel & H. Saner, Eds.). Vittorio Klostermann Verlag/Piper Verlag.

Kucuradi, I. (2011). *Human rights, concepts and problems* (2nd ed.). Philosophical Society of Turkey.

Laertius, D. (1925). *Lives of eminent philosophers* (R. D. Hicks, Trans.). Loeb Classical Library, Harvard University Press.

Nietzsche, F. (1921). *Über die Zukunft unserer Bildungsanstalten, Gesammelte Werke* (Vol. 4). Musarion Ausgabe.

Plato. (1973). *The Republic* (F. M. Cornford, Trans.). Oxford University Press.

Plato. (1976). *The last days of Socrates. Euthyphro, The Apology, Crito, Phaedo* (H. Tredennick, Trans.). Penguin Books.

Yusuf M. Örnek was born in Izmir in 1955. After high school, he studied sociology and philosophy at Hacettepe University in Ankara. He graduated from the university in 1977 and travelled to Germany to pursue a doctorate in philosophy. In 1983, he graduated from Johannes Gutenberg University in Mainz with a doctoral thesis entitled *Karl Jaspers' Concept of Existential Freedom* and returned to Turkey. Between 1984 and 1987, he worked as an assistant professor at Hacettepe University, Department of Philosophy. In 1987, he became an associate professor and left the university to start a business life. During the years he worked in the tourism industry, he gave lectures at several universities in Istanbul and Antalya. Between 2009 and 2013, he worked as a consultant to the Antalya Chamber of Commerce and Industry. Since 2014, Ornek has been lecturing in the Department of Philosophy of Law at Antalya Bilim University, Faculty of Law, and is presently working as a full-time professor at this university. Ornek has translated I. Kant, M. Heidegger, and K. Jaspers into Turkish and has published books and articles on the history of philosophy.

The Zen of Music: On the Energetics of Music as a Source of Personal Well-Being

Oliver Peter Graber

Abstract Music and physiology have been closely linked since time immemorial. Using Gustav Mahler's *Symphony No. 8* as an example, the article addresses essential points of such connections—including tension and synchronization, breathing, construction, and spiritualism—and discusses their potential for well-being in terms of arts for health on the basis of evidence-based findings.

1 Introduction

One of the most remarkable documents in musical history is the *Symphony No. 8* ("Symphony of a Thousand") by Gustav Mahler (1860–1911). In the following, we will look at the program and structure of the piece, which had its world première in 1910, as a way of discussing selected central and empirically established facts from research into the effects of music, which can be used for personal well-being in a noninvasive, cost-effective, and enjoyably usable way.

2 The Upbeat

Before Mahler's mighty *Eighth Symphony* begins with a "deep, droning E flat" (Specht, 1912), the conductor in their function as "ruler of the world" (Canetti, 1960) gets it on its way with an *upbeat*.

This moment of preparation, synchronization, and focusing of the joint will to perform the piece is a moment of joint first *inspiration* in which simply *everything* happens and is decided.

O. P. Graber (✉)
Jam Music Lab Private University @ JAM MUSIC LAB GmbH, Vienna, Austria
e-mail: oliver.graber@jammusiclab.com

M. Mühlböck (ed.), *Sustainable Transformation and Well-being*, CSR, Sustainability, Ethics & Governance,
https://doi.org/10.1007/978-3-031-75566-8_7

As experienced conductors know, you can only bring in an ensemble correctly—whether it's a minimal group or a mighty apparatus in the sense of Mahler's *Eighth Symphony*—with a correct upbeat breath, even without any gesture. Without correct breathing, a successful entry is impossible, even with ambitious gesturing.[1]

Music is structured breathing—and nowhere is this more evident than in singing. But instrumental music and ballet also breathe, in so-called phrases, where a phrase by definition covers as many musical elements as can be produced or consolidated in a breath.

Breathing and music are inseparably linked in many aspects. As described above with the conductor's upbeat, breathing can be both a deliberately utilized element of making music as a "driving force" or, entirely under the influence of music, a "musically driven" flow of breath (Laczika et al., 2013).

In the course of so-called entrainment—the influencing of the body's own rhythms (keywords breathing rhythm and synchronization) by external stimuli such as music—breathing patterns can be specifically influenced by music of a particular nature (Haas et al., 1986; Bernardi et al., 2009; Jiménez et al., 2022).

But what breathing pattern qualifies as physiological, and what produces such "musical patterning" of breathing?

For the sake of clarity, we will limit ourselves at this point to a few facts and figures and assume a healthy adult in a state of rest.

In this case, the respiratory rate, or the number of breaths a minute ($f = 60/T$), averages 12–16. In that formula, T is the time taken for a breathing cycle, which comprises the time for inspiration (I) and expiration (E). The ratio of these two (I:E) can be assumed to be 1:1.5 to 1:2 in our case (Schoppmeyer, 2020).

Singing involves substantially extended expiration, which is both obvious and directly understandable, and has been known since before Socrates (Archytas of Tarent, also the oldest text on breathing physiology specific to music; see Diels & Kranz, 1989).

If expiration increases to I:E = 1:3 or more, for example, this also has effects on the vegetative nervous system and cardiovascular system. Specifically, there is a resulting decrease in blood pressure (Kulinski et al., 2022) and heart rate, which can also be exploited for "relaxation" in the field of wellness.

However, in view of the interpersonal synchronization, the *upbeat* goes far beyond joint inspiration. As was discovered in so-called hyperscanning studies (simultaneous recording of data such as brainwaves for multiple people: Lindenberger et al., 2009; Müller et al., 2013), there are synchronization processes in brain activity in the phase of preparing for making music together.

Even in its emerging stages, music accordingly proves to be a medium which is not merely passively perceived but needs to be actively practiced to develop its full magic, just as in the receptive sphere participation in a live event (e.g., attending a

[1] If anyone wants to delve into this for themselves, the documentary "Deutschland, deine Künstler, Staffel 2 Folge 5: Kurt Masur" by Reinhold Jaretzky from 2009, for example, is recommended (www.youtube.com/watch?v=iTxId9hk4U8&ab_channel=SanCosimato, retrieved on July 17, 2023).

concert) cannot be replaced in its effect (Czepiel et al., 2021) by recordings or electronic media.

3 The "Symphony of a Thousand"

In music history, particularly in the nineteenth century, there was a rise in the size of ensembles to huge dimensions—a trend that the ideas and writings of Anton Reicha (1770–1836), the composition teacher of Hector Berlioz, Franz Liszt, César Franck, Charles Gounod, and others, made a decisive contribution to. The reference to Reicha is far from incidental, as his "theory of composition" is also one of the earliest and most central references to music medicine and music therapy in the modern sense (Reicha, 1832).

Mahler's *Eighth Symphony* is impressive not only for the sheer number of participants (which gave it its later name "Symphony of a Thousand") but also for the large number of instruments in the orchestra, including four keyboard instruments—organ, harmonium, celesta, and piano.

But what is the most outstanding instrument of all these in terms of "musical effect," and why? Can "functional brain imaging" be used to identify in the performer's brain who is playing which instrument?

If someone *actively* and regularly practices a specific instrument, for instance, this results in a process of dynamic adaptation in the brain, which we describe as neural plasticity. We distinguish between "positive" neural plasticity in the form of growth and a "negative" one in the form of a decrease in specific areas of the brain. Studies show that this neural plasticity has an impact which is specific to the instrument (Bangert & Schlaug, 2006). With string instruments (violin, cello, etc.), this significantly correlated positive neural plasticity is only apparent in areas which represent the left (fingering) hand, while the right (bowing) hand does not induce this. However, with keyboard instruments, and particularly the piano, corresponding effects are apparent for both hands and also in both hemispheres of the brain. Pianistic neural plasticity peaks in early years (Bengtsson et al., 2005; Vaquero et al., 2016) but can be triggered lifelong, which results in positive effects, for example, for piano playing for maintaining mental fitness (Worschech et al., 2023; Jünemann et al., 2022; James et al., 2020; Hudak et al., 2019).

The piano differs from other instruments in a number of ways. Its sound produces particularly intensive reactions in the brain (Lütkenhöner et al., 2006), it demands extreme fine motor skills and formation of sequences of movements which increasingly emphasize the role of the cerebellum for playing the piano (Tavor et al., 2020), and piano playing launches almost total networking throughout the brain, resulting in a "full brain workout." This is particularly evident when playing/reading music from sight (so that scores are being read while playing) (Fan et al., 2022). Generally, research shows increasingly fine differentiation of the effects of specific musical activities on the brain (Slater et al., 2017), and it also becomes clear that skills acquired or refined in the course of musical activities can also be used in other areas of life (e.g., language) and can, for example, have decisive benefits in interpersonal dialogue (Strait et al., 2009). Experience shows that improvements in communication and dialogue capacity contribute to "well-being" in every life situation.

The brain dynamic reflected in neural plasticity also results in phenomena which are directly associated with well-being—for example, reduced perception of pain. People who dance together in a rhythmic structure ("synchronized") subsequently also experience less pain (Tarr et al., 2016; Tarr et al., 2015), an effect which can also result from multiyear practice of Zen meditation (Grant et al., 2010). Here, music carries within it the seed of healing, as chronic pain leads to a decline in creativity (Gubler et al., 2022).

4 A Song Is a Song Is a Song

Mahler's *Eighth Symphony* is a choral symphony—a musical genre that began with the *Choral Fantasy op. 80* and *Ninth Symphony op. 125* by Ludwig van Beethoven (1770–1827). Few works followed this initial triumph, including Mendelssohn-Bartholdy's *Second Symphony "Lobgesang" op. 52*, Franz Liszt's *"Faust" Symphony*, which is very relevant to Mahler's *Eighth Symphony*, as it also uses Goethe's *Faust* as the basis for its text, and Mahler's *Second*, *Third*, and *Eighth Symphonies*.

In these pieces, the spoken word is added to the purely instrumental "symphonic language" of the orchestra, adding the dimension of concrete meaning to the orchestral score, although the freedom of the purely musical creation is restricted by the text-based setting.

The difference between text-based and purely instrumental music is a fundamental one in many ways, and its effect can be seen at the level of the order of primates (Videan et al., 2007). Above all, this calls for decisively opposing the trend in so-called playlists of calling any piece of music a song, regardless of whether it has lyrics or not. In fact, there should also be a distinction made within actual songs with lyrics between various kinds of song—art song, folk song, ballad, pop, etc.

This sign of musical ignorance and total indifference is disastrous in many ways, disrupting not only the success of interdisciplinary cooperation in the fields of music-specific neuroscience, music medicine, music therapy, performance medicine, arts for health, and research into the effect of music but also and increasingly the correct selection of music for personal well-being, given the differentiated effect of music established by research.

Dingle et al. (2021) list the following effects:

> Receptive and intentional music listening were found to reduce pain through changes in physiological arousal in some studies but not others. Shared music listening (e.g., concerts or radio programs) enhanced social connections and mood in older adults and in hospital patients. Music listening and carer singing decreased agitation and improved posture, movement, and well-being of people with dementia. Group singing supported cognitive

health and well-being of older adults and those with mental health problems, lung disease, stroke, and dementia through its effects on cognitive functions, mood, and social connections. Playing a musical instrument was associated with improved cognitive health and well-being in school students, older adults, and people with mild brain injuries *via* effects on motor, cognitive and social processes. Dance and movement with music programs were associated with improved health and well-being in people with dementia, women with postnatal depression, and sedentary women with obesity through various cognitive, physical, and social processes. Rapping, songwriting, and composition helped the well-being of marginalized people through effects on social and cultural inclusion and connection, self-esteem, and empowerment.

There is accordingly an essential difference between choosing purely instrumental music for your personal "balance"—for example, if you want to enjoy low tempo music (around 60 beats a minute) purely receptively—and actively singing music with lyrics as a leisure activity (e.g., songs of all kinds, arias, chorales or motets, etc.), alone or in an ensemble, to list just two particularly typical approaches in the well-being sector.

We should also note at this point the linking of music and light or video/multimedia. Presently, we are seeing the growth of hypervisualization, in which the ability to hear and deliberately listen (not only to music) is being totally suppressed, with some highly relevant social consequences, although these are only getting rudimentary public scrutiny (Kisiel, 2008).

Even if there can be no doubt of the different possibilities of combining and interactive effect of music and other stimuli, it is advisable to protect hearing, deliberately cultivating listening, and training our hearing for the purpose of retaining and extending our personal sensory abilities, in order to strengthen our personal well-being.

5 Musical Energetics

After detailed analysis of the works of Anton Bruckner and Richard Wagner, Mahler contemporary and music theorist Ernst Kurth (1886–1946) arrived at the concept of "symphonic waves," musical energy rising and falling in a wavelike manner. These waves display the dramatic ascent from a lower level of energy (which already has all the potential inherent in it) to the final eruption, followed by the inversion of this process, and every stage between. The crucial concept here is movement energy, and in his approach Kurth (Kurth, 1931) provides us with an ideal way of describing the interplay of music and ballet, music and physiology, recalling Nietzsche's comment that "aesthetics is nothing more than applied physiology" (Nietzsche, 1889).

If we want to identify a musical presentation appropriate to well-being, for example, it is advisable to consider as a first step whether the energetic level of the music matches the physiological condition and can in principle be brought into resonance with this, in the sense of the *entrainment* mentioned earlier.

At this point, it is also appropriate to consider a new approach to music history and literature as a whole. Why is there usually an overture at the beginning of a work of music theater? To introduce the actual action? Certainly, but also to bring everyone in the audience to a common physiological starting level. One audience member has been sitting in the opera café for 2 h and has read through the program; another has rushed in from work and public transport at the last minute. No music in the world can speak to both of them immediately in their respective physiological situation, so it is better to start with a flexible overture that sets the level before embarking properly on the musical development—well-being scenarios also benefit from an overture.

6 Intellect and Emotion[2]

In the course of a composition, both work together. Composition follows rules which have to be learned and studied. Ideally, these rules indicate how a musical system can be optimally coupled with a physiological system. The idea that it is so simple to compose is as naïve as it is misleading. We can cobble together a song with one, two, or three parts just as we can build a primitive shelter from a few slats, some brushwood, and nails. A construction on this scale doesn't need any sketches or documentation. For a symphony like Mahler's *8th*, this is no longer enough—a building like the Burj Khalifa needs structural calculations, knowledge, and planning. To take another analogy, many compositions offer a trip in a *Ferrari P4/5*, but almost all of these will generally be ignored in favor of a *1951 Hofmann*—well-being in a *Ferrari* broadens the horizon!

7 Veni Creator Spiritus

In his *Eighth Symphony*, Mahler contrasts the *Easter hymn* in Part One and Goethe's *Faust* in Part Two, which is also a contrasting duo of "heavenly" and "earthly" love, in which the "heavenly" ultimately triumphs, creating a "great hymn to love" (Specht, 1912). The appeal to the Holy Spirit at the start of the symphony also reflects the desire for inspiration of every artistic figure.

Musical composition and improvisation have the same needs but differ with respect to the neural processes involved (e.g., Barrett et al., 2020).

In the course of composition, spontaneous improvisation can help find and develop ideas, overcome blocks, or become aware of functional aspects of the physiognomy of the performing apparatus. Improvisation—even with comb and paper,

[2] This is also the title of a textbook by Josef Dichler.

an empty bottle, and a couple of glasses[3]—can also be the first step toward developing independent creative experience.

Just as music borrows from rhetoric to shape the result (Krones, 2006), so can interpersonal communication borrow from improvisation to achieve new freedom and facility (Haidet, 2007; Haidet et al., 2017).

Music-specific neurotheology (Arias, 2019; Walter & Altorfer, 2022) is still a young field of research. In the course of comparative study of different approaches to mystic contemplation, the significance of breathing for music appears again (Miyashiro et al., 2021; Graber & Tucek, in print), to emphasize this central aspect as a unified solution in music and physiology.

8 The Eternal Feminine Draws Us Upward

With the last syllables set to music in the *Eighth Symphony*—"the appeal of the creative spirit, now incorporate as the almighty and omnipresent spirit of love, has been answered" (Specht, 1912)—Mahler leads us to our own beginnings.

Music (Oikkonen et al., 2016) and dance (Bachner-Melman et al., 2005) show genetic predispositions which are epigenetically regulated (Kanduri et al., 2015; Nair et al., 2019), but decisive impulses for the subsequent musical journey through life come predominantly from the "prenatal world of sound."[4] Robert Chuckrow and Richard Parncutt even argue for a prenatal origin of music (Parncutt & Chuckrow, 2019), which is shown particularly elegantly in a link between the musical and maternal pulse (Teie, 2016).

If inborn talent and the prenatal world of sound are the decisive basis, it is ultimately the individual himself who decides on music as a way of life and grooves this decision into his brain through practice.

9 Dare to Be Silent

When the last note of the mighty opus has died away, silence follows. People seem to fear silence; it means illness, weakness, and death—this appeared in the history of music, particularly since the theory of musical figures, as a general pause (i.e., silence). If we follow the Parncutt-Chuckrow model, this fear of silence, which actually led to a "loudness war" in the music industry,[5] expresses the desire to return to the continuous music in the prenatal state, whose acoustic continuity is an

[3] For example, staged in the film *The Russia House*, 1990, www.imdb.com/title/tt0100530/?ref_=vp_vi_tt, retrieved on July 17, 2023.

[4] This is also the title of a textbook by Alfred Tomatis.

[5] An overview for initial orientation and data analysis can be found on https://dr.loudness-war.info, retrieved on July 17, 2023.

assurance of life and whose warmth, rhythm, and constancy (Bartl, 1989, 2006) reveal basal fundamental needs in the depth of the music.

Silence must be endured – to counter it, allowing it and accepting it requires strength, and entry into the contemplation that is detached from the prenatal world of sound's promise of life and togetherness requires courage. Silence not only recharges, as a contrast to the concentration of music where it can no longer be taken further it is often the loudest signal, which also needs to be properly listened to (Goh et al., 2023).

10 Application and Outlook

What can we learn from all this in the well-being sector which we can integrate into everyday management?

(I) Breathing is the key. Correct breathing makes a decisive contribution to well-being, and breathing exercises which can be accompanied or supported by music or are based entirely on music are suitable for enhancing personal well-being. In easy cases, blood pressure and heart rate can be lowered in this way as well.

(II) Music and dance should be *actively* practiced to retain physical as well as mental fitness and trigger neural plasticity. The piano is particularly suitable for preventive stabilization of mental health into an advanced age. It is never too late, although ideally people should start before or at 45.

(III) Fixation on just one kind of music and small musical forms is limiting. Singing and instrumental music have different benefits for increasing well-being. Critical review of publications and scientific studies helps identify the right activities for yourself. Professional advice on both medical and musical aspects is recommended.

(IV) Music can only have an effect if you give it a chance! The energetic interaction between people and music must be consistent and allow resonance. This applies not only to tempo and dynamic but also to musical style. Openness is the key! If you haven't already done so, discover Mahler's *Eighth Symphony*, allow yourself to be overwhelmed by musical sounds, and have a new experience of music every day.

(V) Musical activities to promote well-being should address both intellect and emotion and have a high artistic quality.

(VI) Creativity in all areas of life! Don't look for excuses to avoid creative concepts. Practice makes perfect—talent isn't everything. If music and dance are used to promote well-being, above all this should be a pleasant experience. Establish a routine; don't overdo it.

(VII) An ideal management experience is "orchestral management"—conducting. There's a lot of debate about the "purpose" of music and art, but research into the effect of music shows clearly that music is far more than entertainment; it's a fundamental human need going back to the prenatal environment. Music should be listened to and understood in exactly this sense but also selected for hearing. Not all music is equal, and encouraging deliberate preference of musical structures and all kinds of intellectual and sensory processes is a decisive step toward sustainability in the use of personal resources and personal potential for development. Not going with the crowd, leaving your comfort zone occasionally, daring to take a musical adventure, and adding highlights to your life—all this is a central "purpose" of the art, just as it can provide catharsis, create a link to the metaphysical, and be an outlet. In this sense, measures to improve the acoustic environment have decisive advantages in corporate structures! One of the most important principles is to create silence, seek silence, and make use of silence.

(VIII) Black out for technology and society, take time for internal and external silence, and find warmth, rhythm, and consistency in life. In the silence, find the Zen of music, and let breath and melodies dissolve in it.

References

Arias, M. (2019). Neurology of ecstatic religious and similar experiences: Ecstatic, orgasmic, and musicogenic seizures. Stendhal syndrome and autoscopic phenomena. Neurología del éxtasis y fenómenos aledaños: epilepsia extática, orgásmica y musicogénica. Síndrome de Stendhal. Fenómenos autoscópicos. *Neurología, 34*(1), 55–61. https://doi.org/10.1016/j.nrl.2016.04.010

Bachner-Melman, R., Dina, C., Zohar, A. H., Constantini, N., Lerer, E., Hoch, S., Sella, S., Nemanov, L., Gritsenko, I., Lichtenberg, P., Granot, R., & Ebstein, R. P. (2005). AVPR1a and SLC6A4 gene polymorphisms are associated with creative dance performance. *PLoS Genetics, 1*(3), e42. https://doi.org/10.1371/journal.pgen.0010042

Bangert, M., & Schlaug, G. (2006). Specialization of the specialized in features of external human brain morphology. *The European Journal of Neuroscience, 24*(6), 1832–1834. https://doi.org/10.1111/j.1460-9568.2006.05031.x

Barrett, K. C., Barrett, F. S., Jiradejvong, P., Rankin, S. K., Landau, A. T., & Limb, C. J. (2020). Classical creativity: A functional magnetic resonance imaging (fMRI) investigation of pianist and improviser Gabriela Montero. *NeuroImage, 209*, 116496. https://doi.org/10.1016/j.neuroimage.2019.116496

Bartl, G. (1989). Strukturbildung im therapeutischen Prozeß. In G. Bartl & F. Pesendorfer (Eds.), *Ärztliche Praxis und Psychotherapie* (Vol. Bd. 3). Literas: Universitätsverlag.

Bartl, G. (2006). Psychotherapeutische Medizin in der ärztlichen (Land-)Praxis. In *Integrierte Psychotherapie*. Springer. https://doi.org/10.1007/3-211-36889-2_3

Bengtsson, S. L., Nagy, Z., Skare, S., Forsman, L., Forssberg, H., & Ullén, F. (2005). Extensive piano practicing has regionally specific effects on white matter development. *Nature Neuroscience, 8*(9), 1148–1150. https://doi.org/10.1038/nn1516

Bernardi, L., Porta, C., Casucci, G., Balsamo, R., Bernardi, N. F., Fogari, R., & Sleight, P. (2009). Dynamic interactions between musical, cardiovascular, and cerebral rhythms in humans. *Circulation, 119*(25), 3171–3180. https://doi.org/10.1161/circulationaha.108.806174

Canetti, E. (1960). *Masse und Macht. Wesentliche Zusammenhänge zum Verständnis unseres Zeitalters*, 1960 im Claassen Verlag GmbH Hamburg erschienen, Lizenzausgabe veröffentlicht 1980 im Fischer Taschenbuch Verlag GmbH, 27th ed., 2001.

Czepiel, A., Fink, L. K., Fink, L. T., Wald-Fuhrmann, M., Tröndle, M., & Merrill, J. (2021). Synchrony in the periphery: Inter-subject correlation of physiological responses during live music concerts. *Scientific Reports, 11*(1), 22457. https://doi.org/10.1038/s41598-021-00492-3

Diels, H., & Kranz, W. (1989). *Die Fragmente der Vorsokratiker. Griechisch und Deutsch von Hermann Diels*. Herausgegeben von Walter Kranz.

Dingle, G. A., Sharman, L. S., Bauer, Z., Beckman, E., Broughton, M., Bunzli, E., Davidson, R., Draper, G., Fairley, S., Farrell, C., Flynn, L. M., Gomersall, S., Hong, M., Larwood, J., Lee, C., Lee, J., Nitschinsk, L., Peluso, N., Reedman, S. E., Vidas, D., … Wright, O. R. L. (2021). How do music activities affect health and Well-being? A scoping review of studies examining psychosocial mechanisms. *Frontiers in Psychology, 12*, 713818. https://doi.org/10.3389/fpsyg.2021.713818

Fan, P., Wong, A. C., & Wong, Y. K. (2022). Visual and visual association abilities predict skilled reading performance: The case of music sight-reading. *Journal of Experimental Psychology. General, 151*(11), 2683–2705. https://doi.org/10.1037/xge0001217

Goh, R. Z., Phillips, I. B., & Firestone, C. (2023). The perception of silence. *Proceedings of the National Academy of Sciences of the United States of America, 120*(29), e2301463120. https://doi.org/10.1073/pnas.2301463120

Graber, O. P., Tucek, G. (in print). *Vox humana. Das Geheimnis der Gregorianik aus Sicht der Musikwirkungsforschung*.

Grant, J. A., Courtemanche, J., Duerden, E. G., Duncan, G. H., & Rainville, P. (2010). Cortical thickness and pain sensitivity in zen meditators. *Emotion (Washington, D.C.), 10*(1), 43–53. https://doi.org/10.1037/a0018334

Gubler, D. A., Rominger, C., Holtforth, M. G., Egloff, N., Frickmann, F., Goetze, B., Harnik, M., Streitberger, K., Zeiss, S., & Troche, S. J. (2022). The impact of chronic pain on creative ideation: An examination of the underlying attention-related psychophysiological mechanisms. *European Journal of Pain (London, England), 26*(8), 1768–1780. https://doi.org/10.1002/ejp.2000

Haas, F., Distenfeld, S., & Axen, K. (1986). Effects of perceived musical rhythm on respiratory pattern. *Journal of Applied Physiology (Bethesda, Md.: 1985), 61*(3), 1185–1191. https://doi.org/10.1152/jappl.1986.61.3.1185

Haidet, P. (2007). Jazz and the 'art' of medicine: Improvisation in the medical encounter. *Annals of Family Medicine, 5*(2), 164–169. https://doi.org/10.1370/afm.624

Haidet, P., Jarecke, J., Yang, C., Teal, C. R., Street, R. L., & Stuckey, H. (2017). Using jazz as a metaphor to teach improvisational communication skills. *Healthcare (Basel, Switzerland), 5*(3), 41. https://doi.org/10.3390/healthcare5030041

Hudak, E. M., Bugos, J., Andel, R., Lister, J. J., Ji, M., & Edwards, J. D. (2019). Keys to staying sharp: A randomized clinical trial of piano training among older adults with and without mild cognitive impairment. *Contemporary Clinical Trials, 84*, 105789. https://doi.org/10.1016/j.cct.2019.06.003

James, C. E., Altenmüller, E., Kliegel, M., Krüger, T. H. C., Van De Ville, D., Worschech, F., Abdili, L., Scholz, D. S., Jünemann, K., Hering, A., Grouiller, F., Sinke, C., & Marie, D. (2020). Train the brain with music (TBM): Brain plasticity and cognitive benefits induced by musical training in elderly people in Germany and Switzerland, a study protocol for an RCT comparing musical instrumental practice to sensitization to music. *BMC Geriatrics, 20*(1), 418. https://doi.org/10.1186/s12877-020-01761-y

Jiménez, A., Lu, Y., Jambhekar, A., & Lahav, G. (2022). Principles, mechanisms and functions of entrainment in biological oscillators. *Interface Focus, 12*(3), 20210088. https://doi.org/10.1098/rsfs.2021.0088

Jünemann, K., Marie, D., Worschech, F., Scholz, D. S., Grouiller, F., Kliegel, M., Van De Ville, D., James, C. E., Krüger, T. H. C., Altenmüller, E., & Sinke, C. (2022). Six months of piano training in healthy elderly stabilizes white matter microstructure in the fornix, compared to an active control group. *Frontiers in Aging Neuroscience, 14*, 817889. https://doi.org/10.3389/fnagi.2022.817889

Kanduri, C., Kuusi, T., Ahvenainen, M., Philips, A. K., Lähdesmäki, H., & Järvelä, I. (2015). The effect of music performance on the transcriptome of professional musicians. *Scientific Reports, 5*, 9506. https://doi.org/10.1038/srep09506

Kisiel, R. (2008). Internet generation don't have the attention span for jury duty, warns Lord Chief Justice. Retrieved July 16, from https://www.dailymail.co.uk/news/article-1083815/Internet-generation-dont-attention-span-jury-duty-warns-Lord-Chief-Justice.html

Krones, H. (2006). Rhetorik und Musik, in: Oesterreichisches Musiklexikon online, begr. Von Rudolf Flotzinger, Ed. von Barbara Boisits. Last change of content: 15.5.2005. Retrieved July 17, 2023 from https://doi.org/10.1553/0x0001df23

Kulinski, J., Ofori, E. K., Visotcky, A., Smith, A., Sparapani, R., & Fleg, J. L. (2022). Effects of music on the cardiovascular system. *Trends in Cardiovascular Medicine, 32*(6), 390–398. https://doi.org/10.1016/j.tcm.2021.06.004

Kurth, E. (1931). Musikpsychologie. Hesse.

Laczika, K., Graber, O. P., Tucek, G., Lohninger, A., Fliri, N., Berka-Schmid, G., Masel, E. K., & Zielinski, C. C. (2013). "Il flauto magico" still works: Mozart's secret of ventilation. *Multidisciplinary Respiratory Medicine, 8*(1), 23. https://doi.org/10.1186/2049-6958-8-23

Lindenberger, U., Li, S. C., Gruber, W., et al. (2009). Brains swinging in concert: Cortical phase synchronization while playing guitar. *BMC Neuroscience, 10*, 22. https://doi.org/10.1186/1471-2202-10-22

Lütkenhöner, B., Seither-Preisler, A., & Seither, S. (2006). Piano tones evoke stronger magnetic fields than pure tones or noise, both in musicians and non-musicians. *NeuroImage, 30*(3), 927–937. https://doi.org/10.1016/j.neuroimage.2005.10.034

Miyashiro, S., Yamada, Y., Muta, T., Ishikawa, H., Abe, T., Hori, M., Oka, K., Koshikawa, F., & Ito, E. (2021). Activation of the orbitofrontal cortex by both meditation and exercise: A near-infrared spectroscopy study. *PLoS One, 16*(2), e0247685. https://doi.org/10.1371/journal.pone.0247685

Müller, V., Sänger, J., & Lindenberger, U. (2013). Intra- and inter-brain synchronization during musical improvisation on the guitar. *PLoS One, 8*(9), e73852. https://doi.org/10.1371/journal.pone.0073852

Nair, P. S., Kuusi, T., Ahvenainen, M., Philips, A. K., & Järvelä, I. (2019). Music-performance regulates microRNAs in professional musicians. *PeerJ, 7*, e6660. https://doi.org/10.7717/peerj.6660

Nietzsche, F. (1889). Nietzsche contra Wagner.

Oikkonen, J., Kuusi, T., Peltonen, P., Raijas, P., Ukkola-Vuoti, L., Karma, K., Onkamo, P., & Järvelä, I. (2016). Creative activities in music – A genome-wide linkage analysis. *PLoS One, 11*(2), e0148679. https://doi.org/10.1371/journal.pone.0148679

Parncutt, R., & Chuckrow, R. (2019). Chuckrow's theory of the prenatal origin of music. *Musicae Scientiae, 23*(4), 403–425. https://doi.org/10.1177/1029864917738130

Reicha, A. (1832). *Traité de haute composition musicale,* dt. Übersetzung von Carl Czerny, Diabelli, Wien.

Schoppmeyer, M. (2020). Atemfrequenz. Retrieved July 17, 2023, from https://www.pschyrembel.de/atemfrequenz/K033R/doc/

Slater, J., Azem, A., Nicol, T., Swedenborg, B., & Kraus, N. (2017). Variations on the theme of musical expertise: Cognitive and sensory processing in percussionists, vocalists and non-musicians. *The European Journal of Neuroscience, 45*(7), 952–963. https://doi.org/10.1111/ejn.13535

Specht, R. (1912). *Gustav Mahlers VIII. Sinfonie. Thematische Analyse mit einer Einleitung und dem Portrait Mahlers*. Universal Edition Leipzig, Wien.

Strait, D. L., Kraus, N., Skoe, E., & Ashley, R. (2009). Musical experience and neural efficiency: Effects of training on subcortical processing of vocal expressions of emotion. *The European Journal of Neuroscience, 29*(3), 661–668. https://doi.org/10.1111/j.1460-9568.2009.06617.x

Tarr, B., Launay, J., Cohen, E., & Dunbar, R. (2015). Synchrony and exertion during dance independently raise pain threshold and encourage social bonding. *Biology Letters, 11*(10), 20150767. https://doi.org/10.1098/rsbl.2015.0767

Tarr, B., Launay, J., & Dunbar, R. I. (2016). Silent disco: Dancing in synchrony leads to elevated pain thresholds and social closeness. *Evolution and Human Behavior: Official Journal of the Human Behavior and Evolution Society, 37*(5), 343–349. https://doi.org/10.1016/j.evolhumbehav.2016.02.004

Tavor, I., Botvinik-Nezer, R., Bernstein-Eliav, M., Tsarfaty, G., & Assaf, Y. (2020). Short-term plasticity following motor sequence learning revealed by diffusion magnetic resonance imaging. *Human Brain Mapping, 41*(2), 442–452. https://doi.org/10.1002/hbm.24814

Teie, D. (2016). A comparative analysis of the universal elements of music and the fetal environment. *Frontiers in Psychology, 7*, 1158. https://doi.org/10.3389/fpsyg.2016.01158

Vaquero, L., Hartmann, K., Ripollés, P., Rojo, N., Sierpowska, J., François, C., Càmara, E., van Vugt, F. T., Mohammadi, B., Samii, A., Münte, T. F., Rodríguez-Fornells, A., & Altenmüller, E. (2016). Structural neuroplasticity in expert pianists depends on the age of musical training onset. *NeuroImage, 126*, 106–119. https://doi.org/10.1016/j.neuroimage.2015.11.008

Videan, E. N., Fritz, J., Howell, S., & Murphy, J. (2007). Effects of two types and two genre of music on social behavior in captive chimpanzees (Pan troglodytes). *Journal of the American Association for Laboratory Animal Science: JAALAS, 46*(1), 66–70, und Snowdon, C.T. Animal Signals, Music and Emotional Well-Being. *Animals* 2021, 11, 2670. https://doi.org/10.3390/ani11092670

Walter, Y., & Altorfer, A. (2022). Physiological foundations for religious experiences in devotional worship practices with music using heart rate and respiration rate analyses. *European Journal of Investigation in Health, Psychology and Education, 12*(2), 127–143. https://doi.org/10.3390/ejihpe12020011

Worschech, F., James, C. E., Jünemann, K., Sinke, C., Krüger, T. H. C., Scholz, D. S., Kliegel, M., Marie, D., & Altenmüller, E. (2023). Fine motor control improves in older adults after 1 year of piano lessons: Analysis of individual development and its coupling with cognition and brain structure. *The European Journal of Neuroscience, 57*(12), 2040–2061. https://doi.org/10.1111/ejn.16031

Oliver Peter Graber is internationally active as a composer, pianist, and author. His main focus is on interdisciplinary projects at the intersection of music, ballet, and science. In addition to this artistic work, he taught at the University of Music and Performing Arts, Vienna, and was the dramaturge of the Vienna State Ballet (Wiener Staatsballett). Presently, he is the artistic head of the "Research Institute for Music Medicine with a Special Focus on Arts for Health" at the JAM MUSIC LAB Private University, Vienna.

The Workplace as a Challenge and a Resource for Adult Development

Rainer von Leoprechting

Abstract Well-being in organizations is different for different people. A major factor is the individual's maturity in life when facing organizational challenges. Scholars in adult development have identified various stages of maturity, each stage offering wider and structurally differentiated capabilities and limitations in dealing with life's challenges. Such maturity phases develop over an adult's lifetime. Their development occurs as a shift in structural qualities such as the nature of one's worldview or identity. In this perspective, well-being occurs when and especially after an individual experiences a stretch in their development and can subsequently live into it and grow. This chapter discusses ways how organizations can support individuals developmentally, either through a deliberately developmental culture or at scale by inherently presenting developmental opportunities in reach. Such inherent methods draw on digital intelligence such as the StoryMatcher developed by the author.

1 Introduction: On Adult Development

The field of adult development was established in the 1900s with the seminal works of Jean Piaget. This constructivist approach to understanding the epistemological evolution of knowledge attested that children construct their own version of reality through a combination of ideas and experiences embedded in social contexts (Piaget, 1963). Many researchers demonstrated that the cognitive and social-emotional development of individuals continues in well-structured and observable patterns beyond the age of 25, as was originally assumed (e.g., Kegan, 1980; Wilber, 2000; Laske, 2011).

R. von Leoprechting (✉)
Ehrenhausen, Austria

M. Mühlböck (ed.), *Sustainable Transformation and Well-being*, CSR, Sustainability, Ethics & Governance,
https://doi.org/10.1007/978-3-031-75566-8_8

1.1 Robert Kegan's Adult Developmental Model

Among the most widely recognized approaches to adult development theory is Robert Kegan's model of the evolving self (Kegan, 1982; 1994).

The model has been thoroughly analyzed and applied for approximately 50 years (Bauger et al., 2021). Kegan's approach to adult development is widely recognized as effective for assessing change processes in adult development and to ascertain the level of capabilities of individuals.

Its five distinct development stages can be summarized as follows:

- Stage 1: Impulsive Mind
  The impulsive mind makes sense of the world through interaction with content that is narrowly limited to their impulses, desires, and perceptions. Individuals at this development state are unable to consider oppositional views and approaches and thus require frequent reminders of rules and expectations.
- Stage 2: Instrumental Mind
  The instrumental mind is motivated by its interests, needs, and desires but has an increased capacity to bring multiple data and things into cognitive order to build models of how the world functions. Individuals in this stage tend to be self-oriented as they seek to instrumentalize the world to serve their needs, desires, and interests.
- Stage 3: Socialized Mind
  The socialized mind has matured enough to see beyond narrow self-interests and is oriented to maintaining connection within one's social groups. Instead of being exclusively self-oriented, individuals at this stage are foremost motivated to nurture interpersonal relationships and act in a manner that serves the needs of others. Instead of trying to model the behaviors of others, as in the case of stage 2, stage 3 individuals can both empathize and respond to the needs of others directly, without a mediating mental model.
- Stage 4: Self-Authoring Mind
  Building on the capacities from stage 3, the self-authoring mind has developed its own principles and values that guide its decisions, irrespective of the felt needs or interests of others. Individuals at this stage are motivated to provide meaningful contributions to interpersonal relationships. Critical at this stage is the ability to empathize and relate to others objectively without confusing one's own needs, wants, or agendas with those of others. The self-authoring mind is highly reflexive, evaluative, and responsible.
- Stage 5: Self-Transforming Mind
  The self-transforming mind has learned how to decouple attachment to their own values and principles so that they can hold various worldviews at the same moment. This increased capacity for dialectic inquiry and systemic thinking increases complexity tolerance, particularly at scales ranging from self to collective and from collective to systemic. The self-transforming mind is oriented

toward a system of decision-making that aligns with what one perceives as beneficial to the wider collective or even mankind as a whole.

- Liminal Spaces in Transition
  Between each of these stages are micro-stages—liminal spaces where the mind is influenced by both the stage before and the stage after. As the tension intensifies, a conflict ensues between the stages. It is not until a significant breakthrough that the old stage will sufficiently weaken its hold and the mind will mature through another phase of cognitive transformation. As the mind enters the new phase, it will reflexively consolidate outdated cognitive orientations. These breakthroughs are often associated with hard, painful decisions, where the individual will have to let go of something or someone valued. This letting go then, in some cases, transforms the meaning-making that had been associated with the matter. This is a form of identity death. Thus, growth most often comes with emotional resistance and distress in the process.

The outcome of adult development is traditionally associated with "wisdom," as we learn to be with what is, including life's messiness and all its limitations.

1.2 A Critical Distinction Between Cognitive and Social-Emotional Development

The author was personally introduced to the art and science of adult development by Otto Laske, who studied with Robert Kegan in the late 1990s. Laske distinguishes between social-emotional and cognitive development as structural movements in their own right, whereas Kegan's model mainly captures the social-emotional development with its guiding question: What shall I do, and for whom? Cognitive development is in contrast concerned with the actual understanding and conceptualizing of the world, under the practical questions (in organizational contexts) of: What is the situation, and what are my options (Laske, 2023a)? The concept of wisdom is made more explicit, as it entails various dimensions, of which the meaning-making and cognition are just two. Laske (2023a) also refers to the psychological dimension with the recognition of needs and partially unconscious drivers or traumata. While all of these dimensions influence each other, one can observe them distinctly.

Cognitive development occurs from a more linear, "simple logical" mind toward a "dialectical," more fully complex mind. To make this explicit, Laske distinguishes 28 thought forms, grouped in four moments of dialectical thinking (Laske, 2023c). As a proxy for cognitive development, one observes the degree of mastery of the full set of thought forms in a dialogue with an experienced practitioner.

As a practitioner concerned with leadership and the development of people with wider responsibilities, the author found this distinction highly valuable for ascertaining the levels of capability of an individual leader. The thought forms provide a rich basis for coaching, mentoring, and consulting (Table 1).

Table 1 Otto Laske's table of thought forms (TFs) as the basis of cognitive development

Process TFs	Context TFs	Relationship TFs	Transformational (meta-systemic) TFs
1. Focus on unceasing movement, hidden dimensions, and negativity	8. Focus on contextualizing a part within a whole; emphasis on part	15. Limits of separation. Focus on existence and value of relationship	22. Focus on limits of stability, harmony, and durability
2. Use of preservative negation (inclusion of antithesis or "other")	9. Focus on equilibrium of whole; emphasis on whole	16. Focus on the value of bringing into relationship	23. Value of conflict leading in a developmental direction
3. Focus on composition by interpenetrating opposites, correlates	10. Focus on structures, functions, and layers defining social systems	17. Critique of reductionism, unrelated discretes, and detotalization (neglecting common ground)	24. Value of developmental potential leading to higher level of functioning, integration, and social change
4. Patterns of ongoing interaction creating patters of movement	11. Focus on the hierarchical nature of structures and layers systems comprise	18. Focus on the relatedness of different value and judgment systems	25. Evaluative comparison of systems in transformation
5. Focus on the active, practical nature of knowledge	12. Focus on stability of system functioning	19. Focus on describing relationships in structural terms	26. Focus on process of coordinating systems
6. Critique of arresting motion and process (reification)	13. Focus on intellectual systems: frames of reference, traditions, and ideologies	20. Focus on describing patterns of interaction in relationships	27. Description of open, self-transforming systems
7. Focus on embeddedness in process, motion	14. Focus on multiplicity of contexts (non-transformational)	21. Focus on describing the constitutive relationship that determines the nature of what some is, its role, and reality	28. Integration of multiple perspectives in order to define complex realities

Note: For a full description and introduction into its use, see Laske (2023c)

The following sections discuss why the workplace has become an increasing challenge to an increasing amount of people and how practices rooted in both social-emotional and cognitive development offer a way forward so that individuals can experience a suitable level of challenge for themselves, which corresponds to their inner, developmental position.

2 The Workplace as a Challenge

In its landmark assessment of the mental demands of modernity, Robert Kegan (1994) discusses the requirements of the late 1900s civilization in many aspects of adulthood and consistently finds that the self-authoring mind is a prerequisite for proper functioning as an adult in various roles with responsibilities for others, like managing, parenting, and counseling, to mention only a few. The problem with this is that most adults don't develop such a self-authoring mind and stay within the reach of a socialized mind. The socialized mind is functional in contexts like traditional villages, where the sense of belonging is essential for the survival of the whole group, and principles and norms don't require constant adaptation.

Pre-modern or indigenous civilizations were much more conducive for survival and social acceptance, where conformity with proven norms as expressed in status quo and stability are expressions of that belonging (Kegan, 1994; Andreotti, 2021).

The challenge is thus produced within modernity itself with its foundational assumption of individually held responsibility and discernment, in its underlying context of dynamic progress and continuous change. The foundational premises of modernity add to the problem: everything is under the rule or control of man. That leads to the exploitation of nature and the living world, including the living humans.

With Andreotti (2021), one might argue that modernity as a civilizational concept is coming into its end phase, where the demands on our collective system can less and less be met by modernity's foundations.

In that sense, we live in a major crisis not just of external factors but foremost in a crisis of our deeply held worldviews and collectively assumed truths. That radicalizes Kegan's (1994) problem, in the sense that modernity not only calls our self-authoring mind to function within modernity. The decay of modernity itself requires more and more of a self-transforming mind so that old and new, more sustainable civilizational patterns can be brought into practice, and this at a planetary scale.

According to Kegan (1994), only 5–7% of adult populations ever reach such a level of maturity, and that often at an age when in modern societies people are expected to have retired from active duties, separated from formal or actual responsibilities.

Contemporary workplaces are a representation of those tendencies, stemming from a variety of factors. To name just a few: the ongoing digitalization and ever faster deployment of new technologies, the globalization of the economy and societies, and increasing stakeholder pressures on business like the pursuit of the Sustainable Development Goals of the United Nations. Underlying those factors is the slow but steady unfolding of climate change. Besides such global trends, any company has to deal with their inner tensions, the introduction of New Work or agile project management together with specific developments of their customers or suppliers.

In brief, the future has become much less predictable, and people live with higher levels of uncertainty and need to observe many more weak signals as managers compared with the generally accepted management standards of the 1990s. A whole generation has seen the rather fast transformation of business and their workplaces in some areas, while their own inner transformation is being challenged to keep the pace. Obviously, changes in contexts outside of us can provoke development. Most often, however, it will generate denial or other forms of resistance so that we can keep our integrity and stay alive in the way we used to be.

An increasing disconnect between the inner and the outer world will manifest itself in various forms of unease or disease. Not surprising from that point of view, the numbers of burnout cases in the western world have been increasing since the early 2000s to date at a staggering rate of 500%: More than six in 1000 workers in Germany received an invalidity pension due to severe burnout in 2022, when in 2005 it was just one per thousand (Radtke, 2024). Among managers, the rates are even slightly higher than among employees in general. This is explained by the higher complexities and conflicts related to the demands of leadership roles (Klinghoffer & Kirkpatrick-Husk, 2023). Kegan's (1994) book title *In Over Our Heads* says it well.

3 Making the Workplace a Resource: Deliberate and Inherent Strategies

3.1 The Workplace as a Resource for Development and Well-Being

If the workplace can stress us out, could it do the opposite and offer us fulfillment, joy, and meaning? As adults, we spend most of our wake time at work. We train and learn for it; we feel our value in society through it and often identify with it fully or in some aspects of it. Developmentally speaking, work is where our capacities play out, where we can demonstrate what we are capable of and perform. We receive positive feedback and recognition. Even faced with obstacles, people can be happy and engaged at work when they know they are pursuing something meaningful. In the early days of his career in the European institutions in Brussels, the author met officials who were about to retire, radiating with optimism, humor, and warmth. What this generation had lived through was the rebuilding of Europe as a peaceful continent after the Second World War. They experienced their work as meaningful and could relate to it. Many of the people who were among the founding generation of the EU reached very old ages and stayed mentally fit for a long time.

While not everyone can connect their work to such historic endeavors, the point here is that the pursuit of meaning at work is generative for health and well-being (see, for instance, the numerous survey results from Gallup in the summary from Harter, 2022). In view of the numerous present challenges, there are still plenty of opportunities to find meaningful work today. For instance, the Sustainable Development Goals offer a succinct summary of them. The challenge in finding meaning at work lies, on the one hand, in the fact that many of today's jobs do not even strive for such noble goals or, on the other hand, that the meaningful output is not visible to the people involved. The developmental challenge is increased when many people at work don't feel the power to find or create such meaning in their work life.

Leaders of organizations or human resources managers concerned with this issue can pursue two main strategies to address this, deliberate and more inherent or covert ones.

3.2 *Deliberately Developmental Strategies and Cultures*

Harvard professors Lahey and Kegan (2016) coined the term "deliberately developmental organization" (DDO) to characterize a New Work approach to support people at work to hold two jobs, their performance job and their developmental job. DDOs recognize the developmental job as part of the work contract with the organization and share part of the developmental responsibility, rather than dumping it on the shoulders of the workers alone.

Such organizations foster humanity at work for everyone and value meaningful exchanges at the workplace. Practices like coaching or developmental conversations are not reserved to a privileged group of senior managers alone but are part of the day-to-day culture at the workplace.

Geveke (2017) gives a summary of popular practices in such organizations:

- Bring purpose into play. One of the breakthroughs in creating organizations where people can be their whole self is aligning the purpose of the organization and individual purpose as a person (Laloux, 2014). If development is part of the purpose of a DDO, then alignment between 'what is my personal desired future and what is the development to that' and the 'desired development for the organization' should be part of a continuous conversation. A supporting question to share could be: 'how can I serve the organization in its development, and how can the organization serve me in my development?'.
- Start experimenting with positive, strength-based practices for development. It is not only our limitations or 'screwing up' where growth is possible. More and more we come to an understanding that by focusing our attention on our strengths and on 'what is working well already and do we want more of', we unleash the energy for positive development of ourselves and the organization. [...] A first practice to consider might be adding a 'success inquiry' in your team meetings. Ask each member 'what has been working really well for you in the last month' or 'what are you really proud of right now'. These types of check-in questions support building positive relationships and energy within the team and for each individual. It creates space to bring our authentic self into the conversation.
- Create a learning environment 'by and for the people', giving opportunities to grow by teaching others. You might call that an in-house academy, more importantly it is a sustainable learning environment dedicated to create space for individual and collective learning moments. Linked into the corporate strategy, the academy connects into all stakeholders in the organization through a networked way where every employee is a co-creator of that academy.

The academy facilitates conversation around all three dimensions of the DDO:

- what are our development aspirations and how can we support those?
- where can development take place in a participatory environment?
- what practices work well here or should we add, and how can we reinforce those?

In the light of primary and secondary processes, we might even say that 'development' becomes the primary process of the organization. Through development we create new products, services or innovation in general.

- In a DDO, managers and leaders are coaches and mentors in the complex, ongoing process of organizational adaptation of which everyone in the DDO is a part. They are instead shapers of the conditions and structures that will allow crew members on the front lines to continually participate in improving the way they work. The DDO dimension 'Home' involves lessons around a more participatory workplace community.

What can we do to start experimenting with more participatory approaches? An evolving practice is the Art of Hosting approach. It blends a suite of powerful conversational processes to invite people to step in and take charge of the challenges facing them. … Participants create and manage their own agenda of parallel working sessions where they share their questions and work together towards possible steps forward, suggestions, solutions. As said, leaders shape the space for conversation and idea generation, and are an equal participant during the open space process.

Ten years after Laloux's (2014) bestselling study about Next-Stage organizations and 8 years after the DDO concept, some sobering has happened among the early enthusiasts. By far not as many managers became successful coaches. The wish to introduce a wider purpose into the work of everyone has generated higher levels of confusion rather than higher levels of job satisfaction. Overall, the work implied in developing a developmental culture feels so countercultural to our dominant modern lifestyles and general norms that the actual realization of DDOs has been limited to few companies. Furthermore, the DDO concept relies on the support of the power hierarchy in organizations, mainly strong individual champions in leadership roles who support the introduction and running of the "academy" as outlined by Geveke above. Gluzdov (2020) shares the painful experiences a well-intentioned leader faces when introducing the DDO approach in the organization they lead and that many of the middle managers left the organization as a consequence.

The main difficulty in implementing a new work culture is that it creates an even more complex work situation for both the people introducing it and for most of the staff involved, as well. In brief, the developmental demands for people risk increasing even further as one tries to consciously address them.

What is not easy does not mean it shouldn't be done or pursued. The author is a co-founder of the consultancy group Fraendi. The group's work is to support and accompany leaders in and of organizations and social initiatives to develop themselves, their ventures, and society. However, the main consideration is not to impose a new standard system on the existing system but to support the development of new patterns from within the organization itself.

These principles seem helpful to avoid some of the most painful pitfalls:

3.2.1 Distinguish Between Business, Organizational, and Individual Purpose

If a group of leaders adopts a meaningful, social purpose, this should not become the purpose of the organization but of the business model. It is in the interactions between stakeholders involved in the business of an organization that a wider purpose is achieved. The organization itself always has a much more modest purpose, which is to serve that business model with appropriate processes, structures, and talent. Individual purpose is found or challenged at the level of individual jobs, where the person's talent may or may not fully unfold in a given span of time (this is inspired by Jaques, 1998).

3.2.2 Support Culture to Develop Gradually and from Within, One Step After Another

Many attempts to "Go Teal"[1] or DDO have been made too fast and too big at a time. Culture is among the slowest things to change, as it is the glue between people and minds that orients everyone without even showing itself. It can't become more human or more evolved by imposition but more by its own unfolding where the necessities of work align with more human sides of the organization, society, politics, and so on. The way the author and his colleagues from Fraendi work is to make important features of the present culture more explicit and to listen to emergent norms, support them by example, and connect them with the processes that support the actual business model. Culture and the behavior that enacts it are then indeed purpose-driven as they support the actual purpose of the organization.

3.2.3 Use Developmental Leadership Assessments for Critical Functions

Building on the insights discussed above, it is unavoidable to observe where people are at in their developmental stages and their cognitive capacity in comparison to the demands of the jobs that are offered to them. The analysis of job requirements can be made in a rather objective way through assessing the degrees of complexity a decision-maker needs to be able to manage, the number of meaningful human relationships to nurture, and the responsibilities toward others.

---

[1] More popular than DDO has been the TEAL movement inspired by Frédéric Laloux's book *Reinventing Organizations* (2014). Laloux argued that organizational cultures follow similar developmental patterns as individuals toward more inclusion and differentiation. "Teal" is a metaphor for "Next-Stage" organizations that foster distinct qualities, like holding participatory or "self-management" principles and having a social-purpose orientation. Many managers, founders of start-ups, and consultants have adopted Teal principles in their work and practice. "Going Teal" became a shorthand for attempts to realize a self-managed organization in existing companies.

To understand the levels of capacity of a person, the author together with a trained team of expert collaborators uses Otto Laske's cognitive interview and assessment (Laske, 2023b) as improved by Iva Vurdelja (2011) and the Subject-Object Interview (see Lahey et al. 2011) to ascertain the developmental position in the Kegan stages of development, as improved by Otto Laske (2011, 2023d). This consists of an in-depth analysis of the language they use when in dialogue with a skillful interviewer. The developmental leadership assessments are peer-reviewed by developmentally trained consultants. The developmental feedback is shared with the person and their hiring managers and gives a highly qualitative assessment of the possible performance of the person in the given work environment.

3.3 *Inherently Developmental Strategies and Processes at Scale*

Inherent strategies will bring developmentally informed services to bear, inside or toward organizations, without the need for visible support or developmental training for leaders or managers. There is no visible champion required to change the organization, nor heavy organizational development initiatives to visibly change the culture.

Instead, the service toward development is embedded in practices, tools, and services that are regularly applied to renew the organization. An obvious application is the recruitment for vacant positions, internal talent development to more senior functions, or horizontal job moves of individuals.

The principle that candidates for senior roles undergo a leadership assessment is in itself informed and done by developmental experts. The world of assessment experts is not there, however. Many assessment professionals will need to learn and apply developmental assessment principles rather than purely psychological practices.

The principle can be adhered to even more inherently, when job portals or in-house learning and talent portals are developmentally informed through a developmental digital intelligence. This is the route the author has been pursuing while developing the StoryMatcher service.[2]

Through that tool, individuals demonstrate their own levels of development in a self-test online when choosing from preferred success moments at work from a wide portfolio of scenarios which are relevant to their workplace. The scenarios have been formulated in such a way that the choices people take tend to correspond to their levels of development according to Kegan's scale. The text analysis of a Subject-Object Interview to ascertain the developmental position of an individual operates in the same logic. (A more detailed discussion is in von Leoprechting, 2023).

[2] https://storymatcher.com/

Thus, when the StoryMatcher tool is embedded in a job portal or learning portal, people will find career or learning opportunities where their present level of development is mildly stretched so that their chances of succeeding from a developmental perspective are high. Hiring managers or colleagues are the ones who select scenarios to fit the vacant position and their desires for a work culture.

Through their selection of desired success scenarios, both sides find each other and their mutually desired culture around the vacancy. As such, a very important element of potential friction has been addressed early on without big fanfare.

If both mentors/coaches and mentees/coachees use the StoryMatcher service for mentoring or coaching services and create a profile, the match for coaches and mentors can be arranged through the coaching portal so that mentors or coaches are systematically a developmental step ahead of their proposed coachee or mentee. A very important principle for developmental coaching is to ensure that coaches should themselves have already experienced the structural shift in developmental terms, which the coachee is just about to get into. All of this happens through online platforms. Individual users don't even need to be aware of the developmental nature of the service, as it is a simple part of their search or recruitment process.

While such tools are still in their early stages, their further development and application in more and more organizations seems to be the most promising strategy to date to address the workplace gap between how people think and what the work requires.

4 Conclusion

We live in times that challenge our inherited worldview and beliefs about ourselves and how organizations work. The ever-accelerating pressure to change, reinvent, and successfully lead an organization has been driving more and more managers and employees to being overwhelmed or even burnt out. This article argued that the underlying crisis of modernity is a deeper level cause. In today's modern organizations and society, humans are reaching more and more limits with regard to what the dominant modern worldview can hold. It is thus unavoidable and necessary that complementary worldviews and holistic practices supplement the modern management systems in operation. The insights about the development of adults over a lifetime offer such a complementary perspective as the requirements of leaders can be weighed against the capabilities of the worldview they hold.

Even in the light of the many, mostly not lasting, attempts to introduce deliberately developmental organizations, there is no shortcut to build both visible and invisible processes in organizations to continuously balance the requirements of the workplace with the developmental capabilities of every worker, identifying and supporting the potential in humans to self-author or self-transform.

The good news is that inherent strategies offer a promising new way to complement or even replace deliberate approaches. Inherent strategies embed developmentally informed software tools in recurring processes like recruitment or professional development and training. Such approaches are much easier to maintain than deliberate initiatives that depend on a sustained effort from enlightened leaders or even go against the need of organizations to support their business model in the first place. As the software tool records and manages the developmental information, only a few experts need the special training and dedication required to maintain their functioning, whereas deliberate approaches need the commitment of everyone. Therefore, inherent models should be further developed and applied as the most promising way to address the developmental challenge in organizations.

References

Andreotti, V. (2021). *Hospicing modernity. Facing humanity's wrongs and the implications for social activism by Vanessa Machado de Oliveira*. Penguin Random House.

Bauger, L., Bongaardt, R., & Bauer, J. J. (2021). Maturity and well-being: The development of self-authorship, eudaimonic motives, age, and subjective well-being. *Journal of Happiness Studies, 22*(3), 1313–1340. https://doi.org/10.1007/s10902-020-00274-0

Geveke, O. (2017). *Practices of a deliberately developmental organisation*. Enlivening Edge. https://enliveningedge.org/tools-practices/potential-practices-towards-deliberately-developmental-organization/

Gluzdov, N. (2020). *Deliberately developmental: How I helped my company upgrade, while also becoming a better person*. Medium. https://medium.com/@gluzd/deliberately-developmental-28e2dc7594c

Harter, J. (2022). *Employee engagement vs. employee satisfaction and organizational culture*. Gallup. https://www.gallup.com/workplace/236366/right-culture-not-employee-satisfaction.aspx

Ives, C. D., Schäpke, N., Woiwode, C., & Wamsler, C. (2023). IMAGINE sustainability: Integrated inner-outer transformation in research, education and practice. *Sustainability Science*. https://doi.org/10.1007/s11625-023-01368-3

Jaques, E. (1998). *Requisite organization. A total system for effective managerial organization and managerial leadership for the 21st century*. Cason Hall & Co.

Kegan, R. (1980). Making meaning: The constructive-developmental approach to persons and practice. *The Personnel and Guidance Journal, 58*(5), 373–380. https://doi.org/10.1002/j.2164-4918.1980.tb00416.x

Kegan, R. (1982). The Evolving Self. Problem and Process in Human Develoment. Harvard University Press.

Kegan, R. (1994). *In over our heads. The mental demands of modern life*. Harvard University Press.

Klinghoffer, D., & Kirkpatrick-Husk, K. (2023). More than 50% of managers feel burned out. *Harvard Business Review*. https://hbr.org/2023/05/more-than-50-of-managers-feel-burned-out

Lahey, L., & Kegan, R. (2016). *An everyone culture: Becoming a deliberately developmental organization*. Harvard Business Review Press.

Lahey, L., Souvaine, E., Kegan, R., Goodman, R., & Felix, S. (2011). *A guide to the subject-object interview: Its administration and interpretation*. CreateSpace Independent Publishing Platform.

Laloux, F. (2014). *Reinventing organizations: A guide to creating organizations inspired by the next stage in human consciousness*. Nelson Parker.

Laske, O. (2011). *Measuring hidden dimensions. The art and science of fully engaging adults* (2nd ed.). Interdevelopmental Institute Press.

Laske, O. (2023a). *Advanced systems-level problem solving* (Approaching real-world complexity with dialectical thinking) (Vol. 1, 2nd ed.). Springer.

Laske, O. (2023b). *Advanced systems-level problem solving* (How to measure and boost thought maturity) (Vol. 2, 2nd ed.). Springer.
Laske, O. (2023c). *Advanced systems-level problem solving* (Manual of dialectical thought forms) (Vol. 3, 2nd ed.). Springer.
Laske, O. (2023d). *Potenziale in Menschen erkennen, wecken und messen. Handbuch der Entwicklungsorientierten Beratung* (3. erweiterte und aktualisierte Auflage). Deutsche Übersetzung von Laske (2011) von Leoprechting, R. Pabst Science Publishers.
Piaget, J. (1963). The attainment of invariants and reversible operations in the development of thinking. *Social Research, 30*(3), 283–299.
Radtke, R. (2024). Arbeitsunfähigkeitsfälle aufgrund von Burn-out-Erkrankungen in Deutschland in den Jahren 2004 bis 2022 (je 1.000 AOK-Mitglieder). https://de.statista.com/statistik/daten/studie/239872/umfrage/arbeitsunfaehigkeitsfaelle-aufgrund-von-burn-out-erkrankungen/
Von Leoprechting, R. (2023). Digital orientation for life choices: Introducing StoryMatcher. *Integral Review, 18*(1). 422–430
Vurdelja, I. (2011). *How leaders think: Measuring cognitive complexity in leading organizational change*. Antioch University Libraries.
Wilber, K. (2000). *Integral psychology: Consciousness, spirit, psychology, therapy* (1st pbk. ed.). Shambhala.

Rainer von Leoprechting (*1965 in Germany) studied economics and its principles in Witten/Herdecke. He served as an EU official in Brussels in managerial roles in budget, evaluation, career guidance, strategy making, design of the EU's personnel portal, and leadership development and learning and started and led the internal consulting service of the European Commission. Relating to the EU's founder, Jean Monnet, he learned and applied any process that could enliven a bureaucracy. He studied adult development and Laske's Constructive Developmental Framework with Otto Laske and translated Laske (2011) into German (Laske, 2023d). Since 2012, he has been working as an independent consultant and companion for leaders to build new systems in light of the challenges of our times. He co-founded Fraendi in 2019 and has developed StoryMatcher as a developmental service for social platforms and portals.

Cornerstones of the School of the Twenty-First Century in the Context of Sustainability and Well-Being

Andreas Salcher

Abstract The vision of a "learning person" is at the heart of overcoming challenges in connection with the digitalization of the world of work, climate change, and social justice. In this context, schools play a central role as a place of teaching and learning. One aspect in particular is necessary to ensure that schools are able to meet the demands of their time: the willingness to engage in ongoing school development. Based on findings from various research disciplines, in particular learning research, sociology, and psychology, as well as an ongoing analysis of best practice examples over many years, the author of this chapter presents principles, forms of work, and methods for a twenty-first century school that provide fertile ground for "learning people." Those are closely linked to the acquisition of skills that make a significant contribution to increasing sustainability and well-being for people, society, and our planet.

Keywords Twenty-first-century school · Learning person · School development · Skills for sustainability · Skills for well-being · Best practices in education

1 Introduction

The UN's 17 Sustainable Development Goals (SDGs) provide us with a guideline for tackling the challenges of our time. If we ask ourselves what role the education system can and should play in this context, we end up in a debate about the skills it should consistently teach. The aim is to support people in achieving these socially necessary goals and in shaping a self-determined and meaningful life, far beyond

A. Salcher (✉)
Vienna, Austria
e-mail: andreas@salcher.co.at

M. Mühlböck (ed.), *Sustainable Transformation and Well-being*, CSR, Sustainability, Ethics & Governance,
https://doi.org/10.1007/978-3-031-75566-8_9

their school career. The 4C model of the twenty-first-century skills developed by the OECD and academics, for example, provides a practical framework, with the main areas of competence being communication, collaboration, creativity, and critical thinking. Both the SDGs and the 4C model open up a variety of starting points for concrete projects for learners. The author of this chapter has evaluated his vision of the holistic embedding of a "learning person" in a "learning community" in the project "Towards a learning nation." The central idea here is not to think of school development primarily from the perspective of individual schools but to form learning regional communities in order to strengthen quality of life and social cohesion (Salcher, 2023). In the following, he presents the pillars outlined in this context for a school of the twenty-first century and supplements them with aspects from Theory U, a method by Otto Scharmer that aims to encourage a holistic transformation of people and organizations.

There is no one right model of school, just as there is no one best way to run a business or educate a child. There are many exemplary schools that show how tried and tested experiences and scientific findings can be successfully implemented for the benefit of curious pupils, committed teachers, and responsible parents. Despite all the differences in content and methodology, innovative school projects have one essential thing in common: the willingness to engage in permanent systemic school development (Salcher, 2023). This builds on a certain basic pedagogical attitude of the school management and teachers, which must be developed and constantly consolidated so that it is not lost in everyday school life. That means passionate and persistent commitment over long periods of time. There are no shortcuts or magic recipes for success along the way. Nevertheless, principles, forms of work, and methods have crystallized that support such a path.

The following selection does not claim to be exhaustive but should be seen as an exemplary guideline for schools in the twenty-first century. It is based on findings from various research disciplines, in particular learning research, sociology, and psychology, as well as an ongoing analysis of best practice examples over many years.[1]

[1] The principles presented as well as the selected forms of work and methods of successful school development are based on the work of Peter Fratton, Bettina Fuhrmann, Daniel Goleman, Markus Hengstschläger, Reinhard Kahl, Harald Katzmair, Michael Meyer, Manfred Prenzel, Margret Rasfeld, Hartmut Rosa, Otto Scharmer, Peter Senge, Michael Schratz, Johannes Steyrer, and Klaus Zierer, among others.

2 Central Principles

2.1 Learning Takes Place Through Relationships: Everything Is Subject to the Law of Appreciation

During the pandemic, the Benedictine monk David Steindl-Rast supported the "Presencia" education project in Argentina, in which the co-founder of "Schule im Aufbruch," Margret Rasfeld,[2] was also involved. The aim was to create happy teachers for the school system, because happy teachers would mean happy children and parents. This concept has worked so well that tens of thousands of Argentinian teachers have now been trained accordingly and are implementing this mindset in their schools. David Steindl-Rast's task was to support the project with his doctrine of grateful living: "For Plato, all learning was a recollection, the unfolding of one's own views from within. Teaching is therefore above all the promotion of self-development. This requires teachers who know their students well. It would be helpful to always be aware of the ideal of the personal relationship between students and teachers, because it is often not expected at all."[3]

This thesis by David Steindl-Rast, formulated from his spiritual perspective, is highly consistent with the following findings from learning research: For example, in the often-cited John Hattie study "Visible Learning," appreciative relationships between teachers and learners are a key factor of sustainable learning (Hattie, 2009).

Hattie is concerned with the mutual visualization and understanding of the learning process between learners and teachers. As long as the channels between people are blocked by incomprehension, indifference, and mistrust, they cannot perceive each other. As soon as learners and teachers realize that it is not primarily about learning for exams and adhering to curricula but first and foremost about the human relationship, mutual trust can develop to such an extent that a teacher can venture into open learning without immediately risking their entire authority if a new method does not immediately work as desired. Establishing this attitude plays an important role, for example, in the change in teaching practice brought about by the introduction of digital devices in schools. If a teacher is integrated into a team with other teachers and is not left to their own devices, they will make faster progress in implementing new forms of learning.

This principle also becomes clear in the case study of the Gebrüder Grimm School in Hamm, Germany (North Rhine-Westphalia): that elementary school with major social challenges, where one in ten children has special educational needs, manages to ensure that all children can recognize and show their talents and

[2] As a principal, Margret Rasfeld established the Evangelische Schule Berlin Zentrum (ESBZ), which encourages school administrators, teachers, parents, and pupils across national borders to rethink their own schools. In 2012, together with Gerald Hüther and Stefan Breidenbach, she founded Schule im Aufbruch to initiate a grassroots movement for a new school culture (https://schule-im-aufbruch.de/).

[3] Interview with David Steindl-Rast on November 18, 2021, at the Gut Aich monastery.

strengths. It starts with the "morning dance" in the school foyer. Pupils can decide for themselves whether they want to dance or prefer to start the school day a little more quietly in the classroom. Learning time then begins. During this quarter of an hour in the morning, the pupils work independently on their individual learning goals, which they have written down on small cards, in stimulatingly structured classrooms while soft music plays in the background. After this learning time, the teachers discuss the school day together with their pupils. Praise and appreciation play a very important role at the school. Small mirrors hang at children's eye level in the stairwell with sayings such as: "You are wonderful!" or "You shine like the sun!" There are small boxes with compliment cards in various places around the school to give away as a way of practicing positive feedback.

All children start each morning in the same "good weather position." Those who follow the class rules and work quietly and carefully can improve their position in the "class weather." If the pupils land at the top of the rainbow 15 times, they receive a reward. The central element of the praise culture is the praise letters. In those letters, the children are praised for special talents or for good behavior in everyday school life, for example, when they help new classmates to find their way around. They are read aloud at the monthly school assemblies. A decisive criterion for the success of the school is evidently the consistently implemented strategy of "learning happens through relationships," both in the relationship between teachers and pupils and within the peer group of learners.

The leading design thinking company IDEO,[4] together with PowerSchool,[5] interviewed teachers, their coaches, administrative staff, and education experts about their changed teaching behavior due to the pandemic (Young & Daniel, 2022). The principle of "humanity first" emerged as one of the most important commonalities among innovative teachers in the study. Teachers developed creative ways to better understand the emotional needs of their students. They led meditations, organized weekly social "campfires" in their classes, were more tolerant of deadlines, and allowed tests to be repeated. This is a quote from a teacher: "As innovators, we need to be flexible with human fallibility and find ways to be more humane in our leadership roles. We need to prioritize the health and well-being of the people we work with over everything else."

2.2 *Learning Needs Resonance and the Acceptance of Unavailability*

The German sociologist Hartmut Rosa distinguishes between competence and resonance (Rosa & Endres, 2016): "Competence means the secure mastery of a technique, the ability to dispose of something at any time that I have acquired as a

---

[4] www.ideo.com

[5] www.powerschool.com

possession. Resonance, on the other hand, means entering into a processual relationship with something [...] Resonance contains a moment of openness and unavailability that distinguishes it from competence. Competence is appropriation, resonance means the transformation of the world: I also transform myself in the process."[6]

Rosa defines a mode of relationship that can be determined by four characteristics. This theoretical concept is explained here using the example of the best practice case study on the "district opera" at the Bremen-Ost Comprehensive School (Gesamtschule Bremen-Ost):

1. The moment of contact: Suddenly something calls to us, moves us from the outside, and gains meaning for us for its own sake.

   Case study Bremen East Comprehensive School: The city council provided the Chamber Philharmonic Orchestra Bremen with rehearsal rooms in the Bremen East Comprehensive School, which is a "Brennpunkt Schule," which is a school in a district in which factors leading to social disadvantage such as poverty and unemployment occur with above-average frequency. There was no intention, and certainly no plan, to bring the professional musicians into contact with the students. The pupils listened to the musicians rehearsing through the walls and found it amazing that music could not only be "downloaded" from the web but that one could make it oneself. They learned from the musicians that mastering a classical instrument was not based on an innate ability but that the orchestral sound could only be achieved through hard professional work.
2. The moment of self-efficacy: On the other hand, we can only speak of resonance when there is an active response to this contact (or invocation).

   Case study Bremen East Comprehensive School: The musicians developed a curiosity for the target group of educationally disadvantaged children who had little or no experience with classical music. Mutual trust developed in everyday encounters and musicians began to take on mentorship for individual classes without switching from the role of members of a world-class ensemble to that of a music teacher.
3. The moment of transformation: There are encounters that we say "transformed us into a different person," and there are transformations that bring about a barely noticeable and perhaps only temporary change, for example, in our mood.

   Case study Bremen East Comprehensive School: The idea of a future lab was born, in which joint projects were developed between the musicians, teachers, and students. Every 9 months, the children, supported by the musicians, presented their dreams, longings, fears, and fantasies in self-written songs to a large audience under the motto "Melody of Life" ("Melodie des Lebens").
4. The moment of unavailability: First of all, this means that there is no method and no seven- or nine-step guide to ensure that we can enter into resonance with people or things. Above all, it means a rejection of the so beloved master plans of school policy.

[6] Translated from German.

Case study Bremen East Comprehensive School: At some point, the spark was ignited to create something really big that would have an impact far beyond the boundaries of the school. In a huge tent surrounded by high-rise buildings far away from the high culture of the city center, up to 700 participants performed the "district opera," which has taken place eight times so far. Children learned that their vocabulary was not sufficient for a cultural project and began to expand it on their own initiative. When they had to communicate with a Swedish conductor, their command of English suddenly made sense because it was necessary. The positive learning effect can also be seen in the school reports and the graduation rate: In the school years from 2011 to 2014, all but two seriously ill pupils graduated in year ten. "Despite 'district opera' and 'melody of life,'" says principal Annette Rüggeberg and immediately corrects herself: "Not despite, but because of, because of that!" Many stay at Bremen-Ost Comprehensive School and complete their high school diploma there.[7]

Hartmut Rosa's conclusion is as follows: Such stories cannot be planned. The actors are not the implementers of external concepts. Every good teacher knows that the decisive learning processes often take place outside of the curriculum, often precisely in those phases in which the curriculum, which attempts to define when what must be learned and mastered, has no meaning. That means that the phases of curriculum learning should be reduced, which takes a lot of pressure off both teachers and learners because there are fewer "planned objectives" to achieve.

This finding is also confirmed by the study "The Talented Teenager" by happiness researcher Mihaly Csikszentmihalyi et al. (1993): 75% of the things that students had to do in formal lessons were perceived as demotivating. It is interesting to note that both motivation and attentiveness were quite high in voluntary exercises that took place outside of formal lessons. It was therefore by no means a lack of interest in learning but obviously the nature of traditional school lessons that so repelled learners. The vast majority felt that the time spent in regular lessons was boring—time that was seen as lost because there was obviously a lack of resonance. This was particularly true for highly talented pupils.

2.3 A Learner-Centered Attitude Leads to a Shared Responsibility for the Design of Learning Processes

Learning is the opposite of being taught, so innovative schools do not see themselves as teaching schools. Schley and Schratz (2021) describe this as a learner-centered attitude. Learning is the highly personal expression of development, the appropriation of the world, and the internalization of insights, methods, and knowledge. Learning is as personal as the different personalities within a learning space.

[7] https://www.youtube.com/watch?v=D4_1AFPZAG8, https://www.youtube.com/watch?v=z8s6Pb4x6Hs

A learner-centered approach aims to open up individual approaches for students to achieve their goals in self-determined learning from an ego perspective. That makes the resonance experiences described by Hartmut Rosa possible, and learning success is understood as a shared responsibility.

That is illustrated by a case study at "Schule am See" in Hard, a town in the westernmost Austrian state of Vorarlberg: The jointly managed primary and secondary school consistently implements its positioning with "individualization and multigrade classes" from the school building to the pedagogical concepts. The structure alternates between independent learning and short, frontal learning impulses for the whole class or group. The establishment of free learning phases in learning offices offers learners the opportunity to solve tasks individually and in a competence-oriented manner and to receive support as required. The acquisition of skills is the top priority. Teachers discuss and plan the learning objectives with each child, taking into account personal gifts and talents. The aim is self-responsible and meaningful learning from planning to performance feedback. The basis is an annual plan or 3-year plan with the involvement of the pupils. They are an active part of the planning and create individual logbooks, learning paths, and learning maps with the help of the teachers. They practice taking responsibility for their learning through learning goal or tutorial discussions.

Social skills and integration are very important at "Schule am See." By mixing the age groups, learners experience diversity and feel that being different is normal and right—the inclusion of children with special needs, those with a first language other than German (around 41% of pupils), and those with intellectual giftedness happens completely naturally at all school levels. Students with giftedness and students with learning difficulties learn together.

The prerequisite for the implementation of a learner-centered approach as described in this case study are teams of teachers with clear responsibilities and ownership in their pedagogical work. Pupils are also self-motivated to work toward their learning goals within clear structures and supported by learning guidance. To ensure that such models do not become too complex, the appropriate "human scale" must be found. That refers to the relationship between the individual and the necessary organizational structure that ensures that goals can be achieved.

Elementary schools have the advantage that the teachers usually know their pupils very well over 4 years and can therefore assess and develop them well in terms of both their personal and professional qualities. With the transition to secondary school, the complexity of administration increases enormously. In the following 5–9 years, all of a sudden up to 25 teachers assess each child. Inevitably, they can only do this from the narrow perspective of their subject and not from the overall personality of the child. That leads to alienation and fragmentation of the learning experience. For that reason, joint teams of five to a maximum of eight teachers, who are responsible for the entire learning process of 50–80 learners, ideally for 4 years, are suitable for the secondary school. Reducing the usual 12–21 subjects to significantly fewer but larger learning areas takes a lot of stress off teachers and learners.

Conclusion: Pupils do not look forward to objects but to people who are role models for them. Learning needs community. Teachers and students perform best in teams. When a young person has finished school, they should be open to the world, not in the sense that they have to be interested in everything, but science, society, the environment, music, art, etc., should mean something to them.

2.4 *Learning Requires Equal Promotion of Social and Cognitive Skills*

In contrast to the intelligence quotient (IQ), which can only be increased slightly, people can significantly improve their social skills. A large number of recent studies prove that the degree of individual self-discipline, for example, is not a matter of fate but can be trained like a muscle over the course of a lifetime. Characteristics such as frustration, tolerance, perseverance, diligence, and patience can be significantly improved regardless of IQ and socioeconomic background (Baumeister, 2022). The systematic promotion of social and communicative skills at school would thus be one of the most sensible tasks. The introduction of the subject "Communication and Social Competence" as part of the high school diploma exam was a central element for the success of the "Sir Karl Popper School," an integral part of a public school named Wiedner Gymnasium in Vienna with a special branch for highly gifted pupils comprising grades nine to 12. Its focus is on promoting talent from the very beginning.

Every bit as important as teaching the cultural techniques of reading, writing, and arithmetic as well as a comprehensive general education is the systematic recording and continuous development of sporting, artistic, and social talents. Harvard psychologist Daniel Goleman (1996) convincingly demonstrated the validity of this assumption in his book *Emotional Intelligence: Why It Can Matter More Than IQ*. He defines the term by means of the following skills: knowing your own emotions, influencing emotions, putting emotions into action, empathy, and dealing with relationships. Thus, if you want to be successful in life, you need to master this emotional alphabet. Goleman makes it clear that formal education and a corresponding intelligence quotient are necessary to get a certain job but that emotional intelligence is the most important prerequisite for being able to do this job successfully. The higher you rise in an organization, the more important the emotional intelligence factor becomes. This is why six out of seven assessment criteria used by companies to select new team members now have to do with emotional intelligence in the broadest sense.

Emotional intelligence is best taught integratively with an understanding of systemic thinking. In their book *Triple Focus: A New Approach to Education*, Daniel Goleman and Peter Senge call for systemic thinking to be taught much more often in schools, as this is the only way to align education with global challenges in the twenty-first century. They define three competencies for that: understanding oneself, understanding others, and understanding complex systems. Even if the two authors do not explicitly refer to the twenty-first-century skills, their statements are very much in line with the models described: "The core dilemma of the Anthropocene era is to understand the systemic consequences of our own actions on a global scale. This work reminds us that the real challenge is not to become even better at the non-systemic ways of thinking that have enabled the accelerated change of the industrial age - but to develop deeper intelligences of ourselves, of others, and of today's systems when we really need them"[8] (Goleman & Senge, 2014).

3 Forms of Work and Methods of the Twenty-First-Century School

An analysis of those schools that have won the German School Award and the Austrian State Prize for Innovative Schools reveals that they contain many elements that the reform pedagogue Peter Fratton developed and practically implemented in his "Houses of Learning" at the Lake Constance School Centre. The trigger for Peter Fratton, who worked many years as a principal of a traditional school, to undertake a massive rethinking was a pupil's question: "Surely that can't be everything after nine years?" (SBW Haus des Lernens AG, 2009).

In an intensive process of change, Fratton, together with his team and academic support, created a school model in which diversity, collaborative learning, and the individual design of learning paths are highly valued. However, Fratton also always emphasizes performance, and the graduates of his schools pass the very demanding Swiss high school diploma well above the national average. In the model, pedagogical practice must not be subordinated to abstract theory. The two dimensions of learning—direct knowledge acquisition and experiential learning—must complement and balance each other. Only practice brings surprise, opens up to the unavailable in the sense of Hartmut Rosa's resonance research, and takes detours, which is why it cannot be planned like a lesson. Fratton formulates pedagogical "primal requests" that are intended to signal a departure from the usual school teaching strategies and curriculum-based thinking:

Don't teach me anything.
Don't explain anything to me.
Don't educate me.
Don't motivate me.

Those "primal requests" are aimed at prospective teachers and are intended to cause irritation among them. Irritated people hold their breath for a brief moment, and an idea can then emerge from this pause. For Fratton, irritation is astonishing surprise. It requires intelligence and understanding. Complex and lively

[8] Translation of the German version.

organizations should be irritable, because allowing irritation means allowing new ideas. The following forms of work and methods inspired by the "Houses of Learning" can be found, sometimes with different names and forms, in many pedagogical concepts of innovative schools and can therefore be seen as paradigmatic. The school of the twenty-first century is based on the values and principles of innovative pedagogy described above and implements them in concrete terms with the following forms of work and learning designs:

- Learning offices, learning labs, or learning studios replace the traditional timetable. Learners acquire and practice basic skills in an open learning format, the so-called learning office, and then apply them in projects with real-life objectives. Many of these projects take place outside of school, usually in collaboration with local businesses and institutions. That ensures that what students learn is relevant and connected to the real world. In the learning offices, there is a wide range of pedagogically excellent teaching materials for autonomous learning, mostly in German, mathematics, and English.
- The 50-minute units are broken up, and instead work is carried out across subjects such as science, mathematics, physics, and biology.
- Each day is divided into clearly structured learning phases, e.g., individual learning in the learning office in the main subjects of German, mathematics, and English, in-depth joint work in workshops, projects with external partners to test practical skills, and, crucially, at least one physical education unit every day.
- Teachers see themselves as learning guides but do not teach in the traditional sense. This philosophy is often expressed in the role designations such as "learning guide" and "learning partner" instead of "teacher" and "pupil."
- High tech: Work is carried out according to learning plans on a digital platform on which learners and teachers transparently document individual teaching successes. Pupils also keep learning diaries.
- High degree of support: Each student has a teacher as a coach for personal support. This coach has an overview of learning progress, learning goals, projects, grades, etc.
- Overall personal development is documented in portfolios instead of being reduced to numerical grades. Instead of traditional exams and schoolwork, learning success is ensured with learning evidence, i.e., learners report when they have understood an area of knowledge; what they have not learned can be practiced again and made up for at a later date.
- The focus is on social learning in heterogeneous, cross-age groups instead of fixed classes.
- Bringing external experts and parents into the classroom.
- Blocks of lessons:[9] For a certain period of time (an epoch), the subjects are taught alongside each other in order to work on a common topic (e.g., "Our district" from a biological, historical, political, artistic, geographical, etc., point of view) from different perspectives.

[9] The German technical term for this special form of structuring lessons is "Epochenunterricht."

- Silence Room: Children who need peace and quiet can work independently here.
- Open Lab: Here students work on projects of their own choosing.
- "Arrival time," e.g., from 8:00 to 9:00 a.m., instead of a rigid start to lessons. Many studies show that adolescents are not really receptive to complex mental tasks before 9:00 a.m., but their performance curve is all the better in the afternoon.
- Marketplace with topics, wishes, and to-do lists: This is where learners are invited to help shape the school.
- Space and time for holistic work on body and mind: This is intended to support the necessary openness for the attitudes of resonance and appreciation described above. In concrete terms, this means, for example, establishing mindfulness exercises, all forms of bodywork, and meditation both in individual practice and in everyday school life.
- Multigrade classes with fixed teams of teachers: In the secondary school, teams of ideally five to eight teachers are formed to cover the areas of science and mathematics, history and geography, languages, philosophy, art, personal development, and sport.

4 Theory U as a Basis for Initiating Transformation Toward More Sustainability and Well-Being

The cornerstones of a twenty-first-century school presented above already contain numerous interfaces in the context of sustainability and well-being. Both the aforementioned principles and the forms of work and methods are aimed at teaching skills that enable people to keep an eye on their own well-being as well as that of society and the environment and to take responsibility for them, regardless of what role they will one day play in society as adults. Being competent and equipped for transformation is one thing; getting involved in change and ultimately taking concrete steps yourself is another. Daring to engage with the aforementioned "unavailable" is easier if it is done from a secure base—that applies to all people and, in the school context, therefore, also to learners and teachers. One factor in this context that has not yet been given much importance in the school context is the inclusion of space and the body in the learning process. One concept that gives importance to the nonmental factors of the body and emotions under the heading of "embodied learning" is Theory U, developed by Otto Scharmer (2009).

Scharmer assumes that our belief that we can change something far exceeds our ability to do so. Anyone who believes that they can escape their comfort zone through willpower alone unfortunately overlooks the fact that all unresolved problems, inhibiting beliefs, and dark stains from the past will accompany them as loyal companions. We have a history that we cannot erase from the beginning and simply rewrite in a better direction. Whenever we try to reshape our lives without taking into account what we already have, reality will catch up with us faster and harder than we would like. That also applies to changes sought by organizations that want to break out of their comfort zone into a hopeful future. The traditional approach to this usually involves four phases: (1) identifying the problems in an organization, (2) finding an external expert, (3) telling everyone in the organization what they need to do differently and better in the future, and (4) fighting all the resistance that has been triggered in stages one to three, especially in stage three.

It is precisely because of this neglect of the complex reality of organizations that well-intentioned goals to change organizations in a positive direction usually fail. "Organizational culture eats every strategy for breakfast" is a common management wisdom.

Theory U, developed at the Massachusetts Institute of Technology (MIT) by Otto Scharmer, aims to develop a deeper and more holistic perspective that views a system as a whole and makes the relationships and connections tangible and comprehensible for everyone involved. Sustainable transformation of individuals, organizations, and the world requires intensive processes that always take the individual and their own willingness to change as the starting point, while traditional change strategies are often based on seemingly rational assumptions.

Theory U is a process with the aim of reconciling the world as we see it in the present reality and as it could be in the future. With our fears and desires, we stand precisely between the two areas of tension of the present and the future. How do we deal with situations of upheaval when the future will be different from the present and we have to open ourselves up to it? Making the necessary knowledge accessible and usable is the content of special individual and group exercises along a U-shaped transformation process (Fig. 1).

The U-curve consists of three phases: downloading, presencing, and performing.

In the first phase, the downloading of old patterns, we consciously focus our attention on the present situation and the existing challenges. That involves questioning preconceptions and beliefs in order to gain new perspectives. In the second phase, called presencing, listening, sensing, and anticipating enables us to include other people's perspectives in order to gain a deeper understanding of the situation and give space to our own intuition. In the third phase, performing, visions are developed and condensed in order to test them in prototypes. That generates positive energy for change among those involved with the aim of creating a new, positive future together.

Theory U is based on the assumption that change must not only take place externally but above all internally. That requires an expansion of awareness and a sensitization of one's own perception and actions. This is why presencing, the visualization of one's own self, is so crucial at the turning point of the U-curve. Meditation, mindfulness exercises, yoga, fasting, martial arts, etc., prove to be helpful here in order to get out of the intellectual comfort zone and feel better again. In that way, we can build a bridge from our present self to a new one.

Otto Scharmer is optimistic that research findings in the fields of brain plasticity, the proven effect of mindfulness exercises, and the further development of emotional intelligence in particular will significantly promote the widespread implementation of Theory U. In times of increasing disruption, the only thing we can

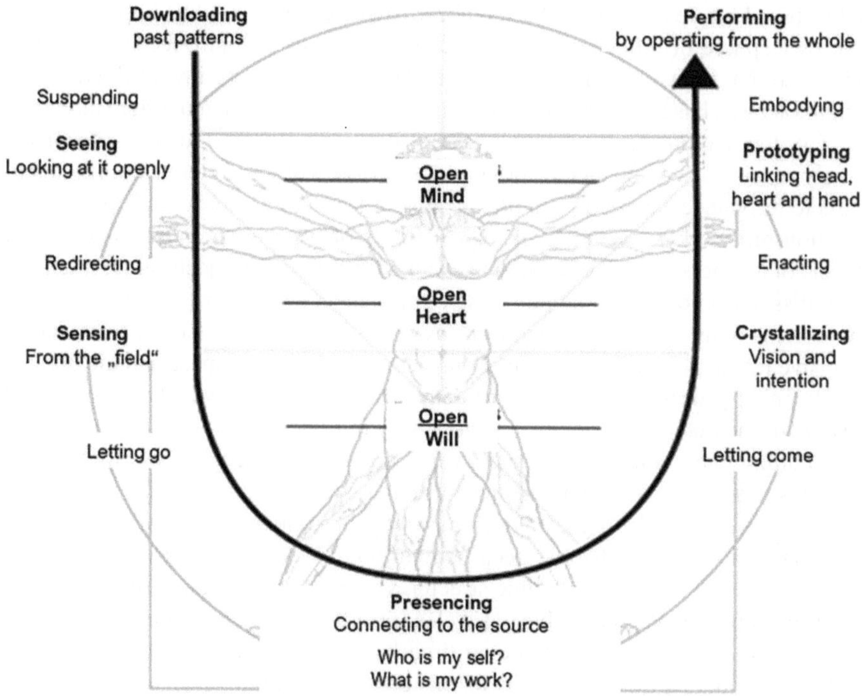

Fig. 1 Theory U (based on Scharmer, 2009)

influence is our inner reaction to it: Do we freeze up with fear, close ourselves off, and persist in habitual behaviors, even when they are demonstrably counterproductive? Or do we manage to let go of the old and open ourselves up to the new? Only those who understand the strong opposing forces within themselves and in social systems can take them into account and overcome them. For many of the challenges of a sustainable transformation of organizations, the process described in Theory U would be helpful in moving from the mindsets of today to those of tomorrow. The future belongs to people who can think in systems. The school of the twenty-first century must therefore successfully teach these skills.

5 Conclusion

The question of the school of the future is often raised in public debate. The many examples of real schools show that the school of the future already exists in the present. School systems only need to orient themselves toward the principles and working methods of the school of the twenty-first century as described. Openness to this type of learning has been increasing enormously and points the way to how we should design learning spaces. A decisive step is therefore to go to where the educational future has already been realized, experience it, and then implement it in our own sphere of influence. Systemic control processes require feedback mechanisms so that the whole of the system can be felt by all those involved. Once this common basis has been created, innovative steps can be taken in the right direction.

The foundations for sustainable school development are the well-known but rarely consistently implemented principles such as a learner-centered attitude, openness to resonance experiences, a high degree of learner autonomy, and appreciative relationship work. Concrete pedagogical forms of work such as learning offices, teams of teachers instead of lone wolves, interdisciplinary projects, at least temporary dissolution of the 50-minute lesson and subject separation, student coaching, new forms of assessment through learning portfolios, etc., are derived from that. A decisive factor is the often completely neglected inclusion of space and the body in the learning process. The concept of "embodied learning" gives these nonmental factors of the body and emotions the necessary importance.

Building the foundation of the twenty-first-century school described above requires time, expertise, and resources through external and internal process support. This work has the highest priority at every innovative school and requires the willingness to give up what has been familiar up to now and to invest time in these issues. Only the more flexible time structures within the innovative school models described above make it possible to set clear priorities for relationship and development work and thus avoid additional time burdens for teachers and students. In order to strengthen this willingness to change in a system in which change is often perceived as a threat, successful intermediate steps are needed that are radical enough to change something structurally and at the same time adaptive enough to be digestible in the system despite initial defense mechanisms. If we want to achieve social change in line with the SDGs and enable people to lead self-determined and meaningful lives, we need to take these intermediate steps. The school of the twenty-first century thus becomes a supporting pillar in the vision concept of the "learning community," which does not elevate learning as an end in itself but enables the population to enjoy a better quality of life through new social relationships and more meaning in life.

References

Baumeister, R. (2022). *Die Macht der Disziplin: Wie wir unseren Willen trainieren können*. Campus Verlag.

Csikszentmihalyi, M., Rathunde, K. R., Whalen, S., & Wong, M. (1993). *Talented teenagers: The roots of success and failure*. Cambridge University Press.

Goleman, D. (1996). *Emotional intelligence: Why it can matter more than IQ*. Bantam.

Goleman, D., & Senge, P. (2014). *Triple focus: A new approach to education*. More Than Sound DBA Key Step Media.

Hattie, J. (2009). *Visible learning: A synthesis of over 800 meta-analyses relating to achievement*. Routledge. https://doi.org/10.4324/9780203887332

Rosa, H., & Endres, W. (2016). *Resonanzpädagogik: Wenn es im Klassenzimmer knistert*. Beltz.

Salcher, A. (2023). *Aufbruch zu einer lernenden Nation*. Andreas Salcher [Report].
SBW Haus des Lernens AG. (2009). Der Lernvirus. Film by Reinhard Khal. https://www.youtube.com/watch?v=5Evd5k9ObJ8
Scharmer, O. (2009). *Theory U: Learning from the futures as it emerges*. Berrett-Koehler.
Schley, W. & Schratz, M. (2021). Führen mit Präsenz und Empathie: Werkzeuge zur schöpferischen Neugestaltung von Schule und Unterricht.
Young, R., & Daniel, M. (2022). Lessons on innovation from teachers. *IDEO Journal*. https://www.ideo.com/journal/lessons-on-innovation-from-teachers

Dr. Andreas Salcher is a management consultant, bestselling author, and critical thought leader on educational issues. He is the co-founder of the "Sir Karl Popper School" for particularly gifted children in Vienna. In 2004, he initiated the "Waldzell Meetings" at Melk Abbey, which were attended by seven Nobel Prize winners and the Dalai Lama. With nine #1 bestsellers and over 250,000 books sold, he is considered one of Austria's most successful nonfiction authors. His latest book was published in October 2023 and is titled *Unsere neue beste Freundin, die Zukunft—Was die Jungen wissen und wir noch nicht (Our new best friend, the future—what the young know and we don't yet).*

Impactful Ventures, Burned-Out Founders? Insights from Two Studies of Social Entrepreneurs

Peter Vandor and Michael Meyer

Abstract Burnout is one of the most consequential forms of low well-being among entrepreneurs. Those who experience symptoms of burnout are more likely to curb growth ambitions, make mistakes, and eventually quit the venture altogether. Founders' burnout therefore creates negative ripple effects for themselves, their teams, their investors, and their customers and beneficiaries. These consequences are even more dire in the case of social entrepreneurs: their beneficiaries are often socially vulnerable groups who are not served by other sectors of society.

Yet, the prevalence and the drivers of burnout among social entrepreneurs are poorly understood in research. In this chapter, we present insights from two international studies: a large quantitative survey of social entrepreneurs in the Impact Hub network (n = 1024) and a qualitative study of young social entrepreneurs at Social Impact Award (n = 51). Results shed light on levels of burnout prevalence, antecedents on the level of the individual and the venture, and coping strategies of entrepreneurs and highlight the importance of a better understanding of social entrepreneurs' well-being among researchers and practitioners.

1 Introduction

> It's overwork, stress and constant pressure that you have to deliver and do this, and people are waiting on your products that could save lives and you are not doing enough about it. It's a constant thing on your mind, the fear that someone could die tomorrow because you took a break.

Though the mental well-being of entrepreneurs has gained some attention over the last years (De Mol et al., 2018; Stephan, 2018; Stroe et al., 2018), research has so far predominantly focused on commercial entrepreneurs. However, an estimated

P. Vandor (✉) · M. Meyer
Vienna, Austria
e-mail: Peter.Vandor@wu.ac.at; Michael.Meyer@wu.ac.at

M. Mühlböck (ed.), *Sustainable Transformation and Well-being*, CSR, Sustainability, Ethics & Governance,
https://doi.org/10.1007/978-3-031-75566-8_10

25% (and in some countries even up to a 50%) of entrepreneurs are social entrepreneurs, who prioritize social or environmental value creation over financial profit and often address pressing societal issues such as environmental challenges, education, or poverty relief by means of innovative business models (Bosma et al., 2016; Dacin et al., 2010; Dees, 1998; Kelley et al., 2016; Leitner et al., 2023).

The concept of burnout was introduced by psychoanalyst Herbert J. Freudenberger (1974) as a state of mental and physical exhaustion caused by one's professional work. Since the 1970s, research has emerged focusing on burnout in social service professions such as social work, nursing, law enforcement, and teaching (Shepherd et al., 2010). Most prominently, Christina Maslach and colleagues have coined burnout as a three-dimensional concept consisting of emotional exhaustion, depersonalization, and a reduced sense of personal accomplishment (Maslach et al., 1997, 2001): (1) emotional exhaustion, which refers to feelings of loss of energy and fatigue; (2) depersonalization, which describes an attitudinal change toward a negative, distant, uncaring, and sometimes even cynical attitude toward others; and (3) reduced personal accomplishment, which refers to a perceived decline in one's own competence performance. Burnout is associated with a number of negative health outcomes including depression, low self-esteem, substance abuse, decreased work performance, and the inability to perform work tasks for a long period of time (Freudenberger, 1974; Maslach et al., 2001; Taris, 2006).

In 2019, burnout was defined as a syndrome in the WHO's ICD-11 (International Classification of Diseases) (Berg, 2019). According to the ICD-11, burnout is a syndrome that occurs due to chronic stress in the workplace. Very similar to the Maslach Burnout Inventory, the main signs of burnout are described as (1) loss of energy, (2) a feeling of exhaustion, (3) psychological distancing from work, (4) increased negative and pessimistic thoughts about work, and (5) reduced professional efficiency. Accordingly, the burnout syndrome is used in a professional context and cannot be used to describe experiences in other areas of life.

The professional context of entrepreneurs is characterized by many challenging working conditions, higher levels of uncertainty, responsibility, and complexity, and more time pressure and longer working hours (Cardon & Patel, 2015; Jamal, 1997; Kim & Ling, 2001). This context is often perceived as stressful and overwhelming (De Mol et al., 2018; Shepherd et al., 2010; Stroe et al., 2018; Wincent et al., 2008). Empirical studies on entrepreneurial well-being have confirmed the harmful effects of such factors. Role stress and negative affective events such as conflicts with co-founders or investors contribute to burnout among business entrepreneurs (Fernet et al., 2016; Shepherd et al., 2010; Wincent et al., 2008), whereas a perceived fit between the job and personal expectations reduces experiences of the burnout symptom (De Mol et al., 2018).

The experience of burnout can have far-reaching effects in entrepreneurship. Burnout syndrome lowers commitment, satisfaction, and work performance for entrepreneurs (Shepherd et al., 2010) and forwards the intention to alter or quit the venture (Wincent et al., 2008). Moreover, when entrepreneurs are suffering from burnout syndrome, it is not only harmful for individuals' health but also for ventures' success and survival (Shepherd et al., 2010; Stephan, 2018). In the case of social entrepreneurs, this is particularly consequential, as such negative effects also cause harm to beneficiaries, vulnerable groups, or the environment. As shown in the quote in the first paragraph (stemming from the interview material introduced in the next section), maintaining founders' mental health and efficacy may thus be of vital importance for beneficiaries.

Despite the growing interest in the phenomenon, we know little about the prevalence of burnout in the field of social entrepreneurship. Among the 140 empirical studies collected in a comprehensive review of research on entrepreneurial well-being (Stephan, 2018), none deals with burnout or any other form of well-being of social entrepreneurs. Some authors have argued that social entrepreneurs face particularly high levels of work demands and stress and are thus more inclined to burnout. In an analysis of popular autobiographies of social entrepreneurs, Dempsey and Sanders (2010, p. 437) point out that these descriptions imply "a troubling account of work/life balance centered on self-sacrifice, underpaid and unpaid labor and the privileging of organizational commitment at the expense of health, family and other aspects of social reproduction." Others have made similar claims and suggested that overwork and burnout are very common in this field (Guclu et al., 2002; Symon and Whiting, 2019). So far, little empirical research has revealed data on the prevalence and drivers of burnout among social entrepreneurs.

Only recently, some studies have scrutinized this "dark side" of social entrepreneurship. Kibler et al. (2019) reveal that entrepreneurs' prosocial motivation is associated with increased levels of work stress and will hence deteriorate life satisfaction. However, this effect dissipates when perceived autonomy at work is high compared to when it is low. Another study has identified the attributions of social entrepreneurs' disappointment and their relation to depression by analyzing social media posts (Drencheva et al., 2021, 2023; Williamson et al., 2022). Posts with internal, stable, and global disappointment attributions (e.g., not fitting entrepreneurial norms) are, on average, from entrepreneurs with stronger depression symptoms than posts with external, unstable, and specific disappointment attributions (e.g., firm performance). This is not surprising, as external unstable attributions of failure rather protect the entrepreneurial identity. Finally, studies of three samples of German and Swiss social entrepreneurs show that (1) the higher the belief in a venture's social value, the higher the job satisfaction and (2) the lower the risk of burnout and (3) that a perception of work meaningfulness also contributed to the belief in social value and thus work-related well-being (Brieger et al., 2021).

In this chapter, we will present data on the prevalence of burnout symptoms among social entrepreneurs, the causes that social entrepreneurs report for these symptoms, and the coping strategies they apply. We investigate two international samples of social entrepreneurs. From this dominantly qualitative evidence, we finally derive implications for mitigating the risk of burnout in the field of social entrepreneurship. Of course, strengthening individual entrepreneurs' resilience is a promising perspective (Korber & McNaughton, 2017), but we rather aim at systemic changes. Therefore, we conclude our chapter suggesting how the ecosystem should respond to these risks by modifying and tightening support structures for social entrepreneurs.

2 Insights from Social Entrepreneurs

2.1 Method and Sample Description

We report findings from two different empirical studies of social entrepreneurs, both of which are cross-sectional and international and have been conducted at WU Vienna's Social Entrepreneurship Center: first, a large-scale dataset of survey responses from over 1000 social entrepreneurs providing insights into the prevalence of burnout and, second, a qualitative study which sheds light on burnout prevalence, drivers, and coping among young social entrepreneurs.

The first, large-scale dataset consists of survey responses collected among members of the Impact Hub network, one of the world's largest support networks for social entrepreneurs.[1] Impact Hub serves as coworking space, service provider, and intermediary in the entrepreneurship ecosystems of many cities across the globe (Vandor et al., 2019; Vandor & Leitner, 2018). The data was collected via an online survey across the network in February and March 2021. After excluding Impact Hub staff, non-founders, and respondents indicating a preference for commercial value creation over social impact, our sample includes 1024 social entrepreneurs from over 47 countries in Africa, Asia, Europe, North America, and South America. In this sample, we applied the 16-item Oldenburg Burnout Inventory (Demerouti & Bakker, 2008; Reis et al., 2015), which measures burnout as a two-factor construct of exhaustion and disengagement (Demerouti and Bakker, 2008). Results of the OLBI have a high convergent validity with the established Maslach Burnout Inventory while offering a more balanced wording that includes eight positively and eight negatively worded items (Halbesleben & Demerouti, 2005). Respondents were on average 36.2 years old, with 48.2% reporting female, 49.0% male, and 1.6% other gender identities.

The second dataset consists of qualitative interviews with former participants and winners of the Social Impact Award, a well-established educational program and incubator for young social entrepreneurs in several European, Asian, and African countries.[2] Respondents were contacted through local implementers of the award and asked for an interview. Between December 2018 and May 2019, 51 interviews were conducted with respondents from Austria, Czechia, Romania, Serbia, and Slovakia. Respondents had participated in the Social Impact Award program on average 3.5 years before the interviews. Given the Social Impact Award's focus on ideation and the earliest stages of entrepreneurship, it is reasonable to assume that most entrepreneurs had founded their organizations during or shortly before participation. Thus, the responses can be regarded as characteristic of early-stage social entrepreneurship (Lepoutre et al., 2013). Respondents were on average 28.0 years old and had founded their organizations at the age of 24.5. Roughly 40% were female and 60% male.

[1] https://impacthub.net/

[2] https://socialimpactaward.net/

The interviews were conducted in semi-structured form with shared guidelines, including questions related to the founders' experience with the support program, the development and performance of the social venture over time, and the founders' individual experience with the entrepreneurship process. Five open-ended questions focused on the well-being of founders, exploring subjective experiences of stress, burnout, and other effects of their entrepreneurship journey on mental health, the perceived causes of such experiences, and founders' coping actions. Respondents were also presented a single-item scale for measuring burnout (Dolan et al., 2015), which is frequently used in the medical field and correlates reasonably highly with the Maslach Burnout Inventory (Gardner et al., 2019; Knox et al., 2018; Puffer et al., 2017).

Out of the 51 respondents, 43 cases provided answers to well-being questions. The interviews were conducted and transcribed in the local languages. All responses were then translated into English and subsequently coded and analyzed by university staff as well as student researchers (see also Hütter, 2018; Katterl, 2019; Ultsch, 2020). The major analytical method was qualitative content analysis (Mayring, 2003). The interview content was interpreted and coded by at least two individuals, partially with help of MAXQDA, and refined in several iterations for the development of a category framework (see again Hütter, 2018; Katterl, 2019; Ultsch, 2020). Next, we present selected highlights and further extensions of these analyses.

2.2 *Prevalence of Burnout*

2.2.1 Prevalence in Large-Scale Dataset

Results from the international dataset (n = 1024) provide a first insight into the prevalence of burnout symptoms among social entrepreneurs. The average reported values for the 16-item Oldenburg Burnout Inventory reach 2.17 for exhaustion and 1.76 for disengagement, marking reasonably low values in the continuum from 1.00 (strong disagreement with a statement indicating burnout) to 4.00 (strong agreement). These values are lower than in comparable studies of burnout prevalence across a wide range of professions, including groups that are at particular risk of experiencing burnout (e.g., healthcare workers) (Table 1).

Of course, comparisons of absolute values between studies of different origins must be interpreted with caution as they may be affected by different sampling strategies, return rates, language, and cultural context in many ways. Nonetheless, they suggest at least that there is no particularly high prevalence of burnout among social entrepreneurs. Instead, especially their level of disengagement seems in fact to be considerably lower than among workers in any other profession.

Table 1 Comparison of average burnout scores between social entrepreneurs and other professional groups

Type of sample	Country context	Source	Ø Exhaustion	Ø Disengagement
Social entrepreneurs	47 countries	Own research in Impact Hub network	2.17	1.76
Hospital nurses and caretakers for the elderly	Germany	Reis et al. (2015)	2.58	1.96
University students	Germany	Reis et al. (2015)	2.28	1.99
University students	Greece	Reis et al. (2015)	2.59	1.90
Hospital nurses	Ireland	Chernoff et al. (2019)	2.58	2.29
Hospital physicians	Ireland	Chernoff et al. (2019)	2.59	2.27
Psychotherapists	United Kingdom	Delgadillo et al. (2018)	2.40	2.20
Refugee resettlement workers	United States	Akinsulure-Smith et al. (2018)	2.59	2.29
Forensic interviewers working with abused children	United States	Perron and Hiltz (2006)	2.26	2.21
Workers at a cement factory	South Africa	Lekutle and Nel (2012)	2.30	2.16
Construction workers	South Africa	Demerouti et al. (2008)	2.17	2.07

2.2.2 Prevalence Among Young Social Entrepreneurs

Findings from our second dataset regarding young social entrepreneurs (n = 51) complement this picture and provide insights into the performance of their social enterprises. Regarding the latter, our data tell a veritable success story. About two-thirds of respondents were still active as social entrepreneurs 3.5 years after having participated in the program. Among winners, i.e., those who received funding and nonmonetary support, the survival rate was even above 75%, which is remarkable considering the early stage of organizations and the young ages of the participants. On average, their social enterprises employed 5.14 paid employees and had 12 volunteers. More than half (58%) of respondents reported that they had also been successful in securing external funding for their venture.

However, this success seems to have come at the expense of the well-being of some founders. Answers to the Dolan Burnout Scale reveal that 43% of respondents experienced symptoms of burnout at least temporarily, and 5% reported having had to discontinue their work due to these symptoms. This prevalence of burnout experiences is higher than what other studies reveal for medical professionals, which report rates between 24.5% and 39% in a sector that is notorious for its high rates of burnout (Dolan et al., 2015; Gardner et al., 2019; Knox et al., 2018; Puffer et al.,

2017). This contrasts with our findings in the large sample and seems to indicate a higher level of burnout among young social entrepreneurs.

We can only speculate about the reasons for the particularly high rate of burnout in this sample. A possible reason might be a specific awareness for burnout in the interview situation, which invited participants to reflect about their well-being more intensely than they might have done in a quantitative survey. It could, however, also reflect the challenges faced particularly by young social entrepreneurs, since both age and low work experience have previously been associated with a higher risk of burnout (Gómez-Urquiza et al., 2017; Marchand and Bearegard, 2018).

2.2.3 Qualitative Experiences of Stress and Burnout

Qualitative evidence from our study of young social entrepreneurs allows a deeper look into participants' experiences. Their answers contain detailed descriptions of extended periods of stress which are frequent precursors of burnout. Descriptions include symptoms often associated with burnout, such as feelings of exhaustion, negative thoughts about work, and reduced personal efficacy (Maslach, 2001), e.g.:

> I cried a few days ago, because I spent so much time and energy and I feel overwhelmed.... and I am not able to find a solution.

> The expectations were very high and maybe it was a bit too much for me. Maybe we could've been…. It was always expected to perform and manage everything. Two of my team members left because they were working too much. One just left, giving up everything.

> We definitely experienced symptoms [of burnout]... insomnia, lack of sleep, constant worrying, fear that someone could get hurt if we didn't do something perfectly.

In seven cases, respondents mentioned explicitly that they believe to have had a burnout. In one case, the answer seems to indicate that this assessment was also validated by a medical professional:

> Yes, I had stress. I was finishing school. I was founding this company, I felt broke down and I had to go to the doctor. The result was that I went to a psychologist. He found that it is from the stress.

Some respondents also described negative physical symptoms, which are often associated with chronic stress experiences and burnout (Redondo-Flórez et al., 2020):

> We had debts that we were not sure we would be able to pay. [This affected] my colleague, who is more stressed than I am. [...] She had a quite serious breakdown, and developed gastritis, because the fund did not come through when it was supposed to.

> Our team of volunteers at that time fell apart and I almost broke down and wanted to return the award for the fear that I will not be able to do it well enough. I was overstressed and tense and my back broke down... for a week I could not move.

2.3 *Antecedents of Burnout*

Analyses of our qualitative data provide interesting cues as for potential drivers of burnout among social entrepreneurs. Asking to assess the potential causes of their stress and burnout experiences, respondents identified many reasons, ranging from exaggerated expectations regarding their own capabilities, lack of financial resources, the challenges associated with balancing entrepreneurship with other obligations in family, work, or education, to the discomfort with the risks of entrepreneurship. Two groups of antecedents emerged as particularly relevant and frequently mentioned:

2.3.1 Role Overload and a Lack of Resources

Several responses point toward the high work demands of the entrepreneurial job description. With their nascent organizations suffering the "liabilities of smallness and newness" (Stinchcombe, 1965), the founders simultaneously face many tasks, from developing services and hiring a team to building trust in the marketplace, navigating legal challenges, and creating organizational structures.

This broad range of tasks can create conflict, role overload, and chronic stress, which are frequently precursors to entrepreneurial burnout (Shepherd et al., 2010; Wincent et al., 2008). Respondents frequently talked about such challenges and mentioned the lack of financial means and disagreements within founding team as sources of stress. Many identified this as a contributor to low well-being and burnout:

> I think social impact projects come with a shitload of stress, we had many things to manage, and we were a very small team.
>
> It's just an overload of work without getting paid and so many decisions.
>
> It is a huge challenge... all the responsibility that comes with it.

Such role overload is further amplified when the level of identification with and the perceived responsibility for the organization is high. As illustrated in one response, founders can experience responsibility to the degree of "becoming" the venture themselves:

> You are a startup and you want to make sure that the things will work. You are investing so much into it. You are investing yourself and your time.

2.3.2 Difficulties to Distance from Beneficiaries and Mission

While the experience of role overload is common among entrepreneurs of any stripe, many respondents also mentioned a second potential driver of burnout which might be more typical for social entrepreneurship. They described a difficulty to set boundaries between themselves and their work for the very reason of their high commitment to the social purpose.

> It's overwork, stress, and constant pressure that you have to deliver and do this, ...people are waiting on your products that could save lives and you are not doing enough about it. It's a constant thing on your mind, the fear that someone could die tomorrow because you took a break.

> You are simply aware of many things going on in the participants' lives. You can't leave that knowledge at work as in a usual 9/5 job.

Their responses suggest that the high levels of motivation associated with prosocial work (e.g., Grant, 2008) can become a boomerang in a setting such as social entrepreneurship, in which the responsibility to act is often felt particularly intensely by founders.

2.4 *Coping Mechanisms*

Results also provide insights into how social entrepreneurs cope with their experiences of stress and burnout. Building on the work of Folksman and Lazarus (1988), we differentiate between coping mechanisms that focus on the problem and try to modify the external source of stress (e.g., changing job demands, enacting change in the organization) and those that focus on the internal emotional experience.

2.4.1 Problem-Oriented Coping

Among the 26 respondents providing insights into their coping, 16 employed mechanisms that can be understood as problem-oriented coping (Katterl, 2019). Examples include changes in the organizational processes and structures, such as hiring new staff or involving external consultants to reduce role overload. In other cases, entrepreneurs aimed to improve communication by introducing new meeting structures, embedding discussions about well-being into staff meetings, or increasing the frequency of informal exchanges in the team. In some cases, entrepreneurs also described restructuring their own work routines, e.g.:

> I have recently reorganized my working time, after 8 p.m. I don't touch my phone and don't receive calls. During the weekend I try to relax and be with my family and my dog. Now, I care much more about myself and I am trying not to carry my work home.

2.4.2 Emotion-Oriented Coping

Another 12 respondents reported emotion-oriented coping (Katterl, 2019). In many cases, this meant distancing themselves from their work and the experience of stress created by it. Some entrepreneurs reported achieving this distance by reappraising and normalizing stress as part of their job as entrepreneurs. Others reported coping through sport, mediation, or taking up healthy dietary habits; several responses also referred to the use of recreational drugs:

> How to cope? Simply get used to it.

> We changed our lifestyle into a healthier and more active one, and it helped.

> Yes. Overworking was the cause of all the stress. It's still ongoing. We try to drink and forget about it.

The use of distancing as a coping mechanism points to an interesting ambivalence in the burnout experience: while experiences of distance and disengagement are a central element of burnout syndrome (Berg, 2019; Maslach et al., 1997, 2001), professional distancing can also be a constructive coping mechanism and is often recommended by work psychologists in social professions (Andersen & Dybbroe, 2017; Green et al., 2006; Sonnenberg, 2021). In this sense, the distancing experience in burnout contains a dysfunctional version of a coping mechanism that—if applied willfully and as part of professional introspection—can be part of a healthy way of managing stress.

3 Discussion

This chapter hopefully contributed to the research on the well-being of social entrepreneurs in a threefold way. First, our quantitative evidence from a large-scale sample of social entrepreneurs showed that burnout is not significantly more prevalent in this field than in other fields. Thus, we see no empirical support for excessive worries about social entrepreneurs' well-being. Second, we nevertheless found indicators of burnout symptoms in a considerable part of the young social entrepreneurs of our qualitative sample. The discrepancy between the rather low prevalence of burnout in the large quantitative sample and the higher prevalence in the sample of young social entrepreneurs should not be overinterpreted but might point to a higher vulnerability among young social entrepreneurs. Such a vulnerability could be caused by their lack of experience and not-yet elaborated coping mechanism (Gómez-Urquiza et al., 2017; Marchand and Bearegard, 2018).

Third, our findings point to specific drivers and coping mechanisms related to burnout among social entrepreneurs. Stress linked to a lack of resources, conflicting commitments to funders and beneficiaries, or conflicts in the founding team as well as the perception of responsibility for beneficiaries are the main drivers of burnout. They are often amplified by founders' difficulties to set boundaries and distance themselves from the beneficiaries' needs and the venture's mission. For coping with these stressors, social entrepreneurs apply functional and dysfunctional, problem-oriented, and emotional mechanisms.

Still, research on the well-being of social entrepreneurs is in its infancy, and our chapter leaves many important questions open. Even though prevalence of burnout is not alarming in our data, we do not know how many former social entrepreneurs have terminated their ventures and left the field because of burnout syndrome (and were thus also missing in our dataset). Moreover, the results from young social entrepreneurs suggest that entrepreneurs with specific characteristics and in specific contexts might be vulnerable to burnout.

Moreover, the findings presented here are widely descriptive and suffer from a couple of limitations. Participation in the surveys was voluntary and not strongly incentivized, which might bias our findings into both directions. Furthermore, our exploratory work elucidates only some of the burnout causes, drivers, countermeasures, and coping mechanisms, which calls for more systematic, theory-based, quantitative, and longitudinal research that is based on more sophisticated sampling techniques. Likewise, we suggest an in-depth analysis of former social entrepreneurs to learn about the causes of exit.

4 Implications for the Social Entrepreneurship Ecosystem

We hope that social entrepreneurs and actors of the ecosystem will find some insights and implications in this chapter. We therefore purposefully avoid recommendations on an individual level and do not add to the vast literature on personal stress management (see, e.g., Hargrove et al., 2011). Instead, we seek to point to systemic reasons and responses to burnout, i.e., in the ecosystem of social entrepreneurs, which has turned out to be pivotal for nurturing social startups (European Commission, 2014; Millner und Vandor, 2014; Roundy, 2017).

First, ecosystem actors such as impact investors, educational and training providers, coworking spaces, and umbrella organizations have to acknowledge the "dark side" of social entrepreneurship: fueled by the very real experiences of stress, risk, and workload typical for (social) entrepreneurship, burnout can be a real problem for founders, their teams, and beneficiaries. Burnout prevention should therefore be regarded as a necessity, not a luxury, starting with raising realistic expectations about the risks and hardships of entrepreneurship. Ecosystem actors should thus avoid portraying entrepreneurs as infallible "heropreneurs" (Papi-Thornton, 2016) and acknowledge the challenges as well as the often collective nature of the entrepreneurial journey. Obviously, such clarifications might shake and destroy naïve visions and overoptimistic assumptions, not only for social entrepreneurs, but also for impact investors and other players in the field.

Second, numerous measures can and should be taken by ecosystem actors to directly contribute to burnout prevention by (a) helping founders to anticipate and prepare for common stressors, e.g., team conflicts, goal conflicts, tight fundings, and providing potential remedies to these problems (access to funding, talent, know-how); (b) providing access to professional coaching and help for individuals and teams; (c) encouraging and supporting peer-help and self-help; or (d) advocating for a better institutional environment (e.g., umbrella organizations, lobbying platforms) and a tailored access to financial markets for social entrepreneurs.

Finally, incubators, investors, and accelerators should reflect whether they themselves are part of the problem by making unrealistic demands or creating pressure on entrepreneurs and include entrepreneurial well-being within their success metrics.

A brief example of how some of these actions can be implemented is provided by the case of the Social Impact Award. Already in the design of each year's incubation program, the Social Impact Award reflects on its role in the field and tries to avoid creating dysfunctional stressors. During their incubation stage, founder teams do not only focus on the development of their business models and impact models and operations but are also encouraged to surface and discuss tensions in the venture and teams through a guided process. Joint events and peer activities (hiking, karaoke evening, etc.) aim to intensify social relations beyond the single venture and facilitate peer coaching. Moreover, the program offers dedicated workshops on well-being and burnout prevention, as well as individual coaching sessions for individual and team-related tensions with external experts. All these activities seem to have a positive impact: in a survey among participants of the program's incubation process, only 14.6% indicated that "the high expectations of SIA staff, experts and mentors have created a lot of stress and pressure," whereas 70.4% responded that "overall, participating in SIA contributed to my ability to deal with conflict and stress when starting a social enterprise." This suggests a clear association between utilizing all these support measures and the participants' inclination of burnout.

5 Conclusion

Impactful ventures do not necessarily leave burned-out founders. Compared with other professional groups, especially those in the wider field of social services, the burnout prevalence of social entrepreneurs is lower, as their autonomy and purposeful projects strengthen intrinsic motivation and performance. However, burnout among social entrepreneurs can be particularly consequential by negatively affecting not only the entrepreneurs but also their teams and often vulnerable beneficiaries. Ecosystem actors have a high responsibility for young social entrepreneurs in particular, as they are more vulnerable to illusionary goals, over-engagement, and then sobering disappointment and exhaustion.

References

Akinsulure-Smith, A. M., Espinosa, A., Chu, T., & Hallock, R. (2018). Secondary traumatic stress and burnout among refugee resettlement workers: The role of coping and emotional intelligence. *Journal of Traumatic Stress, 31*(2), 202–212.

Andersen, L. L., & Dybbroe, B. (2017). Introspection as intra-professionalism in social and health care. *Journal of Social Work Practice, 31*(1), 21–35.

Berg, S. (2019, July). *WHO adds burnout to ICD-11. What it means for physicians*. American Medical Association, Physician Health.

Bosma, N., Schøtt, T., Terjesen, S. A., & Kew, P. (2016). *Global entrepreneurship monitor 2015 to 2016: Special topic report on social entrepreneurship*. Available at SSRN 2786949.

Brieger, S. A., De Clercq, D., & Meynhardt, T. (2021). Doing good, feeling good? Entrepreneurs' social value creation beliefs and work-related well-being. *Journal of Business Ethics, 172*, 707–725.

Cardon, M. S., & Patel, P. C. (2015). Is stress worth it? Stress-related health and wealth trade-offs for entrepreneurs. *Applied Psychology, 64*(2), 379–420.

Chernoff, P., Adedokun, C., O'Sullivan, I., McManus, J., & Payne, A. (2019). Burnout in the emergency department hospital staff at Cork University Hospital. *Irish Journal of Medical Science (1971), 188*, 667–674.

Dacin, P. A., Dacin, M. T., & Matear, M. (2010). Social entrepreneurship: Why we don't need a new theory and how we move forward from here. *Academy of Management Perspectives, 24*(3), Article 3.

De Mol, E., Ho, V. T., & Pollack, J. M. (2018). Predicting entrepreneurial burnout in a moderated mediated model of job fit. *Journal of Small Business Management, 56*(3), 392–411.

Dees, G. (1998). Enterprising nonprofits. *Harvard Business Review, 76*(1), 54–69.

Delgadillo, J., Saxon, D., & Barkham, M. (2018). Associations between therapists' occupational burnout and their patients' depression and anxiety treatment outcomes. *Depression and Anxiety, 35*(9), 844–850.

Demerouti, E., & Bakker, A. B. (2008). The Oldenburg Burnout Inventory: A good alternative to measure burnout and engagement. *Handbook of Stress and Burnout in Health Care, 65*(7), 1–25.

Dempsey, S. E., & Sanders, M. L. (2010). Meaningful work? Nonprofit marketization and work/life imbalance in popular autobiographies of social entrepreneurship. *Organization, 17*(4), 437–459.

Dolan, E. D., Mohr, D., Lempa, M., Joos, S., Fihn, S. D., Nelson, K. M., & Helfrich, C. D. (2015). Using a single item to measure burnout in primary care staff: A psychometric evaluation. *Journal of General Internal Medicine, 30*, 582–587.

Drencheva, A., Stephan, U., Patterson, M. G., & Topakas, A. (2021). Navigating interpersonal feedback seeking in social venturing: The roles of psychological distance and sensemaking. *Journal of Business Venturing, 36*(4), 106123.

Drencheva, A., Au, W. C., & Li Yew, J. (2023). Working for impact, but failing to experience it: Exploring individuals' sensemaking in social enterprises. *Business & Society, 62*(7), 1458–1495.

European Commission. (2014). *A map of social enterprises and their eco-systems in Europe*.

Fernet, C., Torrès, O., Austin, S., & St-Pierre, J. (2016). The psychological costs of owning and managing an SME: Linking job stressors, occupational loneliness, entrepreneurial orientation, and burnout. *Burnout Research, 3*(2), 45–53.

Folkman, S., & Lazarus, R. S. (1988). Coping as a mediator of emotion. *Journal of Personality and Social Psychology, 54*(3), 466.

Freudenberger, H. J. (1974). Staff burn-out. *Journal of Social Issues, 30*(1), 159–165.

Gardner, R. L., Cooper, E., Haskell, J., Harris, D. A., Poplau, S., Kroth, P. J., & Linzer, M. (2019). Physician stress and burnout: The impact of health information technology. *Journal of the American Medical Informatics Association, 26*(2), 106–114.

Gómez-Urquiza, J. L., Vargas, C., De la Fuente, E. I., Fernández-Castillo, R., & Cañadas-De la Fuente, G. A. (2017). Age as a risk factor for burnout syndrome in nursing professionals: A meta-analytic study. *Research in Nursing & Health, 40*(2), 99–110. https://doi.org/10.1002/nur.21774

Grant, A. M. (2008). Does intrinsic motivation fuel the prosocial fire? Motivational synergy in predicting persistence, performance, and productivity. *Journal of Applied Psychology, 93*(1), 48.

Green, R., Gregory, R., & Mason, R. (2006). Professional distance and social work: Stretching the elastic? *Australian Social Work, 59*(4), 449–461.

Guclu, A., Dees, J. G., & Anderson, B. B. (2002). The process of social entrepreneurship: Creating opportunities worthy of serious pursuit. *Center for the Advancement of Social Entrepreneurship, 1*, 1–15.

Halbesleben, J. R., & Demerouti, E. (2005). The construct validity of an alternative measure of burnout: Investigating the English translation of the Oldenburg Burnout Inventory. *Work & Stress, 19*(3), 208–220.

Hargrove, M. B., Quick, J. C., Nelson, D. L., & Quick, J. D. (2011). The theory of preventive stress management: A 33-year review and evaluation. *Stress and Health, 27*(3), 182–193.

Hütter, L. M. (2018). *Mission Drift: Junge Social Enterprises zwischen sozialer und kommerzieller Mission*. Master-Thesis. WU Vienna.

Jamal, M. (1997). Job stress, satisfaction, and mental health: An empirical examination of self-employed and non-self-employed Canadians. *Journal of Small Business Management, 35*(4), 48.

Katterl, H. (2019). *Auslöser von Stress und Coping-Strategien bei Sozialunternehmer*innen*. Bachelor-Thesis, WU Vienna.

Kelley, D., Singer, S., & Herrington, M. (2016). *Global entrepreneurship monitor*. Global Report, Global Entrepreneurship Research Association, London Business School, Regents Park, London.

Kibler, E., Wincent, J., Kautonen, T., Cacciotti, G., & Obschonka, M. (2019). Can prosocial motivation harm entrepreneurs' subjective well-being? *Journal of Business Venturing, 34*(4), 608–624.

Kim, J. L. S., & Ling, C. S. (2001). Work-family conflict of women entrepreneurs in Singapore. *Women in Management Review, 16*(5), 204–221.

Knox, M., Willard-Grace, R., Huang, B., & Grumbach, K. (2018). Maslach burnout inventory and a self-defined, single-item burnout measure produce different clinician and staff burnout estimates. *Journal of General Internal Medicine, 33*, 1344–1351.

Korber, S., & McNaughton, R. B. (2017). Resilience and entrepreneurship: A systematic literature review. *International Journal of Entrepreneurial Behavior & Research, 24*(7), 1129–1154.

Leitner, K. H., Zahradnik, G., Einsiedler, J., Raunig, M., Dömötör, R., & Vandor, P. (2023). *Austrian startup monitor 2022*. AIT Austrian Institute of Technology, 106 p.

Lekutle, M., & Nel, J. A. (2012). Psychometric evaluation of the Utrecht Work Engagement Scale (UWES) and Oldenburg Burnout Inventory (OLBI) within a cement factory. *Journal of Psychology in Africa, 22*(4), 641–647.

Lepoutre, J., Justo, R., Terjesen, S., & Bosma, N. (2013). Designing a global standardized methodology for measuring social entrepreneurship activity: The Global Entrepreneurship Monitor social entrepreneurship study. *Small Business Economics, 40*, 693–714.

Marchand, A., Blanc, M. E., & Beauregard, N. (2018). Do age and gender contribute to workers' burnout symptoms? *Occupational Medicine, 68*(6), 405–411.

Maslach, C., Jackson, S. E., & Leiter, M. P. (1997). *Maslach burnout inventory*. Scarecrow Education.

Maslach, C., Schaufeli, W. B., & Leiter, M. P. (2001). Job burnout. *Annual Review of Psychology, 52*(1), 397–422.

Mayring, P. (2003). *Qualitative Inhaltsanalyse. Grundlagen und Techniken: Vol. 8229: Pädagogik* (8. Aufl., Dr. nach Typoskript). Beltz.

Millner, R., & Vandor, P. (2014). Neues Unternehmertum: Social Entrepreneurship und die Rolle des Umfelds. *Forschung zu Zivilgesellschaft, NPOs und Engagement: Quo vadis?*, 283–300.

Papi-Thornton, D. (2016). Tackling heropreneurship. *Stanford Social Innovation Review*, p. 23.

Perron, B. E., & Hiltz, B. S. (2006). Burnout and secondary trauma among forensic interviewers of abused children. *Child and Adolescent Social Work Journal, 23*, 216–234.

Puffer, J. C., Knight, H. C., O'Neill, T. R., Rassolian, M., Bazemore, A. W., Peterson, L. E., & Baxley, E. G. (2017). Prevalence of burnout in board certified family physicians. *The Journal of the American Board of Family Medicine, 30*(2), 125–126.

Redondo-Flórez, L., Tornero-Aguilera, J. F., Ramos-Campo, D. J., & Clemente-Suárez, V. J. (2020). Gender differences in stress-and burnout-related factors of university professors. *BioMed Research International, 2020*, 6687358.

Reis, D., Xanthopoulou, D., & Tsaousis, I. (2015). Measuring job and academic burnout with the Oldenburg Burnout Inventory (OLBI): Factorial invariance across samples and countries. *Burnout Research, 2*(1), 8–18.

Roundy, P. T. (2017). Social entrepreneurship and entrepreneurial ecosystems. *International Journal of Social Economics, 44*(9), 1252–1267.

Shepherd, C. D., Marchisio, G., Morrish, S. C., Deacon, J. H., & Miles, M. P. (2010). Entrepreneurial burnout: Exploring antecedents, dimensions and outcomes. *Journal of Research in Marketing and Entrepreneurship, 12*(1), 71–79.

Sonnenberg, K. (2021). *Teaching social work in Germany. Social work at the level of international comparison: Examples from Iraqi-Kurdistan and Germany* (pp. 129–141). Springer.

Stephan, U. (2018). Entrepreneurs' mental health and well-being: A review and research agenda. *Academy of Management Perspectives, 32*(3), 290–322.

Stinchcombe, A. L. (1965). Organizations and social structure. *Handbook of Organizations, 44*(2), 142–193.

Stroe, S., Parida, V., & Wincent, J. (2018). Effectuation or causation: An fsQCA analysis of entrepreneurial passion, risk perception, and self-efficacy. *Journal of Business Research, 89*, 265–272.

Symon, G., & Whiting, R. (2019). The sociomaterial negotiation of social entrepreneurs' meaningful work. *Journal of Management Studies, 56*(3), 655–684.

Taris, T. W. (2006). Is there a relationship between burnout and objective performance? A critical review of 16 studies. *Work & Stress, 20*(4), 316–334.

Ultsch, V. (2020). *Social entrepreneurship and well-being: An analysis of social entrepreneurs' well-being and consequences for their ventures*. Bachelor-Thesis, WU Vienna.

Vandor, P., & Leitner, L. (2018). *The support needs of social entrepreneurs-lessons from the impact hub network*. Available at SSRN 3336713.

Vandor, P., Leitner, L., & Stamatiou Nichols, S. (2019). *Beyond co-working: Drivers of collaboration in shared workspaces* (Working Paper). Social Entrepreneurship Center, WU Wien.

Williamson, A. J., Drencheva, A., & Battisti, M. (2022). Entrepreneurial disappointment: Let down and breaking down, a machine-learning study. *Entrepreneurship Theory and Practice, 46*(6), 1500–1533.

Wincent, J., Örtqvist, D., & Drnovsek, M. (2008). The entrepreneur's role stressors and proclivity for a venture withdrawal. *Scandinavian Journal of Management, 24*(3), 232–246.

Peter Vandor is senior researcher and head of the Social Entrepreneurship Center at the WU Vienna, as well as founder of the Social Impact Award. His research focuses on social entrepreneurs and immigrant entrepreneurship and has been published in journals, including *Journal of Business Venturing*, *Journal of World Business*, and *Harvard Business Review*.

Michael Meyer is a professor at the Department of Management at WU Vienna, head of the Institute for Nonprofit-Management, and a member of the Stanford PACS Civic Life of Cities Lab. Michael's present research focuses on urban civil societies, managerialism, nonprofit governance, civic participation, and social entrepreneurship. His teaching and training activities also embrace organization theory, leadership, and organizational behavior. On all these topics, he has (co-) authored and edited more than 200 scholarly publications.

Understanding the Nuances of Women-Owned Businesses in Emerging and Developing Economies

Yanire Brana, Nayan Mitra, and Mohammad Mostak Al Farhad

Abstract Women entrepreneurs play an essential role in economic growth, helping to provide job opportunities and social improvement on a worldwide scale. However, their journey as an entrepreneur is distinct from those of a male-owned enterprise. This article examines recent research on the status of women entrepreneurs in various regions of the world, particularly in emerging countries, and identifies the elements that determine their functioning. In addition to patriarchal attitudes and challenges in accessing capital for business, this contextual analysis confirms anecdotal evidence that outdated legislation and a lack of assistance targeted to women's needs are the fundamental problems and determinants influencing female entrepreneurs in developing market regions. Although research on the psychological well-being of women in this context is still relatively scarce, this chapter refers to some of its potential challenges and opportunities that need to be further explored.

Keywords Women-led business · Woman entrepreneur · Female entrepreneurship · Entrepreneurial ecosystem · Cultural norms · Grants · Capital · Patriarchy · Developing countries · Asia · Middle East

1 Introduction

Female entrepreneurship, defined as women initiating new endeavors to enhance personal and business development (Morazzoni & Sy, 2022), is crucial to socioeconomic development and poverty reduction. Female entrepreneurs play an essential role in economic growth, helping to provide job opportunities and social improvement on a worldwide scale (Bullough et al., 2021).

Y. Brana · N. Mitra (✉) · M. M. Al Farhad
Kolkata, India
e-mail: yanire.brana@opendeusto.es; mostak@sustainableadvancements.com

M. Mühlböck (ed.), *Sustainable Transformation and Well-being*, CSR, Sustainability, Ethics & Governance,
https://doi.org/10.1007/978-3-031-75566-8_11

Significant evidence shows that women's entrepreneurship participation, access to capital, and ecosystem outcomes differ from their male counterparts (Brush et al., 2019; Becker-Blease & Sohl, 2007). Historically, "entrepreneurship has been a 'man's' domain" (Bird & Brush, 2002, p. 41), and literature on the topic shows a substantial amount of references made in masculine terms (Ahl, 2006).

According to the Global Entrepreneurship Monitor, women-owned enterprises are the fastest-growing demographic globally, accounting for around one-third of high-growth businesses (GEM, 2022/2023). According to the World Economic Forum's Global Gender Gap Show 2023 (WEF, GGR, 2023), while most countries state minimal disparities in start-up rates between men and women, some of the nations report that female entrepreneurs outnumber males on average. However, it was observed that there are only four areas having more women than men establishing new enterprises.

Currently, gender disparities still exist in the world of entrepreneurship. While women entrepreneurs have substantially contributed to innovation, employment, and prosperity, challenges and imbalances still need to be addressed (Brush & Cooper, 2012). The number of challenges women face could explain why most companies that have been in business for at least 3½ years are managed mainly by men (WEF Global Gender Gap Report, 2023).

Over the last four decades, there has been substantial research on female entrepreneurs, which may be separated into three primary research streams: entrepreneurial profile, entrepreneurial process, and gender identity (Santos et al., 2018). The research streams in female entrepreneurship have also been classified in two main groups: "developed countries" and "developing countries" (Poggesi et al., 2016 p. 736).

Most studies on female entrepreneurship have focused on analyzing context, including challenges facing women business owners, drivers, and motivations for starting a business. Literature reviews the relevance and impact of the context in which female business owners operate and how preexisting social inequalities could limit women's abilities to pursue entrepreneurial endeavors. These texts highlight the importance of examining female entrepreneurs' unique challenges and opportunities in different countries and contexts.

Despite the global influence and extraordinary rise of female entrepreneurs, most research on female entrepreneurship has been undertaken in developed nations; hence, the conclusions of these studies may not be immediately applicable to developing economies. This restricts awareness of the specific obstacles, opportunities, and experiences that female entrepreneurs from emerging nations encounter when they embark on their entrepreneurial journeys (Yousafzai et al., 2019; Brush & Cooper, 2012; Rosca et al., 2020).

This article examines recent research on the status of women entrepreneurs in various regions of the world, particularly in developing countries, and identifies the elements that determine their functioning. It aims to broaden a body of research that differentiates between female-owned firms in developing and industrialized nations (Alhajri & Aloud, 2023).

2 Contextual Influence on Female Entrepreneurs

Despite the efforts to promote female entrepreneurship, many studies show that women entrepreneurs need help creating businesses, and the kind of help they need varies depending on the context. Female entrepreneurs may need assistance navigating cultural norms and building community relationships, or they may be grappling with a lack of role models, an unaccommodating culture, or the existing gender gap. Multiple contexts may impact women's motivations to pursue entrepreneurial endeavors.

The study of women's entrepreneurship is constantly developing, and new studies may focus on different aspects of female entrepreneurship to present a more complete picture of the situation. So far, female entrepreneurship research has focused chiefly on difficulties, entrepreneurial motivation characteristics, or behavioral differences between genders. Such research collectively contributes to a deeper understanding of women's entrepreneurship challenges, opportunities, and impacts in various social, cultural, and economic contexts.

However, research interests have shifted in recent years to include financing barriers, venture performance, business characteristics, and familial influences that affect female entrepreneurship in developing countries (Poggesi et al., 2016). There is also interest in investigating the impact of context on female entrepreneurs and their new businesses (Alhajri & Aloud, 2023), contexts such as the sociocultural environment, social entrepreneurship, education, social equality, poverty, and inclusion (Cardella et al., 2020). The challenges and opportunities for female entrepreneurs vary greatly depending on the region and culture.

3 Female Entrepreneurship Ecosystems

Academic study on women's entrepreneurship has traditionally focused on the well-established issue of the problems women encounter while founding and maintaining enterprises conducted a study in this area, examining current gender disparities, women's limited access to funding, and the influence of cultural norms on female entrepreneurs. Gender-related difficulties, while acknowledged and handled to some extent, have to be fully resolved internationally; hence, these themes remain a priority on the academic agenda.

Women-owned businesses continue to encounter various problems, impediments, and obstacles. Even though it is critical to include women in the workforce to drive the economy, research has continually proven that women suffer disadvantages and impediments in this context. Entrepreneurial ecosystems are complex where multifaceted environments play a critical role in fostering growth and innovation. Women face challenges when attempting to enter a new market, access financial resources, find infrastructural and institutional support, or attract, train, and retain staff. The rising importance of addressing entrepreneurial ecosystems has incentivized policymakers, academics, and practitioners in many countries to create programs, strategies, and initiatives to encourage female entrepreneurship and company engagement.

As mentioned, most research on female entrepreneurship ecosystems has focused on women business owners in developed countries. However, outlined the main aspects that can influence a business's performance in emerging and developing countries, which include access to capital, institutional and infrastructural support, the regulatory environment, and sociocultural factors that influence business practices, consumer behavior, and network access.

3.1 Access to Capital

Women entrepreneurs frequently experience difficulties in obtaining collateral and financial resources. One of the most critical components of business growth that might affect the success of new and expanding female entrepreneurs is access to monetary financing (Okoth et al., 2016). The financial system is connected to emerging nations' economic success because it encourages savings, capital accumulation, improved resource allocation, and productive efficiency (Sanguinetti et al., 2011). Financial institutions' specialized duty is to acquire and manage financial resources through savings and encourage investment in entrepreneurial enterprises, company, and family activities through various financial products such as loans and credit.

While, women's savings accounts can provide financial independence and economic empowerment, women entrepreneurs, sometimes require more assistance in obtaining finance and investing capital than their male counterparts. Women entrepreneurs have traditionally faced substantial challenges due to a lack of access to cash, commonly known as the financing gap (Brush et al., 2018). For many years, the gender gap in access to financial services in emerging nations lingered around 9%. However, recent attempts to resolve this inequality appear to have worked, as the financing gender gap in these ecosystems has shrunk to 6%, according to the Global Entrepreneurship Monitor (GEM, 2022/2023). However, considerable effort must be made in underdeveloped nations to provide equitable opportunities for women to flourish and access financing.

According to Brush et al. (2018), the main factors contributing to the venture capital gap are human capital, structural limitations, and entrepreneurial choices. Financial loans are unattractive or unaffordable for women business owners due to the predominance of norms or legislation governing property rights and land ownership as well as a lack of knowledge and financial education (Ojong et al., 2021). Because of this, women rely more on family for financial support—both direct family, such as parents, husbands, or children, and extended family, who could be willing to make investments in or provide loans to support the business. According to Lindvert et al. (2017), family networks and connections are crucial to women's entrepreneurial processes in developing nations since they increase their access to financing.

3.2 *Institutional and Infrastructural Support*

The institutional context impacts female entrepreneurship through cultural norms, political-economic aspects, and policy frameworks (Lecuna et al., 2017; Van de Ven, 1993; Kouriloff, 2000). Organizations designed to alleviate the gender disparities in entrepreneurship operate under several conditions within countries and sectors, including gender culture and gender equality policies and laws, usually called the "gender context." Van der Lippe et al. (2011) have analyzed the link between gender context and institutional structures. Institutionality, an essential element in improving the business environment for women, should include bringing to the fore the gender perspective in the government, formalizing processes that support women-owned businesses, and providing a space for women entrepreneurs to participate in business development policies (ILO, 2014).

For women entrepreneurs to prosper, infrastructure assistance is essential. The availability of water and power, for example, has evident implications on women's capacity to establish a company (Pueyo & Maestre, 2019). Energy consumption has a gendered influence on demand and supply, resulting in varied economic prospects for female entrepreneurs. Policymakers could follow a gender approach to the productive use of energy, increasing the potential consumers for the electricity suppliers and benefiting women business owners (Pueyo & Maestre, 2019). The studies by Pueyo and Maestre (2019) based in Latin America, India, and South Africa show a gendered impact of electrification; electricity improves productivity, allowing the use of more sophisticated machinery in business endeavors, thereby increasing female earnings.

However, not all categories of services and infrastructure are as apparent. Providing infrastructural support also includes delivering types of technical infrastructure, such as transportation and communication, which can greatly impact female start-up activity and business efficiency (Audretsch et al., 2015).

3.3 *Regulatory Framework*

Several studies confirm that the basics of entrepreneurship and the legal and political environment have an intimate link that may considerably affect both male and female entrepreneurs (Weeks & Seiler, 2001; Murdock, 2012; Kim & Li, 2014). Ardagna and Lusardi (2010) examined the influence of gender environment on entrepreneurship and discovered that regulation moderates female entrepreneurship. They found that in highly controlled settings, women are more likely than males to establish a firm only when it is required to meet their financial requirement.

More governments have grown conscious of the need to implement legal reforms that ensure women's empowerment while safeguarding their people and economy. According to the 2022 Women Business and the Law study, the Middle East and North Africa area improved the most from 2020 to 2022, followed by sub-Saharan Africa.

3.4 Sociocultural Factors

Women's entrepreneurship is not a one-size-fits-all concept. The circumstances and dynamics of women's lives can vary significantly based on cultural, societal, and individual factors (Bullough et al., 2021), and sociocultural norms, values, and expectations significantly affect the intent of women to create a business.

According to Bui et al. (2018), there is a need to increase the public's acceptance of female business owners, since patriarchal assumptions of women harm female entrepreneurship. The effects of this macho culture extend outside the family. According to the, women are underrepresented in business and leadership positions and have fewer opportunities for career growth and training.

In addition to patriarchal settings, other sociocultural contexts need to be taken into account. In contrast to the individualistic cultures of established economies that support capitalism and self-interest, many women in emerging and developing economies live in more collectivist societies that place a higher value on the welfare of the group as a whole than the individual. The intricate relationship between individualism, collectivism, and women's entrepreneurship is highlighted by research examining the relationship between women's entrepreneurship and societies with extreme degrees of either collectivism or individualism at the macro-societal level (Bullough et al., 2021).

In these circumstances, women's entrepreneurship can be extremely important for modernizing nations, boosting economic activity, and resolving social issues. Social objectives are more common among women than among businessmen, which is beneficial for people and their communities (Minniti, 2010).

Since women at the bottom of the pyramid (BOP) seem to have a variety of motivations, it is especially important to understand the personal rewards that entrepreneurship brings to them. As Gray and Finley-Hervey (2005) point out, women in developing nations typically pursue entrepreneurship as a means of generating income and escaping poverty, but they are also motivated by a desire for autonomy and personal fulfillment.These women benefited, for example, from Unilever's widely recognized Shakti initiative, which gave rural Indian women "self-esteem, a sense of empowerment, and a place in society," thus providing "dignity to the women entrepreneurs" (Neath & Sharma, 2008).

According to Williams and Shepherd (2016), entrepreneurship can therefore aid in meeting people's basic psychological requirements and enhancing their well-being. The company benefits greatly from this kind of personal well-being because it encourages tenacity and production. On a personal level, though, this well-being inspires entrepreneurs to act and feel empowered (Williams & Shepherd, 2016), which has positive ripple effects on their communities and families (Lepeley et al., 2019).

Consequently, given the challenges they and their communities encounter, personal well-being is crucial for female entrepreneurs at the BOP. Therefore, it's indispensable to investigate how entrepreneurship affects BOP women entrepreneurs' well-being beyond the success of their businesses, rather than assuming that financial gains and company performance inevitably translate into personal well-being.

4 Women-Owned Businesses in Emerging and Developing Economies

It is critical to approach the study of female entrepreneurship from a nuanced and localized viewpoint, given that the problems, possibilities, and cultural dynamics that women entrepreneurs encounter differ significantly from area to region or nation to region or country. Analyzing context-specific tactics might significantly help to improve women's entrepreneurship in emerging and developing nations.

Ecosystems that consistently exhibit high levels of gender parity in entrepreneurial activity are those in which women own firms at a rate equal to that of males (Hill et al., 2022). One noteworthy trend in entrepreneurship are the high levels of gender equality in certain countries, where women are just as likely as men to be business owners, if not more so. Gaining knowledge about the conditions and factors that contribute to gender differences could help promote female entrepreneurship throughout the world.

The World Bank's (2020) regional classification, which divides developing countries into low- and middle-income economies, serves as the basis for the division of countries into "developed" and "developing" categories. Financial institutions and international organizations like the International Monetary Fund (IMF) divide the world into two main categories: advanced countries and emerging markets and developing economies. These divisions are based on economic criteria and common characteristics. This categorization enables the examination and understanding of the characteristics and financial development of women-owned businesses across international borders.

In line with the IMF World Economic Outlook (2023), there are 41 developed or industrialized countries classified as "advanced economies" that have a high GDP per capita, a varied export market, and a financial sector that is linked into the global financial system. The majority of research on women entrepreneurs and their involvement in entrepreneurship has historically been based on this group, which consists of the EURO region, the USA, and the ASEAN-5 countries (Indonesia, Malaysia, the Philippines, Singapore, and Thailand).

Often referred to as "emerging market and middle-income economies," the category of "emerging market and developing economies" includes middle- and low-income countries. This IMF group includes Latin America, the Caribbean, Asia, the Middle East and Central Asia, and sub-Saharan Africa. Ninety-five nations that have increased their incomes significantly but have not yet caught up to advanced economies fall within the subclass of middle-income economies. The Poverty Reduction and Growth Trust, which offers financial support to alleviate economic vulnerabilities and promote sustainable growth, is needed by 59 countries, which make up the low-income/developing economies subgroup, according to the IMF Fiscal Monitor (2023). This collective group of emerging markets and developing economies has made substantial progress in income growth and entrepreneurship-related economic development. However, they still need to achieve the status of advanced economies. They may have more developed financial systems and infrastructure than the broader category of developing economies.

4.1 Asia

As noted, there is an increased interest in the research of female entrepreneurship in emerging markets. Thus far, much of the emerging markets literature has focused on businesses in Asia (especially China), highlighting entrepreneurship and women's empowerment (Yunis et al., 2020; Parvin et al., 2012).

Even though enterprises from rising nations operate in a variety of political, social, economic, and geographic environments, many Asian societies, particularly those in India, have a history of patriarchal norms and societal structures that place women in inferior roles (Shukla et al., 2018; Goyal & Parkash, 2011; Rashid & Ratten, 2020). While significant progress has been made in challenging traditions and promoting gender equality, much research has focused on the challenges that Asian women entrepreneurs continue to face and the impact of patriarchal culture on various aspects of their lives, such as family dynamics and social hierarchies.

The literature on Asian women frequently emphasizes women's economic vulnerability due to a mix of issues such as restricted access to formal education, work, and health care. Gender studies in this region place a strong emphasis on feminine poverty. Feminized poverty is the belief that women have a greater poverty rate than males, that their poverty is more severe, and that poverty among women is growing. As scholars investigate the reasons for women's poverty in Asian nations such as Bangladesh, India, and Sri Lanka, this subject is frequently mentioned (Lucy et al., 2008; Mahmood et al., 2014). Many Asian cultural conventions, such as marriage, contribute to the feminization of poverty (Kishor & Gupta, 2004).

The societal acceptance of women company owners and how choice is curtailed is a prominent topic in studies of Asian female entrepreneurs (Nawaz, 2010; Sharma, 2013; Tripathi & Singh, 2018). Women from upper castes, for example, may suffer discriminatory norms that limit their independence and commercial potential. Local efforts, such as instilling social responsibility in women, have raised awareness of and pushed for women's rights and empowerment across all castes (Anderson et al., 2019), undermining the patriarchal social order. Women are increasingly involved in grassroots movements or civil society organizations that help rural women in their areas (Field et al., 2010; De Vita et al., 2014).

4.2 The Middle East and North Africa (MENA)

The MENA area has the most significant percentage of gender inequality globally, with a 40% gender difference (Bastian et al., 2019). Historically, the region's political and economic models were not inclusive of women, despite no explicit laws preventing women-owned businesses (Hattab, 2012). Many MENA nations' constitutions recognize constitutional equality between men and women, giving women entrepreneurs equal rights and duties (Hattab, 2012). However, due to the diverse sociocultural customs of the countries, the execution of these rights differs greatly (De Vita et al., 2014; Bastian et al., 2019). Traditional patriarchal conventions and social norms, supported by legal, political, and religious institutions, maintain gender inequities, restricting women's entrepreneurial activity and empowerment (Bastian et al., 2018; Caputo et al., 2016).

Female entrepreneurship is among the lowest in MENA, accounting for only 4% of total entrepreneurial activity (GEM, 2022/2023). The MENA region's current gender inequities are affected by women's restricted involvement in and access to economic and political power (Bastian et al., 2019). Women appear to be motivated to start businesses primarily by economic necessity (Terjesen et al., 2010).

Literature on the subject generally notes that women in the MENA region are less likely than men to start businesses in all countries except Morocco, Iran, and Sudan, where women undertake start-up activities more often than men. However, although start-up activity rates for women decreased in five countries during the COVID-19 pandemic period, it also increased in three, including Oman (5.8% to 11.9%), Saudi Arabia (14% to 19%), and Egypt (4% to 5.7%) (GEM Women, 2021/2022).

Countries that traditionally have the strictest laws governing women's rights and behavior have seen an easing of restrictions in recent years. For example, since 2019, Saudi Arabia has quietly reformed repressive laws to allow women more independence: strict dress codes and mandatory gender segregation were abolished, the guardianship system was amended, and women were finally allowed driver's licenses. According to Women Business and The Law (2023), despite economies having enacted 122 reforms, only 14 economies achieved legal gender parity in 2022.

4.3 Sub-Saharan Africa (SSA)

SSA is a diversified region south of the Sahara Desert where women entrepreneurs are affected by a variety of variables like cultural norms, access to education and resources, and government programs that encourage gender equality and economic empowerment for women (Amine & Staub, 2009; Mordi et al., 2010). SSA has the world's most significant number of female entrepreneurs (Kelley et al., 2017), with women starting or running more than a quarter of all firms, mainly in the micro and small company sectors, many of which operate in the informal economy. African women are twice as likely as males to establish a business, accounting for half of all new entrepreneurs (GEM, 2022/2023; De Vita et al., 2014).

The entrepreneurial landscape, on the other hand, differs significantly between African nations and regions. SSA's intrinsic sexism elevates males as home leaders and family caregivers (Woldie & Adersua, 2004; Okafor & Mordi, 2010), restricting women's access to resources, power, and mobility. Such cultural and gender restrictions might make it difficult for female entrepreneurs to manage work and home obligations and establish a firm (Xiong et al., 2018; Quagrainie, 2016). Several societal assumptions around women's primary function as caretakers tend to place women in a subordinate position to males (Aldrich & Cliff, 2003; Webb et al., 2015).

Because of these factors, many women-led companies in SSA are part of the informal sector, including most domestic workers, street sellers, waste collectors, and hawkers. Though definitions vary by nation, casual economy workers are mainly self-employed or wage employees who labor without official contracts or social security. Women who work in the informal economy have less access to credit and are more vulnerable. Because SSA nations frequently have a frail and inadequate institutional framework with restricted market access (Mair & Marti, 2009; Webb et al., 2015), there is a growing economic requirement that drives women into entrepreneurship.

4.4 Latin America and the Caribbean (LAC)

Even though LAC is a region of significant differences in which entrepreneurs confront considerable hurdles in the start-up ecosystem (Buendía-Martínez & Carrasco, 2013), the area is strongly entrepreneurial and a global leader in business ambitions and start-up activity. According to the Global Entrepreneurship Monitor (2022/2023), the Latin American and Caribbean (LAC) area has the most excellent total entrepreneurial activity (TEA) of any international region, with 24.1% TEA for women and 30.4% TEA for men. The LAC area also boasts the highest percentage of female investors globally, with a 12-month rate of company investment of 13.6% (GEM, 2022/2023). Because of the significant proportion of women-owned businesses in this region, there is a wealth of literature on female entrepreneurship.

Even though social exclusion and gender discrimination in the labor market pose significant challenges for LAC women in business (Rubach et al., 2015; World Bank, 2010, 2011; Khwaja & Mian, 2005; Herranz et al., 2010), LAC women continue to strive for economic independence and freedom from discrimination (Giménez et al., 2017). Many Latin American countries have anti-discriminatory legislation in place to promote gender equality and financial autonomy (Giménez et al., 2017), but a widely accepted machismo culture reinforces gender inequality in a variety of ways, including the belief that a woman's primary role in society is to care for others (Tabbush, 2010). As a result, in most businesses, female entrepreneurs are motivated by necessity rather than desire (Terjesen et al., 2010).

The role of Latin American women entrepreneurs is growing in importance; nonetheless, household duties sometimes limit their engagement in the market (Terjesen et al., 2011). Women continue to undertake the majority of unpaid productive labor, such as housework, childcare and parenting, and community or volunteer work, impeding their entrepreneurial ambitions. "Even though women have been substantially absorbed into the economically active population and participate in paid labor, there is no family shared responsibility since home responsibilities are not dispersed. Due to societal norms, these tasks are still entrusted to women". Thus, women's entrepreneurial ventures in LAC are mainly microenterprises and informal (Buendía-Martínez & Carrasco, 2013).

Access to financial funding is one of the most significant problems for Latin American women entrepreneurs. According to Scherr et al. (1993), lenders sometimes discriminate against women based on unfavorable preconceptions. As a result, women prefer to invest their personal and family savings in business. Even though the majority of LAC nations has begun to assist and is working on clear entrepreneurship strategies (Hidalgo et al., 2014), private institutions are more widespread and functional, and many women company owners take the initiative themselves. Networking and personal connections provide critical possibilities for women establishing and expanding businesses; over two-thirds of women in the LAC area personally know at least one other entrepreneur, the highest rate for women globally (GEM, 2022/2023).

Despite the hurdles and disadvantages, the LAC area has the most excellent rates of entrepreneurial intent and activity and the largest proportion of female entrepreneurs and investors (GEM, 2022/2023). According to the Women's Entrepreneurship Report (2022), 43% of Latin American women intend to start a business, the highest percentage in the world. This region is seeing a lively and diversified entrepreneurial landscape, emphasizing the potential for economic development and innovation and emphasizing the significance of ongoing support for entrepreneurship, particularly for women.

5 Conclusions

Female entrepreneurs sometimes face significant obstacles while starting their businesses, which vary depending on the location. It is essential to look at the local entrepreneurship culture and the background factors of women-owned enterprises to gain a deeper understanding of female entrepreneurial experiences (Bui et al., 2018). The experiences of these entrepreneurs in different geographic and cultural contexts, including emerging and developing markets, have not received much attention, even though understanding the unique opportunities and challenges faced by female entrepreneurs in emerging markets is crucial for developing effective support networks and policies.

Amine and Staub (2009), Lindvert et al. (2017), and Katre (2018) have all done research on female entrepreneurs in developing market situations (Rosca et al., 2020). While the majority of research on female entrepreneurship in developing markets and middle-income countries has been conducted in Asia and Latin America, little has been published on countries in sub-Saharan Africa or the Middle East and North Africa region (Hatem, 2012). This contextual study validates anecdotal evidence that the primary issues and drivers impacting female entrepreneurship in developing market regions are outdated legislation and a lack of help tailored to women's needs, in addition to patriarchal and macho attitudes. Although these concepts take many different forms based on cultural, historical, and geographic contexts, they are both based on emerging and rising economies and contribute to the perpetuation of gender inequality (Tlaiss & Kauser, 2019).

Women entrepreneurs face challenges in securing funding, acquiring institutional and infrastructure support, negotiating the regulatory landscape, and conquering societal barriers, according to studies on female entrepreneurship in emerging markets and middle-income countries. In order to gain a deeper understanding of the issue, it is imperative to conduct extensive study on the characteristics of these ecosystems, assess the progress made in assisting women entrepreneurs across different regions, and consider the future trajectory of this kind of assistance (Poggesi et al., 2016). First research indicates that women entrepreneurship might also entail positive effects in regard to women's psychological well-being. Further research is required to explore this field and better understand potential opportunities and challenges in this matter for developing market regions.

References

Ahl, H. (2006). Why research on women entrepreneurship needs a new direction. *Entrepreneurship Theory and Practice, 30*(5), 595–621.

Aldrich, H. E., & Cliff, J. E. (2003). The pervasive effects of family on entrepreneurship: Toward a family embeddedness perspective. *Journal of Business Venturing, 18*, 573–596.

Alhajri, A., & Aloud, M. (2023). Female digital entrepreneurship: A structured literature review. *International Journal of Entrepreneurial Behavior & Research, 30*(5). https://doi.org/10.1108/IJEBR-09-2022-0790

Amine, L. S., & Staub, K. M. (2009). Women entrepreneurs in sub-Saharan Africa: An institutional theory analysis from a social marketing point of view. *Entrepreneurship & Regional Development, 21*, 183–211.

Anderson, A. R., Younis, S., Hashim, H., & Air, C. (2019). Social enterprising informs our concept; of exploring informal micro social enterprise. *Social Enterprise Journal, 15*, 94–110.

Ardagna, S., & Lusardi, A. (2010). Heterogeneity in the effect of regulation on entrepreneurship and entry size. *Journal of the European Economic Association, 8*(2–3), 594–560.

Audretsch, D. B., Heger, D., & Veith, T. (2015). Infrastructure and entrepreneurship. *Small Business Economics, 44*(2), 219–230. https://doi.org/10.1007/s11187-014-9600-6

Bastian, B. L., Sidani, Y. M., & El Amine, Y. (2018). Women entrepreneurship in the Middle East and North Africa: A review of knowledge areas and research gaps. *Gender in Management, 33*, 14–29.

Bastian, B. L., Metcalfe, B. D., & Zali, M. R. (2019). Gender inequality: Entrepreneurship development in the MENA region. *Sustainability, 11*, 6472.

Becker-Blease, J. R., & Sohl, J. E. (2007). Do women-owned businesses have equal access to angel capital. *Journal of Business Venturing, 22*, 503–521.

Bird, B., & Brush, C. G. (2002). A gender perspective on organizational creation. *Entrepreneurship Theory and Practice, 26*(3), 41–65.

Brush, C. G., & Cooper, S. Y. (2012). Female entrepreneurship and economic development: An international perspective. *Entrepreneurship and Regional Development, 24*(1–2), 1–6.

Brush, C., Greene, P., Balachandra, L., & Davis, A. (2018). The gender gap in venture capital-progress, problems, and perspectives. *Venture Capital, 20*(2), 115–136. https://doi.org/10.1080/13691066.2017.1349266

Brush, C., Edelman, L. F., Manolova, T., & Welter, F. (2019). A gendered look at entrepreneurship ecosystems. *Small Business Economics, 53*(2), 393–408.

Buendía-Martínez, I., & Carrasco, I. (2013). Women, entrepreneurial activity and rural development in Latin America and the Caribbean. *Rural Development Notebooks, 10*, 21–45.

Bui, H. T. M., Kuan, A., & Chu, T. T. (2018). Female entrepreneurship in a patriarchal society: Motivation and challenges. *Journal of Small Business & Entrepreneurship, 30*(4), 325–343.

Bullough, A., Guelich, U., Manolova, T. S., & Schjoedt, L. (2021). Women's entrepreneurship and culture: Gender role expectations and identities, societal culture, and the entrepreneurial environment. *Small Business Economics, 58*(2), 985–996.

Caputo, A., Mehtap, S., Pellegrini, M., & Al-Refai, R. (2016). Supporting opportunities for female entrepreneurs in Jordan. *International Journal of Entrepreneurship and Small Business, 27*, 384–409.

Cardella, G. M., Hernandez-Sanchez, B. R., & Sanchez-Garcia, J. C. (2020). Women entrepreneurship: A systematic review to outline the boundaries of scientific literature. *Frontiers in Psychology, 11*, 1557. https://doi.org/10.3389/fpsyg.2020.01557

De Vita, L., Mari, M., & Poggesi, S. (2014). Women entrepreneurs in and from developing countries: Evidences from the literature. *European Management Journal, 32*, 451–460.

Field, E., Jayachandran, S., & Pande, R. (2010). Do traditional institutions constrain female Entrepreneurship? A field experiment on business training in India. *American Economic Review, 100*, 125–129.

GEM Women. (2021/2022). *Women's entrepreneurship report from crisis to opportunity*. Global Entrepreneurship Research Association, London Business School.

Giménez, D., Gabaldón, P., & Seierstad, C. (2017). Women's entrepreneurship in Latin America and the Caribbean: A multidimensional approach. In T. S. Monolova, C. G. Brush, L. F. Edelman, A. Robb, & F. Welter (Eds.), Entrepreneurial ecosystems and growth of women's entrepreneurship: A comparative analysis (pp. 44–71). Cheltenham, UK: Edward Elgar Publishing.

Global Entrepreneurship Monitor. (2022/2023). *Global report: Adapting to a "new normal"*. GEM.

Goyal, M., & Parkash, J. (2011). Women entrepreneurship in India-problems and prospects. *International Journal of Multidisciplinary Research, 1*, 195–207.

Gray, K. R., & Finley-Hervey, J. (2005). Women and entrepreneurship in Morocco: Debunking stereotypes and discerning strategies. *Entrepreneurship Management, 1*, 203–217.

Hatem, O. (2012). *High growth and rapid internationalization of firms from emerging markets: The case of the Middle East and North Africa (MENA) region*. PhD Dissertation, Edinburgh.

Hattab, H. (2012). Towards understanding female entrepreneurship in Middle Eastern and North African countries. *Education, Business and Society: Contemporary Middle Eastern Issues, 5*, 171–186.

Herranz, A. Á., De Lara, P. V., Barraza, S., & Legato, A. M. (2010). Factors determining the entrepreneurial consolidation in Latin America. *African Journal of Business Management, 4*, 1717.

Hidalgo, G., Kamiya, M., & Reyes, M. (2014). *Emprendimientos dinÁmicos en América Latina: Avances en prÁcticas y políticas* (Serie Políticas Públicas y Transformación Productiva 16). CAF.

Hill, S., Bosma, N., Coduras, A., Guerrero, M., Ionescu-Somers, A., Roomi, Muhammad, A., Sahasranamam, S., & Shay, J. (2022). *GEM 2021/2022 global report: Opportunity amid disruption*. GEM.

ILO (International Labour Organization). (2014). *Marco integrado de evaluación de un entorno favorable para el desarrollo de empresas de mujeres*. Organización Internacional del Trabajo.

Katre, A. (2018). Facilitating affective experiences to stimulate women's entrepreneurship in rural India. *International Journal of Gender and Entrepreneurship, 10*(3), 270–288.

Kelley, D. J., Baumer, B.S., Brush, C., Greene, P. G., Mahdavi, M., Cole, M., Dean, M., & Heavlow, R. (2017). *Global Entrepreneurship Research Association (GERA) 2017*. GEM.

Khwaja, A. I., & Mian, A. (2005). Do Lenders Favor Politically Connected Firms? Rent Provision in an Emerging Financial Market. The Quarterly Journal of Economics, 120(4), 1371–1411. http://www.jstor.org/stable/25098774

Kim, P. H., & Li, M. (2014). Seeking assurances when taking action: Legal systems, social trust, and starting businesses in emerging economies. *Organization Studies, 35*(3), 359–391.

Kishor, S., & Gupta, K. (2004). Women's empowerment in India and its states: Evidence from the NFHS. *Economic and Political Weekly, 39*, 694–712.

Kouriloff, M. (2000). Exploring perceptions of a priori barriers to entrepreneurship: A multidisciplinary approach. *Entrepreneurship: Theory and Practice, 25*, 59–80.

Lecuna, A., Cohen, B., & Chavez, R. (2017). Characteristics of high-growth entrepreneurs in Latin America. *International Entrepreneurship and Management Journal, 13*(1), 141–159.

Lepeley, M. T., Kuschel, K., Beutell, N., Pouw, N., & Eijdenberg, E. (2019). *The wellbeing of women in entrepreneurship – A global perspective*. Routledge.

Lindvert, M., Patel, P. C., & Wincent, J. (2017). Struggling with social capital: Pakistani women micro entrepreneurs' challenges in acquiring resources. *Entrepreneurship & Regional Development, 29*(7–8), 759–790.

Lucy, D. M., Ghosh, J., & Kujawa, E. (2008). Empowering women's leadership: A case study of Bangladeshi Microcredit Business. *S.A.M. Advanced Management Journal, 73*, 31–50.

Mahmood, S., Hussain, J., & Matlay, H. Z. (2014). Optimal microfinance loan size and poverty reduction amongst female entrepreneurs in Pakistan. *Journal of Small Business and Enterprise Development, 21*, 231–249.

Mair, J., & Marti, I. (2009). Entrepreneurship in and around institutional voids: A case study from Bangladesh. *Journal of Business Venturing, 24*, 419–435.

Minniti, M. N. (2010). What do we know about the patterns and determinants of female entrepreneurship across countries? *European Journal of Development Research, 22*, 277–293.

Morazzoni, M., & Sy, A. (2022). Female entrepreneurship, financial frictions, and capital misallocation in the USA. *Journal of Monetary Economics, 129*, 93–118.

Mordi, C., Simpson, R., Singh, S., & Okafor, C. (2010). The role of cultural values in understanding the challenges faced by female entrepreneurs in Nigeria. *Gender in Management: An International Journal, 25*, 5–21.

Murdock, K. (2012). Entrepreneurship policy: Tradeoffs and impact in the EU. *Entrepreneurship and Regional Development, 24*(9–10), 879–893.

Nawaz, F. (2010). Nexus between women entrepreneurship development and empowerment: Bangladesh context. *Nepalese Journal of Public Policy and Governance, 26*, 62–74.

Neath, G., & Sharma, V. (2008). The Shakti Revolution: How the world's largest home-to-home operation is changing lives and stimulating economic activity in rural India. *Development Outreach, 10*(2), 13–16.

Ojong, N., Simba, A., & Dana, L. (2021). Female entrepreneurship in Africa: A review, trends, and future research directions. *Journal of Business Research, 132*, 233–248. https://doi.org/10.1016/j.jbusres.2021.04.032

Okafor, C., & Mordi, C. (2010). Women entrepreneurship development in Nigeria: The effects of environmental factors. *BULETINUL Universitatii Petrol-Gaze din Ploie~ ti, Economic Science Series, 62*, 43–52.

Okoth, D., Nambuswa, E., & Namusonge, G. (2016). Determinants of effective women entrepreneurship in Kenya: A survey of trans Nzoia county. *European Journal of Business and Management, 8*(30), 174–182.

Parvin, L., Rahman, M. W., & Jia, J. (2012). Determinates of women micro-entrepreneurship development: An empirical investigation in rural Bangladesh. *International Journal of Economics and Finance, 4*, 254–260.

Poggesi, S., Mari, M., & De Vita, L. (2016). What's new in female entrepreneurship research? Answers from the literature. *International Entrepreneurship and Management Journal, 12*(3), 735–764.

Pueyo, A., & Maestre, M. (2019). Linking energy access, gender and poverty: A review of the literature on productive uses of energy. *Energy Research & Social Science, 53*, 170–181. https://doi.org/10.1016/j.erss.2019.02.019

Quagrainie, F. A. (2016). Institutional approach and competitive behaviours of informal Ghanaian women entrepreneurs. *International Journal of Entrepreneurship and Small Business, 28*, 323–338.

Rashid, S., & Ratten, V. (2020). A systematic literature review on women entrepreneurship in emerging economies while reflecting specifically on SAARC countries. In V. Ratten (Ed.), *Entrepreneurship and organizational change. Contributions to Management Science* (pp. 37–88). Springer.

Rosca, E., Agarwal, N., & Brem, A. (2020). Women entrepreneurs as agents of change: A comparative analysis of social entrepreneurship processes in emerging markets. *Technological Forecasting and Social Change, 157*(120), 067.

Rubach, M. J., Bradley, D. B., & Kluck, N. (2015). Necessity entrepreneurship: A Latin American study. *International Journal of Entrepreneurship, 19*, 126–139.

Sanguinetti, P., Arreaza, A., Rodríguez, P., Álvarez, F., Ortega, D., & Penfold, M. (2011). *RED 2011: Servicios financieros para el desarrollo. Promoviendo el acceso en América Latina* (Reporte de Economía y Desarrollo (RED)). CAF.

Santos, G., Marques, C. S., & Ferreira, J. J. (2018). A look back over the past 40 years of female entrepreneurship: Mapping knowledge networks. *Scientometrics, 115*(2), 953–987.

Scherr, F. C., Sugrue, T. F., & Ward, J. B. (1993). Financing the small firm start-up: Determinants of debt use: Financing the small firm start-up: Determinants of debt use. *Journal of Entrepreneurial Finance, 3*(1), 17–36.

Sharma, Y. (2013). Women entrepreneurship in India. *IOSR Journal of Business and Management, 15*, 9–14.

Shukla, T., Chauhan, G. S., & Null, S. (2018). Traversing the women entrepreneurship in South Asia: A journey of Indian start-ups through Lucite ceiling phenomenon. *Journal of Enterprising Communities: People and Places in the Global Economy, 12*, 50–66.

Tabbush, C. (2010). Latin American women's protection after adjustment: A feminist critique of conditional cash transfers in Chile and Argentina. *Oxford Development Studies, 38*, 437–459.

Terjesen, S., Amoros, J., & E. (2010). Female entrepreneurship in Latin America and the Caribbean: Characteristics, drivers, and relationship to economic development. *European Journal of Development Research, 22*, 313–330.

Terjesen, S., Elam, A., & Brush, C. (2011). Gender and New Venture Creation. In K. Hindle & K. Klyver (Eds.), *Handbook of research on new venture creation* (pp. 85–98). Edward Elgar.

Tlaiss, H. A., & Kauser, S. (2019). Entrepreneurial leadership, patriarchy, gender, and identity in the Arab world: Lebanon in focus. *Journal of Small Business Management, 57*(2), 517–537.

Tripathi, K. A., & Singh, S. (2018). Analysis of barriers to women entrepreneurship through ISM and MICMAC: A case of Indian MSMEs. *Journal of Enterprising Communities: People and Places in the Global Economy, 12*, 346–373.

Van de Ven, A. H. (1993). The development of an infrastructure for entrepreneurship. *Journal of Business Venturing, 8*, 211–230.

Van der Lippe, T., de Ruijter, J., de Ruijter, E., & Raub, W. (2011). Persistent inequalities in time use between men and women: A detailed look at the influence of economic circumstances, policies, and culture. *European Sociological Review, 27*(2), 164–179.

Webb, J. W., Pryor, C. G., & Kellermanns, F. W. (2015). The influence of institutions and family embeddedness. *Africa Journal of Management, 1*, 115–136.

Weeks, J. R., & Seiler, D. (2001). *Women's entrepreneurship in Latin America: An exploration of current knowledge* (Sustainable Development Dept. Technical Papers Series MSM-111). Inter-American Development Bank.

Williams, T. A., & Shepherd, D. A. (2016). Building resilience or providing sustenance: Different paths of emergent ventures in the aftermath of the Haiti earthquake. *Academy of Management Journal, 59*(6), 2069–2102.

Woldie, A., & Adersua, A. (2004). Female entrepreneurs in a transitional economy: Businesswomen in Nigeria. *International Journal of Social Economics, 31*, 78–93.

World Bank. (2010). *Women entrepreneurs: Barriers and opportunities in the formal private sector in Latin America and the Caribbean*. The World Bank.

World Bank. (2011). *An ongoing (R) gender evolution. Expanding economic opportunities for women in Central America. Review of the last decade*. The World Bank.

World Bank. (2020). *World development indicators database*. Available at: https://databank.worldbank.org/data/download/GDP.pdf

Xiong, L., Ukanwa, I., & Anderson, A. R. (2018). Institutional influence and the role of family in poor women's micropreneurship. *International Journal of Entrepreneurial Behavior and Research, 26*, 122–140.

Yousafzai, S., Fayolle, A., Saeed, S., Henry, C., & Lindgreen, A. (2019). The contextual embeddedness of women's entrepreneurship: Towards a more informed research agenda. *Entrepreneurship and Regional Development, 31*(3–4), 167–177.

Yunis, M. S., Hashim, H., & Tayyeb, S. (2020). The gendering of context: A fresh perspective of women social enterprise in Pakistan. *Business and Economic Review, 12*, 25–48.

Yanire Brana is the president and founder of MET Community, a nonprofit organization that supports and promotes female entrepreneurship around the globe through mentoring, training, and networking. She is also a guest lecturer at Georgetown University and several other universities in Latin America and Spain. With over 20 years of experience in ESG, change management, training, and mentoring, Yanire works for the World Bank Group, and her former employers include Inter-American Development Bank, Banco Santander Group, Accenture, and Booz Allen. She received postgraduate certifications from the London Business School, Harvard Business School, and Georgetown University. Prior to that she completed the PhD studies in the University of Complutense. Yanire holds a law degree from the University of Deusto, a master's degree in international relations from the University of the Basque Country and an MBA in eBusiness from IE Business School. She recently completed her studies in Business Sustainability Management at the University of Cambridge.

Nayan Mitra straddles seamlessly between academic, social, and corporate sectors in the area of CSR and sustainability. Her company, Sustainable Advancements, broadly aims to promote the 5Ps of the Sustainable Development Goals (SDGs). She works closely with some of the eminent corporations and civil societies of India. She is a pro-bono member of the Advisory Council and is in the Board of some of the eminent NGOs in India, whose work has received exemplary status in international and national circuits. Recipient of multiple prestigious international and national awards, Dr. Mitra is the series editor of Responsible Leadership and Sustainable Management by Springer and has over 60 peer-reviewed papers in her name.

Mohammad Mostak Al Farhad was a DAAD scholarship awardee in 2018 and worked with Cologne University on peri-urban issues in India and Germany. Having worked with the Centre for Science and Down to Earth as a researcher, his work primarily focused on African and South Asian issues of mining and Environmental Impact Assessment. Besides authoring a technical manual for 11 African and 4 South Asian national governments, he is currently working with Sustainable Advancements.

The Phenomenon of Family Businesses: A Cross-Generational Commitment to Sustainability—Attitudes and Values of the Successive Generation in Family Businesses in the Case of Austria

Tien Thanh Nguyen and Christoph Robinson

Abstract Research shows that the unique intertwining of family and business renders a special commitment of family businesses to long-term, sustainable thinking and at the same time affects the well-being of the individual family members. Sustainability is to be considered a core principle that ensures the survival and success of family businesses. More broadly understood, sustainability is far from being limited to utilising sustainable resources and productions only. It also involves intergenerational thinking and acting.

Especially in Austria, family businesses play a crucial and indispensable role in the economy and society. Austria's overall well-being heavily depends on its family businesses. However, recent estimates suggest that an increasing number of family businesses will not be able to find successors within the family, despite successive generations (the so-called Next Gens) displaying a high degree of awareness regarding sustainability and long-term thinking.

In order to understand and consequently strengthen the future of family businesses, it is necessary to analyse the attitudes and values of Next Gens. A common theme among Next Gens is their desire to be able to define themselves in what they are doing while not neglecting their commitment to their family's business. These efforts to re-think their roles contribute to a generational shift in how family businesses are seen and operated.

T. T. Nguyen · C. Robinson (✉)
Vienna, Austria
e-mail: christoph.robinson@iv.at

M. Mühlböck (ed.), *Sustainable Transformation and Well-being*, CSR, Sustainability, Ethics & Governance,
https://doi.org/10.1007/978-3-031-75566-8_12

1 Introduction

"Sustainability" has become a buzzword in recent years. In times of quick-paced, far-reaching changes and uncertain outcomes, people look for sustainability as means to succeed in an unknown future. Only when provided with at least a specific level of certainty, businesses may be able to plan and look into the future. Sustainable entrepreneurial success relies—in huge parts—on long-term thinking that even exceeds generational boundaries. "Sustainability" is, hence, more than a mere buzzword. It is a core principle that ensures the survival and success of businesses.

In the realms of the business world, thinking sustainably does not only mean the utilisation of sustainable resources and production methods, but it also involves successful intergenerational operations. This is even more so crucial for family businesses. Typically, family businesses are tightly connected to the family that runs and owns it. Inevitably, long-term success requires handing over the business to a willing succeeding generation. For this to occur, successors need to commit to the cause. This includes the business itself. Family businesses are especially embedded in a wide-reaching web of values, commitments and relationships.

In Austria, family businesses appear to be deeply rooted and well-regarded in their local and regional communities while also being successful in their respective industries—domestically as well as internationally. However, a continuously rising number of family businesses, even highly profitable ones, struggle to find successors within the family. Estimates suggest, for example, that until 2030 only about 40% of family businesses will be retained within the family in Austria (KMU Forschung Austria, 2021)—continuing an ongoing downward trend. This, however, seems to not only be the case in Austria but also in many places around the world. In an article in *The Economist* from January 29, 2024, it was pointed out that family-owned businesses of all sizes worldwide struggle to find successors within the family.

Something appears to be in disorder. There seems to be a significant shift happening in attitudes and values but also in life choices among younger generations. In a recent study on the next generation of family entrepreneurs in Austria, we have shed light on the motivations, ambitions and thought processes of young family members. Doing so provides insight not only into how future generations aim to conduct businesses but also into how they think about values and sustainability itself. Ultimately, sustainability also includes the ability to adapt to changes—however fundamental they may be (Fig. 1).

2 Why Are Family Businesses So Important?

Austria is a land of family businesses. In general, family businesses are deeply regionally rooted and play an important economic and social role. According to KMU Forschung Austria (2020), around 157,000 businesses count as family businesses (excluding sole proprietorships), which employ 1.8 million people or

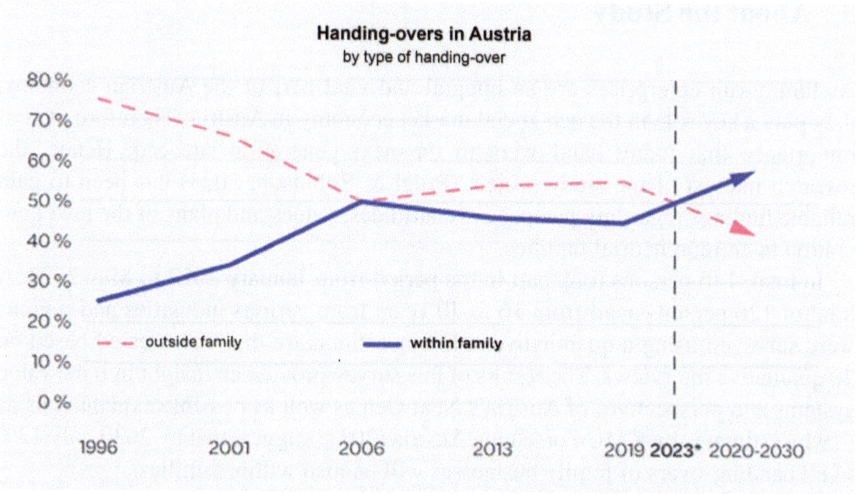

Fig. 1 Handing-overs in Austria (Data source: KMU Forschung Austria, 2021)

two-thirds of the working population on the free market in Austria, whether self-employed or employed. They generate a total revenue of 442 billion euro, which is more than half of all revenue earned by businesses. Furthermore, 80% of the world- and European market-leading companies from Austria are family-owned. The highest share of family-run companies is in the tourism sector (67%), followed by 62% in the manufacturing and construction sector and 50% in retail (KMU Forschung Austria, 2020). The regional importance is underlined by the fact that 46% of family businesses are based in rural areas, 30% in regions with medium-sized population density and only 24% in cities.

A special characteristic of family businesses is how they are defined by a set of specific values in connection with a long-term perspective in decision-making of an entrepreneurially active family ready to take responsibility (see Altenburger & Schmidpeter, 2018). Thousands of businesses are expected to be handed over in the coming years. However, estimates suggest that handing-overs within families are—absolutely and relatively speaking—declining. Successful handing-overs are crucial for a prospering economy and future wealth in Austria. It is, therefore, important to look at the perspective, attitudes and beliefs of the future generation in entrepreneurial families, the so-called "Next Gen". This article is based on findings of a study conducted by the Julius Raab Foundation together with the Friedrichshafen Institute for Family Entrepreneurship at the Zeppelin University in 2022 and 2023—for the first time in Austria—following related study projects for Germany in recent years.

3 About the Study

As family-run enterprises are an integral and vital part of the Austrian economy, they play a key role in the eco-social market economy in Austria. Therefore, it is of importance that many hand-overs to the next generation succeed. Hence, the research interest of the study project (Prügl & Robinson, 2023) has been to gain reliable findings regarding perspectives, attitudes, values and plans of the next generation in entrepreneurial families.

In total, 146 persons took part in the period from January 2022 to May 2023. A total of 126 persons aged from 16 to 40 years from various industries and regions were surveyed using a quantitative online questionnaire that was created based on 20 qualitative interviews. The results of this survey provide an insight into the value systems and perspectives of Austria's Next Gen as well as possible explanations as to why estimates by KMU Forschung Austria (2021) suggest that by 2030 only 42% of all handing-overs of family businesses will happen within families.

4 Sustainability in Terms of Family Businesses

The importance of sustainability for family businesses is attested by most Next Gens as they consider long-term, intergenerational thinking to be a prerequisite in the present—and even more so in the future. Notably, when asked about future opportunities, a Next Gen noted that re-emphasising sustainable thinking in the broadest sense renders an ever-increasing competitive advantage for their family's business. The majority of Next Gens do not pay lip service when considering sustainability important. It seems that sustainability is ingrained in the mindset of family businesses. In fact, studies show that family businesses usually act more ecologically sustainably compared to companies without a relation to a family (Berrone et al., 2010; Campopiano & De Massis, 2015; Dyer & Whetten, 2006). Indeed, Next Gens are even more conscious about sustainability than preceding generations. Then, 61.9% state that strengthening their family business' sustainability efforts is important for the future. In Germany, for example, 75% of Next Gens strongly prioritise sustainability efforts (Stiftung Familienunternehmen, 2023).

However, while most people primarily link the term "sustainability" to environmental issues, it can also be understood more broadly. The United Nations, for example, defines "sustainability" as "meeting the needs of the present without compromising the ability of future generations to meet their own needs" (United Nations, 1987). In the context of (family) business, this means, in order for an enterprise to be "sustainable", it needs to address not only environmental matters but also all the other aspects related to its long-term success. For instance, decisions that might have positive short-term effects but harm the business in the long run cannot be viewed as sustainable. In other words, past and current decisions need to make future endeavours possible, preferably easier. Sourcing environmentally friendly resources or providing vocational training for future employees is as much of a matter of sustainability as deciding to whom and how a business is handed over.

In this context, sustainability requires a certain level of resilience. A business can only be sustainable if it manages to deal with problems accordingly. Moreover, 78.4% of Next Gens say that enhancing the resilience of the business is an important goal for the future. Taking this approach to sustainability, 78.2% of Next Gens believe that the business is stronger when family members are involved. Fifty-eight percent agree with the idea that children have responsibility for their family's business and should therefore be introduced to it early in life. In fact, over 80% imagine themselves participating in the family business in one way or another when turning 40 years old, whether it be as an executive, shareholder or employee. Nonetheless, Next Gens emphasise the importance of being able to make their own experiences and to have the ability to experiment while also receiving proper guidance from preceding generations in a non-intrusive manner. As the handing-over process may be extensively time-consuming, Next Gens stress the need for parents to make the appropriate arrangements for successful succession as soon as possible.

This shows that handing-over is understood as a means to ensure the long-term existence and resilience of the family business. This is a crucial point in the lifespan of a (family-run) business. Thus, all generations involved have to work together to achieve the common goals set. Especially, the handing-over of a family business is also the convergence of generations. For it to render satisfying results, a common ground must be established. Unsurprisingly, what ties a family together is also a key ingredient in building a foundation for a business existing, and ideally thriving, over multiple generations. Far more than the mere existence of family ties, families, and therefore their businesses, need shared values.

In this web of commonalities, the mental well-being of all involved must not be forgotten because any handing-over is likely to fail when family structures are negatively impacted. This is the case due to family dynamics determining how family members operate the business. It also becomes obvious that the family business has a profound influence on family life in general. While research on this issue is still fragmented and limited, existing academic literature points to the special nature of family businesses and their implications on the mental health of family members. According to a meta-analysis of existing research by Diane Arijs and Anneleen Michiels (2021), this can be explained by the "intertwining of the family and the business system" as this "represents a double-edged sword for business families that strive for mental health at individual, family and business levels". In the studies the authors analysed they found unique characteristics chiefly pertaining to family businesses, such as more intense intertwining of personal and business matters but also the lasting effects of parenting on how children operate the family business in the future. In addition, the authors conclude that satisfaction of family members partaking in the family's business is determined by, inter alia, the level "of feeling instrumental, feeling supported, and feeling enabled" as well as "by opportunities for personal growth, birth-to-death care, frequent direct communication, and effective behavior control". This underlines that, among other things, understanding the views and ambitions of Next Gens is crucial for successful handing-overs. Without this, the mental well-being of all involved falls short, posing serious threats to the sustainability of the family business.

5 Trends in Handing-Over of Family Businesses and Ambitions of the Next Generation

5.1 Values and Ambitions

A family business is where family and business matters inevitably intertwine. This combination of family and business requires a set of values that gives orientation for everyone involved. So it is of high interest to know for which values and attitudes the next generation stands. The findings show a high importance of "classic" family business attitudes like family as well as individual entrepreneurial responsibility. Besides leading a life with personal responsibility, Next Gens emphasise the significance of having a trustworthy partner, a good family life as well as diligence and ambition. Even though the study participants highlight that businesses are viewed more favourably when family members are actively involved, they also insist that family and business matters should be clearly separated.

Austria's Next Gens set themselves high standards and insist on their commitment and individual responsibility. Notably, they see themselves connected to their family's business but also want to go their own way. Next Gens do not want to rest on the fruits of the work of preceding generations. Instead, they strive to re-define their role and position. Many express intentions to start and build their own business. Nevertheless, over 80% look favourably at the idea to be actively involved in the family business when turning 40. They aim to do so by incorporating all garnered experience when taking on entrepreneurial responsibilities while also entering new fields of business involving crucial topics of the future such as digitalisation, innovation and sustainability in a panic-proof and resilient manner. However, despite their eagerness to start their own business, once in charge of the family business, the willingness of Next Gens, for example, to cooperate with start-ups is rather low—something that should be challenged in future research projects.

5.2 Generational Shift

An intergenerational element inheres in family-run enterprises, of course. Due to the strong combination of the family and business, the task and process of handing-over is key for the long-term existence of the company. The ways Next Gens can take over their family's business are diverse. They can, inter alia, take on operative leadership, be shareholders, assume an advisory council position or work as an employee with no further responsibilities. It is important for parents to be ready to allow this diversity of options and roles. There are no universal answers for how a business can be handed over ideally. It rather depends on the specific circumstances. Overall, the results indicate room for improvements when it comes to how handing-overs are done. It is therefore necessary to start handing-over procedures as soon as possible and to understand them as a process. Furthermore, there is a need for finding appropriate roles for parents after completing a handing-over. Nonetheless, handing-overs can also be fulfilled outside the family.

5.3 *Digitalisation, Innovation and Sustainability*

Three areas were identified as key for the long-term success of (family-run) businesses based on the qualitative interviews that were conducted at beginning of the study project, and therefore we wanted to explore the attribution of Next Gens to "digitalisation", "innovation" and "sustainability" further in the online questionnaire.

Digitalisation is undisputedly a strong factor of change in all sectors and parts of our life. The ability to deal and make use of it can decide the fate of organisations and enterprises. Since current successor generations are considered "digital natives", they have a leading role in the digital transformation of their family's business. Innovation is a prerequisite for competitiveness and continuing market success. Innovation does not only mean disruption, but also the improvement of products, services, processes and business models (see OECD & Eurostat, 2018). Sustainability in the context of family enterprises is not limited to environmental aspects, but it is, as outlined above, more comprehensive.

In general, great importance is attributed to those three big areas. Digitalisation and innovation, and to a lesser extent sustainability, are viewed as matters providing competitive advantages. In addition, many indicate that they are ready but not sufficiently prepared to pursue innovation. On the whole, there is a general consensus of being satisfied with the current state of their family's business in all three areas. Levels of satisfaction regarding the state of sustainability of the business are the highest and the lowest regarding the status of digitalisation. Generally, most are convinced of being adequately competitive when it comes to digitalisation, innovation and sustainability—if not already being forerunners themselves.

5.4 *Entrepreneurial Responsibility*

Due to the deeply regional roots of the family businesses and their close connection to their social environment, these companies are very active in several areas of societal life in addition to their economic contribution of creating jobs, generating wealth and paying taxes.

The overwhelming majority of Austrian Next Gens declare being happy to commit to philanthropic causes, especially in social matters, sports, environmental issues, art and culture as well as education. This shows a thorough family-entrepreneurial mindset that is at the core of societal and social commitments. Such is underlined by 62.1% of participants that state they provide vocational training. Providing vocational training is a special aspect of entrepreneurial responsibility and sustainability, since not only are future employees trained and ideally retained but an immense contribution to their educational path and success is also made. In a broader sense, vocational training is also a contributor in providing the Austrian labour force with highly skilled workers, hence making the Austrian economy an attractive place to do business.

5.5 *Future Challenges and Opportunities*

After a successful handing-over, Next Gens have to tackle the upcoming challenges and set priorities for the continuation of their family's business. Many Next Gens highly prioritise driving digitalisation and building resilience. Even though family and traditions that are typically related are considered important, Next Gens do not want to "stick to the old ways". In terms of sustainability, 62% of Next Gens see it as important that they embed sustainability strategically into their business.

Furthermore, special attention is paid to finding and retaining employees. Respondents say that appreciating employees is extremely important to their families. In overcoming the general shortage of skilled workers, not only vocational training but also the approach to employee recruitment and retention plays a decisive role. Top priority is given to creating an ideal working environment and a family-friendly atmosphere, followed by financial incentives. Likewise, the flexibilisation of working hours and location as well as the creation of decision-making leeway and opportunities for advancement are receiving increased attention, too.

Regarding the attractiveness of doing business in Austria, there is great agreement on the inevitability of reducing bureaucracy and solving the shortage of skilled workers. For this, it is pivotal to implement appropriate labour market policies but also to provide broader societal and political answers (such as reforming taxes and social welfare). Additionally, Next Gens are worried about the future viability of Austria's education system and infrastructure.

6 Conclusion

There seems to be a global trend in handing-overs of family businesses: Families struggle to find successors within the family. It is expected that only 42% of handing-overs in Austria will occur within the family by 2030. While further research still needs to be done, already existing literature highlights the special nature of family business as in the intertwining of personal and business matters. Therefore, for a family business to succeed, attention must not only be paid to how the business is operated but also to all family members involved, their views and ambitions as well as the dynamics among them.

Austrian Next Gens especially emphasise the importance of sustainability and family businesses generally act more ecologically sustainable compared to companies without significant links to a family. When understanding sustainability more broadly, families owning businesses appear to employ a seemingly inherent long-term, intergenerational thinking. While Next Gens acknowledge the importance of sharing common values and beliefs, it is also important for them to be able to do things to their liking to fulfil their own visions and ambitions.

Still, being part of a family that runs a business means that it is an integral part of the lives of any family member. It is almost impossible to separate them, so much so that the family business' success can affect the well-being of the family as well as its members and vice versa. The study results presented above have delivered first results that point to the two sides of the same coin or, as Arijs and Michiels (2021) have put it, "the double-edged sword" that comes with the strong intertwining of family and business regarding mental health and well-being. Future research should investigate this relationship further. Especially for Austria as well as for any country where family businesses are a key pillar of the economy and society as a whole, it is crucial to understand their commitment to sustainability in the broadest terms—and what is necessary for them to fulfil it.

References

Altenburger, R., & Schmidpeter, R. (2018). Die gesellschaftliche Verantwortung von Familienunternehmen. In *Management-Reihe corporate social responsibility* (pp. 1–15). Springer Berlin Heidelberg. https://doi.org/10.1007/978-3-662-55618-4_1

Arijs, D., & Michiels, A. (2021). Mental health in family businesses and business families: A systematic review. *International Journal of Environmental Research and Public Health, 18*(5), 2589. https://doi.org/10.3390/ijerph18052589

Berrone, P., Cruz, C., Gomez-Mejia, L. R., & Larraza-Kintana, M. (2010). Socioemotional wealth and corporate responses to institutional pressures: Do family-controlled firms pollute less? *Administrative Science Quarterly, 55*(1), 82–113. https://doi.org/10.2189/asqu.2010.55.1.82

Campopiano, G., & De Massis, A. (2015). Corporate social responsibility reporting: A content analysis in family and non-family firms. *Journal of Business Ethics, 129*(3), 511–534. https://doi.org/10.1007/s10551-014-2174-z

Dyer, W. G., & Whetten, D. A. (2006). Family firms and social responsibility: Preliminary evidence from the S&P 500. *Entrepreneurship Theory and Practice, 30*(6), 785–802. https://doi.org/10.1111/j.1540-6520.2006.00151.x

KMU Forschung Austria. (2020). Familienunternehmen in Österreich 2019. Retrieved from https://www.kmuforschung.ac.at/wp-content/uploads/2020/04/forschung-bericht-familienunternehmen-2019.pdf

KMU Forschung Austria. (2021). Unternehmensübergaben und -nachfolgen in Österreich. Retrieved from https://www.kmuforschung.ac.at/wp-content/uploads/2023/07/Nachfolge2021_Endbericht_final.pdf

OECD & Eurostat. (2018). *Oslo manual 2018: Guidelines for collecting, reporting and using data on innovation, 4th edition, the measurement of scientific, technological and innovation activities*. OECD Publishing/Eurostat. https://doi.org/10.1787/9789264304604-en
Prügl, R., & Robinson, C. (2023). Die nächste Generation: Wir übernehmen!: Wie ticken Österreichs nachfolgende Familienunternehmerinnen und -unternehmer? https://assets.julius-raabstiftung.at/2024/10/JRS_WirUebernehmen_OA.pdf
Stiftung Familienunternehmen. (2023). Deutschlands nächste Unternehmergeneration – 6. Studie zu Einstellungen, Werten und Zukunftsplänen der Next Gens. Retrieved from https://www.familienunternehmen.de/media/public/pdf/publikationen-studien/studien/Deutschlands-naechste-Unternehmergeneration_6_Studie_Stiftung-Familienunternehmen.pdf
The Economist. (2024, January 29). Many family firms lack heirs. Unrelated help is at hand. Retrieved from https://www.economist.com/business/2024/01/29/many-family-firms-lack-heirs-unrelated-help-is-at-hand
United Nations. (1987). Report of the World Commission on Environment and Development: Our common future. Retrieved from http://www.un-documents.net/our-common-future.pdf

Tien Thanh Nguyen obtained a Master's degree in Business Law and a Bachelor's degree in Business Administration at the Vienna University of Economics and Business. From 2019 to 2024, he worked as Policy Advisor in the Julius Raab Foundation. In this capacity, he was involved in the study project on Austria's Next Gens.

Christoph Robinson studied Law at the University of Graz and completed a part-time MBA program in Digital Transformation & Data Science at the Vienna University of Economics and Business. He started his professional career from 2013 to 2016 in the Graduate Programme of the Federation of Austrian Industries with postings in Vienna, Brussels and Graz. From 2016 to 2017, he joined the Foreign Ministry as an Adviser in the Strategy & Policy Planning Unit, and in 2018 he became a deputy head of the Unit for Strategy, Analysis & Policy Planning in the Federal Chancellery. From 2019 to 2023, Christoph Robinson was the Managing Director of the Julius Raab Foundation and responsible for the study project on the Next Generation in Austrian family businesses.

How to Achieve a Wellbeing Economy from Macro to Micro

Friedrich Hinterberger

Abstract More than 50 years after the pathbreaking publication of the Club of Rome's first report on the "Limits to growth", the organization published "Earth For All. A Survival Guide For Humanity". This new report suggests complying with the limits set by the Paris Agreement not only to reduce global greenhouse gas emissions and other environmental goals but also and foremost to achieve a "wellbeing economy" all over the world. To attain this, it is crucial to intertwine macro, meso, micro and individual levels of action in policy, business and personal fields. By means of this contribution, I will discuss some framework conditions for more well-being for everyone: How the macro, meso, micro and individual level correlate (and which indicators help us picture this) and which challenges and opportunities derive from that (e.g. planetary boundaries set by nature). Therefore, this article brings together theoretical insights and practical learnings and, drawing on those, proposes elements that are applicable to various domains at every level.

1 Background

Despite the multiple crises in the environment, society and the economy that we are facing today, current studies also show ways out that nurture the hope of escaping them. One of those studies is the "Earth4All" report (Dixson-Declève et al., 2022), which was published by the Club of Rome 50 years after its pathbreaking first report on the "Limits to growth" (Meadows et al., 1972). It basically requires a new social contract based on a systemic view of the world in which five comprehensive

I am grateful to Meike Bukowski, Mathias Kirchner, Marisa Mühlböck, Walter Ötsch and René Schmidpeter for their valuable suggestions while being responsible for all remaining shortcomings.

F. Hinterberger (✉)
Austrian Chapter, Club of Rome, Vienna, Austria
e-mail: fritz@hinterberger.com

M. Mühlböck (ed.), *Sustainable Transformation and Well-being*, CSR, Sustainability, Ethics & Governance,
https://doi.org/10.1007/978-3-031-75566-8_13

turnarounds are described that would enable a giant leap towards sustainability. In such a view, wellbeing serves as an ultimate aim that is strived for in order not to exceed the planet's limits and at the same time to fulfill fundamental social conditions that make such a transformation possible in the first place. This can be described as a threefold target system of ecological, distributive and economic targets as well as wellbeing goals in which the economy can develop without overstressing planetary and social systems and requires a monitoring system at the micro, meso and macroeconomic levels, which are closely intertwined.

When we define wellbeing as a key factor in dealing with the "limits to growth", it needs to be integrated at every level (micro, meso, macro and individual/collective) and requires a framework that allows the intertwinement of every level (in alignment with a systemic worldview). For this, we must define what wellbeing means to us, formulate appropriate goals in this context and identify indicators that qualify to measure and monitor the progress.

1.1 An Earth for Everyone: Dealing with the Limits to Growth

Climate change is unavoidable, and has turned into a veritable crisis, which comes on top of the recent pandemic, wars, economic stagnation and inflation. How can we best deal with these upheavals that seemed impossible just a few years ago? Shall we continue as before with a policy of small steps?

In their report (Dixson-Declève et al., 2022), the Club of Rome describes the efforts made to date in the field of sustainability as "too little too late". This scenario shows that these efforts do not bode well, neither for the climate and the environment nor for society and the economy. It is therefore predicted that global warming will continue, the quality of life will decline despite economic growth, social tensions within society and between regions of the world will continue to increase and the planet will become even more unsafe as a result.

However, the Club of Rome also outlines a way out of this downward spiral: taking the upheavals seriously and actively working on a transformation that the Club describes as a "giant leap", which the report describes as "possible and necessary". To achieve that, five turnarounds in the fields of energy, food, poverty, inequality and "full gender equality in terms of representation, rights, resources and power in law and employment" are crucial. This means nothing more and nothing less than "updating our economic system with the aim of redefining what is really important for economic strategies" (Dixson-Declève et al., 2022). The Club of Rome further writes in the executive summary of that report:

> If we act now, with the largest effort and investment in this decade, then within a single generation we can achieve many of the sustainable development goals, and we can build societies that respect planetary boundaries. (p. 2)

Such a future will be built on the basis of a new social contract between a government and its citizens to upgrade the economic system. In the report, the following is described as an overarching goal in this context:

> Choose wellbeing economies, adopt new economic indicators that deliver better outcomes for people and planet and place them at the centre of policymaking. (p. 12)

This requires action at every level: the individual as well as in business and policies.

The five aforementioned turnarounds refer to poverty, inequality, empowerment, food and energy, and their development is based on:

- Forward thinking—long-term and intergenerational
- Reorganized markets and a new global financial system
- New ways of thinking about property rights so that everyone benefits from the global commons such as the climate system and biodiversity
- Circular economy and regeneration

This can be understood as a contemporary version of the notion of sustainability.

1.2 A Systemic World View

Contributions from individual people and companies (micro level) will only lead to major changes if government policy at the national or supranational macro levels sets appropriate framework conditions. On the other hand, politicians will only take action if they can rely on acceptance from the micro level and companies will only provide sustainable solutions if they expect sufficient appreciation on the market.

Between the micro and macro levels, there are various meso levels, such as regions or sectors or even socially more or less strongly organized groups or communities, such as trade unions, citizens' initiatives or "bubbles" linked by social media, which act as a hinge between the two levels.

Therefore, it is important for decision-makers at every level (local, regional, national and multilateral governments and organizations) as well as businesses and individuals to understand their systemic contributions to current and potential problems and hence to the achievement of comprehensive goals for the future. Or in other words, it is crucial to intertwine macro, meso, micro and individual levels of action in policy, business and personal fields and find suitable ways to measure and monitor the progress.

Viewing the world as a whole systemically (and understanding it with corresponding computer models) has always been one of the main concerns of the Club of Rome. In this article, I want to add a further system aspect to this: the interweaving of the macro level with underlying levels down to the individual and thus the subjective. This brings the notion of wellbeing into focus.

1.3 The Doughnut as a Sustainability Framework at Different Levels

Empirically, rates of economic growth are declining worldwide in the long term. At the same time, crises are lasting longer and longer. Yet, the economic system should be fit to enable a high quality of life while respecting planetary boundaries and minimum social conditions in a sustainable way regardless of whether the economy grows in monetary terms or not. Therefore, governments, businesses and individuals need to set new goals.

The British economist Kate Raworth (2012, 2017) proposed the image of a "doughnut" to illustrate what the economy and society are ultimately about: the simultaneous observance of planetary boundaries and minimum social conditions for a good life for everyone—now and in the future. The economy and society would then be able to develop freely within those guidelines. The graphical representation as a doughnut is a playfully serious approach to framing this challenge, and it serves as a compass for human progress in this century.

The ecological ceiling consists of nine planetary boundaries, as set out in publications in journals such as *Science* and *Nature* (Rockström et al., 2009; Steffen et al., 2015; Richardson et al., 2023), beyond which lie unacceptable environmental degradation and potential tipping points in Earth systems. On the other hand, Raworth derived 12 dimensions of the social foundation from internationally agreed minimum social standards as set out by the world's governments in the 2015 Sustainable Development Goals. Between the social and planetary boundaries lies an ecologically safe and socially just space in which humanity can thrive.

From this follows that the economic system makes a high quality of life possible while respecting planetary boundaries and minimum social conditions in a sustainable way, regardless of whether the economy grows in monetary terms or not. Wellbeing, on the other hand, will have to be decoupled from economic growth in order to allow a flourishing society and at the same time meet nature's needs to be able to support our human needs (i.e. to provide ecosystem services).

If we now take a systemic view of the updated version of sustainability in the sense of an Earth for everyone ("Earth4All"), the wellbeing of people becomes the guiding concept both at an individual (micro) and at higher levels (meso and macro), and systemic analysis shows us how improvements towards wellbeing at one level can influence the other levels.

1.4 The Role of Monitoring

The doughnut can be used to both inform and evaluate policies on the macro level but also the management of businesses as well as individual choices. Respecting planetary boundaries requires a reduction of resource consumption, especially but not only of fossil fuels, to 10–20% of today's level in a few decades (see IPCC,

2023; United Nations Environment Programme, 2024 for details) to achieve the macro goals of climate neutrality and biodiversity loss reduction. This requirement concerns each actor in society at the micro level such as individuals of companies.

Individuals, companies, organizations and governments set goals, make plans and monitor their compliance. Indicators for monitoring need to refer to the goals set. If not, implicitly used indicators define goals. Often, one-dimensional indicators are used, for instance, monetary values such as individual income, a company's profit, regional or gross domestic product (GDP), which measure production or income at a regional or national level.

The concept of measuring progress in terms of GDP was developed about 80 years ago in close coordination with corresponding scientific findings (of Keynesianism as a response to the prevailing neoclassicism) and political programmes (of deficit spending—after the Great Depression) (Lepenies, 2013; Schmelzer, 2016). At the micro level, this refers to income or profit. The great achievement of national accounting is that individual incomes and business profits can be summed up to regional, national and global incomes in terms of GDP.

The importance of GDP hints at economic growth as the implicit goal. Indicators for income do not say much, though, about whether or how this affects wellbeing, as long as the purchasing power behind it is not taken into account. The same applies to other indicators such as poverty risk or inequality parameters: they do not directly say anything about target values (where do we want to go) or how they can be achieved (Becker, 2017; Koch, 2022).

For a transformation towards a more sustainable and distributive society, the measurement of progress beyond the financial (income, GDP) needs to be broader (Binswanger, 2006; Easterlin, 1974). The global Sustainable Development Goals (SDGs) of the UN's 2030 Agenda are scientifically (e.g. Alliance for Sustainable Universities in Austria, 2021) and politically (e.g. United Nations, 2015) justified.

The development of indicators requires a strict metric to determine whether I myself, whether a company, a region or a country is moving towards sustainability or not. For this we need a method that is applicable to all regions, companies, organizations and countries so that it is possible to describe the respective contribution of the individual, the company or a country to the (non-)achievement of the overall goals, in the same way as GDP can be understood as an aggregation of the income of people, companies, sectors or regions.

1.5 Wellbeing as an Ultimate Aim

Every individual has his or her personal goals and every company has genuine corporate goals, which are not exhausted in the maximization of income. It starts with dreams and wishes for the future to achieve personal wellbeing or the company's purpose, which relates to a concrete demand on the part of its customers. Sustainability goals must be integrated with those aspirations before integrating individual and company goals at higher societal levels.

When it comes to concrete issues, such as the construction of a motorway under a nature reserve or the question of how to deal with a pandemic, deep social divides become apparent—not only through society but right through families, companies, organizations and communities. The baby-boomer generation, now aged around 65, is increasingly being asked by young people: what have you actually done to prevent the ecological, social and economic crises?

The COVID-19 pandemic has also worsened the mental health of many people around the world. There can and will only be an "improvement of the world" in terms of ecological, social and economic sustainability if each individual has the feeling and awareness that they are improving their own situation and at the same time get acceptance from the community. This acceptance is in the end necessary because without acceptance within a community, people are not willing to entertain suggestions that should lead to improvements. Therefore, the approaches presented below are oriented towards the goal of a "good" life (or wellbeing) in the sense of an objective and subjective quality of life, as advocated politically by the "Wellbeing Economy Alliance", for example.

The international Wellbeing Economy Alliance defines a "Wellbeing Economy" (WE) as an economy.

> designed to serve people and the planet, not the other way around. Rather than treating economic growth as an end in and of itself and pursuing it at all costs, a Wellbeing Economy puts our human and planetary needs at the centre of its activities, ensuring that these needs are all equally met, by default. (Wellbeing Economy Alliance, 2024, para. 1)

This can be traced back to the model of fundamental basic needs by Manfred Max-Neef et al. (1990). Last but not least, in addition to the issues addressed in the global Sustainable Development Goals (SDGs), political participation is also considered to play an essential role for a "good life" (see also Hare (2017) on the importance of prosociality ("kindness"), Isham et al. (2022) on the issue of material values and Hauser et al. (2016) on the contribution of global and local action to individual wellbeing).

2 How to Develop and Implement the Necessary Transition in a Participatory Way?

How can we (as people, companies, municipalities, regions or states) tackle those challenges? Based on earlier work by the Austrian Chapter of the Club of Rome, the Alliance for Sustainable Universities in Austria, the University of Applied Arts Vienna and the University of Salzburg, we have been working in several projects on schemes to monitor sustainable development derived from goals and involving citizens and stakeholders, in which we applied participatory, i.e. transdisciplinary approaches, which support citizens and/or stakeholders in developing pathways into a positive future in line with their own wellbeing as well as with planetary and societal conditions.

The following part of this article describes the concept developed in those projects along with first experiences at the micro, meso and macro levels. They are all based on participatory workshops in which we accompany citizens and stakeholders to:

- Understand the system at stake
- Formulate collective visions and goals based on individual needs, wants and wishes
- Derive indicators from the very goals of the participant that lead the way
- into developing pathways to achieve the goals together.

The participants are guided and supported to elaborate measurable goals in a co-creative way, the implementation of which can then be communicated. Progress can be documented, for example, in sustainability reports and the programme adapted systematically. This is how a transformation towards more wellbeing in times of multiple crises can succeed.

In such a way, individuals, businesses, regions and countries (in the end: all of us on Earth as a global community) can achieve a common system understanding and problem identification to reach their goals in terms of synergies and trade-offs as well as problems and conflict potentials that may arise.

2.1 Understanding the System at Stake

Firstly, we help the participants to familiarize themselves with possible goals in a systemic view, i.e. understanding possible synergies and trade-offs as well as balancing and reinforcing loops. The result is an imagined systemic picture of the future that also shows how the players (can) work together, but also to develop an understanding of which goals are achievable and need not remain mere wishful thinking. This creates a common language, a common understanding of the dynamics that need to be considered.

The aforementioned ecological and social limits are based on our current scientific understanding and should not be crossed for defined bandwidths if we want to be able to be in charge of our future. Crossing these limits could trigger ecological and social tipping points that would bring us into an unsafe operating space (Steffen et al., 2015; O'Neill et al., 2018), and in order to get to action, we must also develop a shared understanding of them. The main question to be asked is: How do those goals relate to one's own wellbeing and the wellbeing of other parts of the system (human and non-human)?

2.2 *Formulating Collective Goals Based on Individual Needs, Wants and Wishes*

Individual and collective goals and visions are instrumental for real change. As the second step, we therefore develop concrete, personal goals for ourselves and then, in a group process, common goals for the organization, the company, the region, the country—always linked to the question: How can I know whether we are getting closer to the goal? For example: Which kind of poverty do we seek to reduce? How do we reduce global warming, protect ecosystems or contribute to world peace?

In order for all people on Earth to be able to live well in the future, it is necessary to set goals at all three levels, the individual, national and global, which describe qualitatively and quantitatively what kind of future they envision. When it comes to ideas about the future, people have images in their minds that they formulate as stories of how the world should become for them. Breithaupt (2022) states that "[n]arratives make the offer of an end to the resolution of a crisis" (p. 187).

Images are central to cooperation and therefore to a future-oriented, creative policy, especially for the problems of the environment (Ötsch & Horaczek, 2021). Ötsch and Horaczek (2021) speak of the "imaginative person" (homo imaginens) who thinks that the world can be shaped in contrast to "externally orientated people" (as we find this, e.g. in standard economics).

Therefore, we invite them to develop positive images of the future at all levels (from the individual to the collective and ultimately for the world as a whole). Those images can then be developed as communicable stories, which allows participants to relate them to each other.

In addition to scientific methods, artistic strategies and communication formats specifically adapted to the context are used. This way participants can open their minds from the confinement of problems and focus on what is really important to them and combine fact-based rational knowledge with subjective perceptions in order to develop a holistic understanding of which goals are achievable and need not remain mere wishful thinking. Methodologically, artistic actions and media can be used together with intellectual ones (e.g. systemic analyses) in order to look beyond the familiar.

The results of that step are qualitative narratives and collages or even songs in which the goals are described in the form of positive images of the future. The system understanding of the current situation needs to be confronted with the system understanding of the future vision. From that, transition pathways and leverage points (Meadows, 1997, 2008) can be identified in terms of stories for change, focusing on possible actions on how to choose, to start, to continue, to alter and to stop on a pathway and asking questions on how to recognize if one is on the right pathway, for interpreting signs if one is to change or leave the pathway will help to pinpoint the most effective places to intervene so that the future vision may be achieved.

That results in qualitatively and quantitatively formulated statements in which they describe their goals in the form of positive images or stories of the future. In this way, the participants formulate and achieve their goals as a team with everyone involved, without this being at the expense of others.

2.3 Deriving Indicators from People's Very Goals

But how do we measure whether we are moving towards the agreed targets or away from them? Indicators that are important to people should be derived from their goals and not vice versa. Indicators derived in this way must be applicable at different levels (similar to the methodology of the carbon footprint) and make it possible to describe the respective contribution of the individual, a company, an organization or a country to the (non-)achievement of the goals (just as GDP can be understood as an aggregation of the incomes of people, companies, sectors or regions).

This measurement in the three dimensions of sustainability (ecological, economic, social) is combined with the assessment of the personal subjective wellbeing of all those involved, i.e. ourselves and all those affected by our actions.

To this end, a conceptual framework was created within the UniNEtZ project (Hinterberger & Spittler, 2021) based on Kate Raworth's (2012) concept of "doughnut economics" and including personal wellbeing in a comprehensive and at the same time limited set of indicators. It describes a society and economy that make a good life possible for everyone within the planetary boundaries set by nature but also within social boundaries.

The task is now to develop a comprehensive indicator system in the sense of a wellbeing index that describes in a directionally reliable and easily understandable way whether and to what extent individual actors, concrete developments and politicians (as well as entire countries and regions) contribute to the achievement of the global goals.

It is important that the impacts of human activities within the existing environmental boundaries (planetary boundaries; Steffen et al., 2015) contribute to the achievement of social goals and thus ensure human wellbeing. Importantly, the micro-level indicators can be related to the macro level to determine how a person, household, company, product or service contributes to goal achievement at the macro (global, regional) level.

This is related to discourses on the topic of "Beyond GDP", which has recently gained some attention again. The central idea is that GDP is not a meaningful indicator (Kubiszewski et al., 2013). Therefore, various alternatives to GDP have been developed to help move societies towards a "wellbeing economy".

Both objective conditions and subjective wellbeing are important for the assessment. A wellbeing index should therefore capture and visualize both. For the objective factors, threshold values must first be defined for social target values and ecological indicators.

As stated before, to ensure that indicators are used by people and decision-makers and are therefore effective, it is necessary that indicators relate to the goals that people also have. Implementation can subsequently strengthen the self-efficacy of the participants, the organizations, regions or companies involved.

In the framework of the so-called Bellagio STAMP Principles (Shortall et al., 2015), participation was defined as an important principle for the development of assessment frameworks for sustainability, including indicators (Hardi & Zdan, 1997). This requires a process of affected citizens or relevant stakeholders (business, politics, interest groups, media, science and civil society).

Finally, that makes it possible to review the progress in the form of sustainability reports to determine the extent to which the goals have been achieved. In this way, at least ideally, communities can achieve goals as part of the larger system developing towards the goals, without individual goals being achieved at the expense of others' goals.

2.4 Achieving Goals Together

In a final step, all those realizations are used to create pathways and concrete steps at both the individual and collective levels in a way that macro, meso, micro and individual levels of action are intertwined, as are policy, business and personal goals.

Individuals, businesses, regions and countries (in the end, all of us on Earth as a global community) achieve a common systemic understanding and problem identification regarding reaching their goals in terms of synergies and trade-offs between these goals as well as problems and conflict potentials that may arise due to potential trade-offs between the goals. That insight is crucial for creating pathways together, which allows the participants to achieve their respective goals.

3 Some Cases: A Practitioner's Point of View

Since 2020, the Austrian Chapter of the Club of Rome together with the UniNEtZ project at the University of Applied Arts Vienna and the University of Salzburg have been working on the "Wellbeing – a good life in a world full of crises" project, which is essentially about monitoring sustainable development derived from goals and involving citizens and stakeholders. In several projects in association with the University of Natural Resources and Life Sciences, Vienna (BOKU), cooppa and Saphenus Medical Technology GmbH, we have been applying and testing several aspects of this approach with and for concerned people, organizations and regions (Hinterberger et al., 2023; Kirchner et al., 2023; Bukowski et al., 2024). Those applications have included workshops in the Austrian region of Salzkammergut with five start-up companies, students and Austrian stakeholders interested in consolidating goals related to the SDGs 1, 8, 10 and 13. It is in principle applicable to other issues and communities worldwide.

3.1 An Application at the Macro Level

In a project called SDGVisionPath,[1] we combined the application of "Communities of Practice (CoP)" (Wenger, 1999) for stakeholder and expert collaboration with two models specifically suited to address SDG interactions and for integrating stakeholder and expert knowledge at the Austrian national level.

The project's theme focuses on SDG 8 (decent work and economic growth), SDG 13 (climate action) and SDGs 1/10 (no poverty/reduced inequalities). In a first workshop, the stakeholders developed a common system's understanding of these SDGs, applying a method from systems thinking called causal loop diagrams (CLDs) (Sterman, 2000). In a second workshop, stakeholders jointly formulated visions and goals applying an arts-based method called speculative thinking (Dunne & Raby, 2013). In a third workshop, stakeholders applied storytelling to developed pathways that could bring us to their desired futures. In a final workshop, yet to take place at the moment of writing, stakeholders together with the research project team will try to identify a set of policy recommendations based on the findings in the previous workshops and guided by findings from the quantitative simulation model iSDG (Spittler & Kirchner, 2022; Allen et al., 2019) as well as the qualitative Inequality and Poverty Assessment Model (IPAM) (Bukowski & Kreissl, 2022).

In addition to the workshops, we used a survey to identify the usefulness of indicators suggested to measure the degree to which the goals had been achieved and asked for target values. To this end, the project team proposed indicators not only based on the objectives formulated in the SDGs but also by the stakeholders. In this way, we wanted to ensure that these indicators really measure what the stakeholders directly or indirectly involved in the transformation process are really aiming for, rather than imposing concepts developed by experts or bureaucrats.

In addition to typical SDG indicators, such as "be able to heat living space adequately" (SDG 1/10—energy poverty), "job satisfaction" (SDG 8), "real GDP per capita" (SDG 13) and "greenhouse gas emissions" (SDG 13), stakeholders broadened this horizon by including additional indicators for goals that were important in their goals, for example, "hours spent on basic needs such as eating or sleeping and leisure activities" (work-life balance), "gender equality at work" (gender equality), "material footprint" (circular economy) and "soil sealing" (biodiversity). Furthermore, stakeholders suggested to include institutional goals and indicators for education (e.g. education for sustainable development), transparency (e.g. monitoring progress) and governance (e.g. citizen panels) (Bukowski et al., 2024; Hinterberger et al., 2024a, b).

[1] https://sdg.visionpath.at. The work on the research "SDGVisionPath" project (Co-creating future visions and transition pathways for the SDGs climate action, inequality and decent work and economic growth) mentioned in this chapter is funded by the Austrian "Klima-und Energiefonds" ("Climate and Energy Fund") within the Austrian Climate Research Program ACRP (funding number KR21KB0K00001).

3.2 Applications at the Meso Level

In another project (Hinterberger et al., 2023), which consisted of four regional workshops, interested people and decision-makers from the Dachstein-Salzkammergut-Totes Gebirge region developed a shared positive image of how the economy, society and nature should change in the face of the climate crisis and how they can subsequently recognize ("measure") whether the region is moving in that direction.

For this, adaptation to climate change plays a central role thematically. What do we want to adapt to in the face of climate change and, going further/deeper, where do we want to go? Or even more concretely: How do we (people and nature) in the region want to live with climate change? The global sustainability goals provide a globally recognized framework for this. "Adaptation to climate change" represents one of the 169 targets (i.e. sub-goals) of the UN's 2030 Agenda—more precisely, Target 13.1. The measurement of the success of "adaptation to climate change" can thus be placed in relation (synergies and trade-offs) to the other goals of the people in the region: What adaptation is needed to achieve the goals?

Together, these images of the future were developed in line with the goals that were set as a contribution to a good life for everyone, now and in the future. This resulted in communicable stories and strict metrics (ecological and social footprint) that complement each other to form a credible statement on the current progress towards a sustainable future.

More recently, we could test the feasibility of this concept with a group of students interested in the topic of affordable sustainable housing in Salzburg (Bukowski et al., 2024) using a songwriting workshop to develop goals for a "good life for all" in the sense of "leave no one behind" under the conditions of the climate crisis. The success of the corresponding implementation steps can then be evaluated in a holistic sense with the help of a set of comprehensive leading indicators.

3.3 An Application at the Micro Level

In 2022, we supported five companies and start-ups in formulating their sustainability programme, in which companies formulate 5–10 goals with key figures that are accompanied by ambitious stories and a carbon footprint. The sustainability data is then followed up with annual sustainability reports. It is important to us that indicators for monitoring sustainable development be derived from the company's specific goals.

The process comprised four modules, the sequence of which follows the path outlined above. From a footprint perspective, a common, comprehensive and systemic understanding of the challenges and one's own role in the topic of "sustainability" was established and the challenges of one's company within the framework of the 2030 Agenda were clarified.

On this basis, personal and shared goals were formulated in the form of visions of the future and stories. The leading indicators derived from this (similar to the carbon footprint methodology) must be applicable at different levels (from individual people to the world as a whole) and make it possible to describe the respective contribution to the (non-)achievement of the goals. The targets developed were quantified: How can I know if we are getting closer to our goal?

The group provided a wealth of ideas and perspectives. For example, one of the companies was able to report on its plans for the circular economy. Scraps and offcuts from the construction of large yachts were to be used for new products in the tiny house segment.

Another participant was an innovative farm with its comprehensive range of agricultural products (cattle, pigs, bread and cider) for sale in the farm shop and its own restaurant. For them, the purchase of organic products from neighbouring farms as well as a further education programme for young and old were important goals linking sustainability to their own personal and company goals. We encountered a lively need for discussion on the topic of carbon footprint and its reference value: How is the carbon footprint of the entire farm divided up among the various products?

A typical office-based company had to learn how difficult it sometimes is to collect data. What is the annual electricity or heating requirement in the rented office? Where can the carbon footprint be reduced? Employee mobility and nutrition usually had a major impact. Or how do you measure the impact of a company's consulting services?

4 Conclusions

The methods described in this article show promising opportunities based on transdisciplinary research for participatory processes that support people and communities in striving for their aims in terms of wellbeing and sustainability. Wellbeing can be considered as the basis for positive visions and therefore transformation towards what is considered as a "better world" by the people involved.

The combination of arts-based workshops to develop visions, goals and systemic methodologies to judge their feasibility along with monitoring indicators derived from those goals proved to be applicable at the micro, meso and macro levels.

The initiators are looking for more experts and stakeholders, individuals and organizations such as companies or citizen initiatives to apply the tools developed and support them in the further development and implementation of the concepts developed.

References

Allen, C., Metternicht, G., Wiedmann, T., & Pedercini, M. (2019). Greater gains for Australia by tackling all SDGs but the last steps will be the most challenging. *Nature Sustainability, 2*, 1041–1050. https://doi.org/10.1038/s41893-019-0409-9

Allianz Nachhaltige Universitäten in Österreich (Hrsg.). (2021). *UniNEtZ-Optionenbericht: Österreichs Handlungs- optionen für die Umsetzung der UN-Agenda 2030 für eine lebenswerte Zukunft*. UniNEtZ – Universitäten und Nachhaltige Entwicklungsziele.

Becker, I. (2017). Kritik am Konzept relativer Armut–berechtigt oder irreführend? *WSI-Mitteilungen, 70*(2), 98–107.

Binswanger, M. (2006). Why does income growth fail to make us happier? *The Journal of Socio-Economics, 35*(2), 366–381. https://doi.org/10.1016/j.socec.2005.11.040

Breithaupt, F. (2022). *Das narrative Gehirn. Was unsere Neuronen erzählen*. Suhrkamp.

Bukowski, M., & Kreissl, K. (2022). Social and climate (in-)equality perspectives within the SDGs: Introducing the inequality and poverty assessment model for a sustainable transformation of housing. *Sustainability, 14*, 15869. https://doi.org/10.3390/su142315869

Bukowski, M., Hinterberger, F., Palmetshofer, G., & Kirchner, M., 2024. Pathways – How do we get there? (Working Paper No. 3), SDGVisionPath. die Angewandte, Paris Lodron Universität Salzburg, cooppa Mediengenossenschaft eG, Vienna. https://sdg.visionpath.at/wp-content/uploads/SDGVisionPath_WP3.pdf

Dixson-Declève, S., Gaffney, O., Ghosh, J., Randers, J., Rockström, J., & Stoknes, P. E. (2022). *Earth for all: A survival guide for humanity: A report to the Club of Rome (2022), fifty years after "The limits to growth" (1972)*. New Society Publishers.

Dunne, A., & Raby, F. (2013). *Speculative everything: Design, fiction, and social dreaming*. The MIT Press.

Easterlin, & Richard A. 1974. "Does Economic Growth Improve the Human Lot? Some Empirical Evidence." In Nations and Households in Economic Growth, edited by P. David and W. Melvin, 89–125. Palo Alto, CA.: Stanford University Press.

Hardi, P., & Zdan, T. (1997). *Assessing sustainable development: Principles in practice*. The International Institute for Sustainable Development.

Hare, B. (2017). Survival of the friendliest: Homo sapiens evolved via selection for prosociality. *Annual Review of Psychology, 2017*(68), 155–186. https://doi.org/10.1146/annurev-psych-010416-044201

Hauser, O., Hendriks, A., Rand, D., et al. (2016). Think global, act local: Preserving the global commons. *Scientific Reports, 6*, 36079. https://doi.org/10.1038/srep36079

Hinterberger, F., & Spittler, N. (2021). Option 8_01 "Neue Messung des Fortschritts beyond GDP". In Allianz Nachhaltige Universitäten in Österreich (Hrsg.), *Österreichs Handlungsoptionen zur Umsetzung der UN-Agenda 2030 für eine lebenswerte Zukunft*. UniNEtZ-Optionenbericht: UniNEtZ – Universitäten und Nachhaltige Entwicklungsziele.

Hinterberger, F., Bukowski, M., Spittler, N., & Payerhofer, U. (2023). *Klimawandelanpassung umfassend messen*. Endbericht von StartClim2022.D in StartClim2022: Schlüsselmaßnahmen, Messbarkeit und Notfallszenarien, Auftraggeber: BMK, BMWFW, Klima- und Energiefonds, Land Oberösterreich.

Hinterberger, F., Payerhofer, U., Muhr, M., Janu, R., Bukowski, M., Palmetshofer, G., & Fingerlos, I., 2024a. Vision and Goals - Where do we want to go? (Working Paper No. 2a), SDGVisionPath. die Angewandte, Paris Lodron Universität Salzburg, cooppa Mediengenossenschaft eG, Vienna. https://sdg.visionpath.at/wp-content/uploads/SDGVisionPath_WP2a.pdf

Hinterberger, F., Bukowski, M., & Fingerlos, I., 2024b. Indicators - How do we measure whether we are achieving our goals? (Working Paper No. 2b), SDGVisionPath. die Angewandte, Paris Lodron Universität Salzburg, Vienna. https://sdg.visionpath.at/wp-content/uploads/SDGVisionPath_WP2b.pdf

Kubiszewski, I., Costanza, R., Franco , C., Lawn, P., Talberth, J., Jackson, & T., Aylmer, C. (2013): Beyond GDP: Measuring and achieving global genuine progress. Ecological Economics 93, pp. 57–68. https://doi.org/10.1016/j.ecolecon.2013.04.019

IPCC. (2023). Summary for policymakers. In Core Writing Team, H. Lee, & J. Romero (Eds.), *Climate Change 2023: Synthesis Report. Contribution of Working Groups I, II and III to the Sixth Assessment Report of the Intergovernmental Panel on Climate Change* (pp. 1–34). IPCC. https://10.59327/IPCC/AR6-9789291691647.001

Isham, A., Verfuerth, C., Armstrong, A., Elf, P., Gatersleben, B., & Jackson, T. (2022). The problematic role of materialistic values in the pursuit of sustainable well-being. *International Journal of Environmental Research and Public Health, 19*, 3673. https://doi.org/10.3390/ijerph19063673

Kirchner, M., Spittler, N., Hinterberger, F., Bukowski, M., Palmetshofer, G., & Wretschitsch, E. (2023). *Co-creating future visions and transition pathways for the SDGs climate action, inequality and decent work and economic growth* (SDGVisionPath). Erster Zwischenbericht an den Klima- und Energiefonds.

Koch, A. (2022). *Armut? Klare Antworten aus erster Hand* (1st ed.). UVK UTB.

Lepenies, P. (2013). *Die Macht der einen Zahl. Eine politische Geschichte des Bruttoinlandsprodukts*. Suhrkamp.

Max-Neef, M., Elizalde, A., & Hopenhayn, M. (1990). *Entwicklung nach menschlichem Maß. Eine Option für die Zukunft*. Aus dem Spanischen von Norbert Rehrmann und Horst Steigler/CEPAUR.

Meadows, D. (1997). *Feeling our feelings might not be a trivial exercise*. Dear Folks. https://donellameadows.org/archives/feeling-our-feelings-might-not-be-a-trivial-exercise/

Meadows, D. (2008). *Thinking in systems* (D. Wright, Ed.). Chelsea Green Publishing.

Meadows, D., Meadows, D., Randers, J., Behrens, & William, W., III. (1972). *The limits to growth: A report for the Club of Rome's project on the predicament of mankind*. Potomac Associates. ISBN 0-87663-165-0.

O'Neill, D. W., Fanning, A. L., Lamb, W. F., et al. (2018). A good life for all within planetary boundaries. *Nature Sustainability, 1*, 88–95. https://doi.org/10.1038/s41893-018-0021-4

Ötsch, W. O., & Horaczek, N. (2021). *Wir wollen unsere Zukunft zurück! Streitschrift für mehr Phantasie in der Politik*. Facultas.

Raworth, K. (2012). *A safe and just space for humanity: Can we live within the doughnut?* Oxfam International.

Raworth, K. (2017). Why it's time for doughnut economics. *Progressive Review, 24*(3), 216–222. https://doi.org/10.1111/newe.12058

Richardson, K., Steffen, W., Lucht, W., Bendtsen, J., Cornell, S. E., Donges, J. F., Drüke, M., Fetzer, I., Bala, G., von Bloh, W., Feulner, G., Fiedler, S., Gerten, D., Gleeson, T., Hofmann, M., Huiskamp, W. N., Kummu, M., Mohan, C., Nogués-Bravo, D., Petri, S., Porkka, M., Rahmstorf, S., Schaphoff, S., Thonicke, K., Tobian, A., Virkki, V., Wang-Erlandsson, L., Weber, L., & Rockström, J. (2023). Earth beyond six of nine planetary boundaries. *Science Advances, 9*(37), eadh2458. https://doi.org/10.1126/sciadv.adh2458

Rockström, J., Will, S., Noone, K., Persson, Å., Chapin, F. S., Lambin, E. F., Lenton, T. M., et al. (2009). Safe operating space for humanity. *Nature, 461*(7263), 472–475. https://doi.org/10.1038/461472a

Schmelzer, M. (2016). *The hegemony of growth. The making and remaking of the economic growth paradigm and the OECD, 1948–1974*. Cambridge University Press.

Shortall, R., Davidsdottir, B., & Axelsson, G. (2015). A sustainability assessment framework for geothermal energy projects: Development in Iceland, New Zealand and Kenya. *Renewable and Sustainable Energy Reviews, 50*, 372–407. https://doi.org/10.1016/j.rser.2015.04.175

Spittler, N., & Kirchner, M. (2022). iSDG_KlimAT: Ein SDG Modell für Österreich – Erfassung der Easterlin, R. A. (1974). Does economic growth improve the human lot? Some empirical evidence. In *Nations and households in economic growth* (pp. 89–125). Elsevier. https://doi.org/10.1016/B978-0-12-205050-3.50008-7

Steffen, W., Richardson, K., Rockström, J., Cornell, S., Fetzer, I., Bennett, E. M., Biggs, R., Carpenter, S. R., de Vries, W., de Wit, C. A., Folke, C., Gerten, D., Heinke, J., Mace, G. M., Persson, L. M., Ramanathan, V., Reyers, B., & Sörlin, S. (2015). Planetary boundaries: Guiding

human development on a changing planet. *Science, 347*, 1259855. https://doi.org/10.1126/science.1259855
Sterman, J. D. (2000). *Business dynamics: Systems thinking and modeling for a complex world, management*. Irwin McGraw Hill.
United Nations. (2015). *Transforming our worlds: The 2030 Agenda for Sustainable Development.*
United Nations Environment Programme. (2024). *Global resources outlook 2024 summary for policymakers: Bend the trend – Pathways to a liveable planet as resource use spikes*. International Resource Panel. https://wedocs.unep.org/20.500.11822/44902
Wenger, E., 1999. Communities of practice: Learning, meaning, and identity. Cambridge university press.
Wellbeing Economy Alliance. (2024). Retrieved May 10, 2024, from https://weall.org/what-is-wellbeing-economy

Friedrich Hinterberger is the Vice President of the Austrian Chapter of the Club of Rome. He studied economics and public finance at the universities of Linz and Giessen with research stays in Rome, Florence and New York. His research focuses on ecological economics, future and quality of life research and the measurement of progress towards sustainability. He is the author of several books, including "Wachstumswahn. Was uns in die Krise führt - und wie wir wieder heraus kommen" (with Christine Ax). He was the founding president of the Sustainable Europe Research Institute (SERI) in Vienna and head of the Ecological Economics and Ecological Economic Policy working group at the Wuppertal Institute for Climate, Environment and Energy and Senior Scientist at the University of Applied Arts Vienna and the University of Salzburg.

Eudaimonic Well-Being: A Foundation for Public Administration Systems Fit to Build a Sustainable Future

Fredrik Lindencrona

Abstract Despite global efforts to achieve the UN Agenda 2030 and the Sustainable Development Goals (SDGs), progress remains insufficient. Embracing the concept of eudaimonia, which prioritizes practical wisdom and excellence as core to human well-being, is deemed essential to achieve the significant transformations necessary for a sustainable future for people and the planet. Based on this assumption, a framework for enhancing public administration systems (PAS) guided by the principles of eudaimonia is developed throughout this chapter. Overall, the framework aims to clarify how the administrative direction, organization, and skills in public administration systems can be built on a foundation of core eudaimonic values such as equity, innovation and stewardship. Based on this analysis, the chapter sketches a viable transformation agenda for making PAS develop in the intended direction in order to become much more responsive to the imperative for immediate action and to develop innovative strategies to drive sustainable development. By designing PAS based on principles of eudaimonia, a future is envisioned where such systems are able to steward the needed societal transformation at the speed and magnitude required to address the multifaceted challenges of the twenty-first century.

1 Introduction

"Despite impressive engagement around the Sustainable Development Goals, the world is far off track. Much more effort, investment and systemic change are required" (Foreword by UN General Secretary Guterres in the Global Sustainable Development Report 2023 by the Independent Group of Scientists appointed by the Secretary-General).

F. Lindencrona (✉)
Inner Development Goals, Stockholm, Sweden
e-mail: fredrik@innerdevelopmentgoals.org

M. Mühlböck (ed.), *Sustainable Transformation and Well-being*, CSR, Sustainability, Ethics & Governance,
https://doi.org/10.1007/978-3-031-75566-8_14

Well into the 2020s, amid the decade for action for Agenda 2030 as declared by the United Nations, we face significant challenges: during 2022 UNDP reporeds a reversal in the human development index for the first time, 2023 marked the hottest year on record, and new conflict zones have emerged globally (UNDP, 2022; Independent Group of Scientists appointed by the UN General Secretary, 2023). In the preface to the 2023 Global Progress Report on Sustainable Development, UN Secretary General Guterres reminds us of the commitment made by all 193 country leaders to "transform our world for people, planet, and prosperity" (Independent Group of Scientists to UN General Secretary, 2023). Accelerating progress in human development while maintaining a safe and fair operating environment is the paramount challenge for the global community in the decades ahead.

The independent group of scientists behind the 2023 Global Sustainable Development Report, tasked with reviewing UN Agenda 2030, asserts that while transformations are inevitable, their trajectory, direction, and pace hinge on human resolve. Human determination and agency for transformation must not only be robust but also resilient in our current landscape where "complexity and uncertainty have become the norm – they are contexts, not merely risks" (OECD 2017b). In this era of interconnected meta-crises (UNDP, 2022), successful outcomes demand comprehensive transformations that address an entangled interconnected web of social, economic, and ecological sustainability dimensions. Now, more than ever, it is imperative for humanity not only to adapt but also to cultivate the willingness and capacity to steer our world toward a sustainable future for both people and the planet. While the specific paths to be taken may certainly vary across countries and regions, successful navigation and stewardship of transformation necessitate a firm foundation in core guiding values. This sentiment is eloquently expressed by UN Secretary General Guterres, who underscores that these transformative shifts require us to engage our "values, hearts, and minds" (Independent Group of Scientists to UN General Secretary, 2023, foreword). Navigating and stewarding through this new terrain demands that we think more effectively and act more swiftly as a collective human endeavor. The imperative to comprehend and address human determination, along with the capabilities and conditions that foster it, has now firmly entrenched itself as a central lever for policy transformation across all sectors of society (e.g., Independent Group of Scientists appointed by the Secretary-General, 2019; UNDP, 2020, 2022; IPCC, 2022). In this context, I advocate for the Aristotelian concept of eudaimonia as particularly pertinent, providing a solid ethical foundation for steering the necessary transitions amid the meta-crises we face. Drawing upon the conceptual framework of systems thinker Donella Meadows (1999), we can view eudaimonia as a critical deep leverage point guiding the human determination required to drive integrated transitions for the betterment of both people and the planet. However, as emphasized in discussions surrounding transformation, it is imperative not only to grasp the challenge better but also to swiftly and collaboratively address it to have any hope of altering our trajectory.

According to Aristotle, eudaimonia is the ultimate human good pursued for its own sake, representing the highest virtue to aspire to. In practical terms, this entails a focus on cultivating and applying our best human potential, encapsulated in our practical wisdom. Maximizing eudaimonia necessitates fostering conditions that support mutual flourishing and recognizing the interconnectedness of our qualities, capacities, and skills. To thrive in the Anthropocene, we must adhere to values that foster human flourishing while respecting planetary boundaries. UNDP 2020 articulates three fundamental principles—equity, innovation, and stewardship—as foundational elements for living a good life, suggesting a contemporary expression of eudaimonic principles for the twenty-first century. Despite the challenges facing our planet and global communities, our collective efforts to address these issues often fall far too short.

Among the myriad of actors essential for driving sustainable transformation, public administrations—across all levels and sectors—are undeniably pivotal. However, despite their significance, public administration systems (PAS) often fall short in serving societal needs. This chapter delves into PAS from a eudaimonic perspective, posing the question: How can those systems realize their full potential in steering societal development toward a sustainable future? While the focus is on internal structures, processes, and cultures within PAS, it is essential not to isolate them from the broader societies they serve. PAS are integral subsystems of a larger societal ecosystem, in constant coordination with civil society, academia, business, citizens, and communities. PAS should be considered a part of such integrated living systems, systems that always take place in distinct bioregions with their own cultural history and linkages to other neighboring and more distant systems. Progress toward sustainability hinges on the effective functioning of these interconnected systems at all levels (Carayannis et al., 2012).

UN Secretary General Guterres has called for a global plan to rescue the Sustainable Development Goals (SDGs), recognizing the urgency of intensifying and rethinking our approaches. As a researcher and public servant, I recognize the pressing need to confront complex challenges at scale and want to contribute to a positive development. My aim is that this chapter can serve as an initial attempt to sketch a coherent approach to explore how eudaimonic principles for PAS can help us in our precarious situation. I believe that, like in any innovation process, providing a starting point by bringing together previously separate fragments of specific theory and empirical findings into a some coherent whole is important to help catalyze progress in new ways to approach these challenges and find better approaches to drive necessary transformation. Through this exercise I hope to invite further exploration and debate.

2 Sketching an Analytical Framework for Eudaimonic Public Administration Systems

The subsequent sections of this chapter delve into what a eudaimonic perspective might entail for directing, organizing, and supporting PAS suitable for their twenty-first century role. It's important to note that the literature reviewed, along with my own experiences, are not exhaustive and primarily draw from European and Commonwealth countries, potentially providing a somewhat limited view that may not be universally applicable to all PAS worldwide. The approach and necessary steps will likely vary based on the starting point and specific intentions of each jurisdiction. I hope that practitioners and scholars from diverse regions will collaborate to explore these topics from their unique perspectives, contributing to our collective understanding.

2.1 Initial Considerations

Given humanity's inseparable connection to all living beings and our interdependence with nature, the flourishing of both current and future generations is contingent upon thriving biodiversity and healthy ecosystems (intergovernmental science-policy platform on biodiversity and ecosystem services' framework, as presented in Díaz & Demissew, 2015). Achieving sustainable development entails bringing long-term well-being to both people and nature, transcending technical challenges to encompass the aspirations and rights of individuals worldwide (UN Secretary General Guterres to UNGA, 2023).

A central challenge of eudaimonia in this era is thus how to cultivate constructive synergy through relationships at all levels, spanning from individuals to global communities always living in relation to nature and the planet. In a recent examination, Allen and Malekpour (2023) summarize literature on methods to expedite the required transitions toward a sustainable world for both people and the planet. They focus on positive tipping points and deep leverage points, which they believe can indicate "opportunities to redesign systems so that critical feedback mechanisms foster acceleration" and present a useful model categorizing levers for systems change, inspired by Donella Meadow's 12 levers, condensed into the categories of "Intent," "Design," "Feedbacks," and "Parameters."

The analytical framework hones in on four interconnected areas, offering practical responses to levers proposed by Allen and Malekpour. Each section covers one of these tangible approaches for a eudaimonic PAS:

1. Policy directionality: This area concerns intent and its tangible representation in parameters, specifically metrics. It focuses on the guiding beliefs behind eudaimonic PAS, and how can they be reflected in relevant metrics.
2. Governance and organization: This area focuses on design and feedbacks, primarily exploring how PAS should be structured and managed to embody eudaimonia.
3. Fundamental skills necessary for public servants in the eudaimonic PAS: Here, attention is given to the people working within PAS at all levels, investigating the fundamental values and skills required for effective leadership, teamwork, and performance in a eudaimonic PAS.

The chapter concludes by synthesizing lessons from the previous sections and presents early ideas for a transformation agenda outlining steps to facilitate the emergence of eudaimonic PAS and proposing promising examples of new practices.

2.2 Thinking by Analogy

To inject a bit of levity and relevance into the theoretical discussions, I will employ a simple analogy throughout this chapter. Imagine a group of individuals embarking on a journey, whether on foot or in wheelchairs, to a picturesque lakeside camping spot for a weekend of leisure activities. They know where they want to go but are unsure of the route, anticipating challenges along the way. As we progress through the chapter, we'll revisit this analogy to explore how each level of the analytical framework aligns with choices they will need to take during their journey.

2.3 Roots in Ancient Greece

Eudaimonia traces its roots back to the Greek philosophers, most prominently Aristotle, who viewed it as an ongoing process requiring conscious action based on practical wisdom and moral virtues. For Aristotle eudaimonia means living well, which consists of activities exercising the rational part of the psyche in accordance with the virtues or excellency of reason. These virtues, neither innate nor easily acquired, develop through habituation, reflection, and social experience. Eudaimonia can also be represented in a thing and it will then refer to the performance of the thing's characteristic function. The virtue or excellence of a thing consists of whatever traits or qualities enable it to perform its intended function well (Eudaimonia, 2024; more references on eudaimonia can be found in Lavecchia's (2025) contribution in this book).

The philosophical principles of eudaimonia for humans as well as human-made things remain relevant today and find support in modern psychological theories. One example is Daniel Kahneman's work on human reasoning that distinguishes between two systems of thinking: System 1, which functions fast with automatized thinking but is at the same time prone to biases, and System 2, which functions much more slowly, but provides deliberate reasoning aligned with practical wisdom in Aristotelian terms (Kahneman, 2011). Similarly, Mick Cooper emphasizes the importance of directionality and synergy in human behavior to meet higher-order needs and wants, a phenomenon lying the foundation for human motivation (Cooper, 2021; 2023).

Applying eudaimonia as a guiding principle for PAS underscores the importance of core values in setting direction, shaping outcomes and quality criteria for organizational functioning. Scholars advocate for different values to represent the Aristotelian notion of flourishing at the present moment. Some of these alternatives include dignity, community, sustainability, beauty, and innovation, among others (see, e.g., Bednar, 2023; Raworth, 2017). For all of these alternatives to be materialized, conditions conducive to eudaimonia must be designed and delivered. In any eudaimonic PAS, the purpose and governance of PAS must be intentionally developed and driven by both public representatives and public servants who embody certain virtues and excellences.

I believe that equity, stewardship, and innovation, initially proposed by UNDP as personal and cultural values for sustainable futures for people and planet (UNDP, 2020), can serve as helpful summary dimensions integrating most of the core aspects that one could envision as fundamental organizing principles for eudaimonic PAS. PAS demonstrating these virtues prioritize opportunities for all to flourish sustainably, continuously striving for improvement. These values, intertwined and indispensable, unify PAS ethics toward the goal of maximal flourishing for all involved.

Returning to our analogy, when the journey and destination prioritize long-term benefits (stewardship), address challenges equitably for all participants (equity), and create conditions for innovation and problem-solving (innovation), the criteria for a flourishing journey are met for the entire group now and for the longer run.

3 Three Levels of a Eudaimonic PAS

3.1 Policy Directionality

> How would the world change if we made policy, resource and other decisions not based solely on short-term targets, like growing the economy or preventing specific diseases, but to achieve long-term well-being for all people, communities and the environment? (Acharya & Plough, 2020, p. xi).

That question, formulated out of insights from well-being policymakers, practitioners, and academics, sets out a fundamental question regarding policy directionality of a eudaimonic PAS.

In this section, we delve into the overarching intents and beliefs guiding eudaimonic PAS and how they can be reflected in relevant metrics. A crucial aspect of any robust system is its directionality, indicating a clear path forward (Cooper, 2021). The highest-order virtues necessary for a thriving public administration system must align with the fundamental priorities of the people it serves, a topic that should be actively debated in any democracy. However, for a PAS to be eudaimonic, it must prioritize the well-being and flourishing of individuals and communities, both present and future. Therefore, as previously argued, it inherently requires a sustainability perspective.

For PAS, overarching directionality is most clearly articulated in foundational documents such as constitutions, treaties, and similar legislative frameworks. These formal expressions of higher-order needs and aspirations for the system establish a unique framework for shared directionality within a PAS (Table 1).

Shared higher-order needs and wants serve as the bedrock for PAS cohesion, akin to rock formations shaping the course of a river. Those establish boundaries aligning with overarching system goals within which actions can fluctuate,

Table 1 Examples of foundational documents expressing policy directionality in line with eudaimonic principles

Swedish constitution, SFS 1974:152:
2 § Public power shall be exercised with respect for the equal value of all people and for the freedom and dignity of the individual. The personal, economic, and cultural welfare of the individual shall be the fundamental objective of public activities. In particular, the public sector shall:
Safeguard the right to work, housing, and education and promote social care and security and good health conditions
Promote sustainable development that leads to a good environment for present and future generations
Endeavor to ensure that the ideas of democracy guide all areas of society and protect the private and family life of the individual
Work to ensure that all people can achieve participation and equality in society and that the rights of children are realized
Combat discrimination against people on the basis of gender, skin color, national or ethnic origin, linguistic or religious affiliation, disability, sexual orientation, age, or other circumstances relating to the individual as a person.
The opportunities of the Sami people and ethnic, linguistic, and religious minorities to maintain and develop their own cultural and community life shall be promoted.
EU constitution: The Treaty establishing a Constitution for Europe, Lisbon 2004
The Union's aim is to promote peace, its values, and the well-being of its peoples. It shall:
Provide its citizens with an area of freedom, security, and justice without internal frontiers and an internal market where competition is free and undistorted
Work for the sustainable development of Europe based on balanced economic growth and price stability, a highly competitive social market economy, aiming at full employment and social progress, and a high level of protection and improvement of the quality of the environment
Combat social exclusion and discrimination
Promote social justice and protection, equality between women and men, solidarity between generations, and protection of the rights of the child
Promote economic, social, and territorial cohesion and solidarity among Member States
Respect its rich cultural and linguistic diversity
Ensure that Europe's cultural heritage is safeguarded and enhanced

(continued)

Table 1 (continued)

Political declaration of the high-level political forum on sustainable development convened under the auspices of the General Assembly, UNGA 2023
Our shared commitment 1. We, the Heads of State and Government and high representatives, have met at United Nations Headquarters in New York on 18 and 19 September 2023, at the Sustainable Development Goals Summit, to review progress and accelerate the implementation of the 2030 Agenda for Sustainable Development 2. We reaffirm our commitment to effectively implement the 2030 Agenda and its SDGs and uphold all principles enshrined in it. The 2030 Agenda remains our overarching roadmap for achieving sustainable development and overcoming the multiple crises we face. We will act with urgency to realize its vision as a plan of action for people, planet, prosperity, peace, and partnership, leaving no one behind. We will endeavor to reach the furthest behind first (….) 9. We commit to bold, ambitious, accelerated, just, and transformative actions, anchored in international solidarity and effective cooperation at all levels. We will promote a systemic shift toward a more inclusive, just, peaceful, resilient, and sustainable world for people and planet, for present and future generations 10. We will devote ourselves collectively to the pursuit of sustainable development including through international cooperation and partnership on the basis of mutual trust and the full benefit of all, in a spirit of global solidarity, for the common future of present and coming generations

Adhering to democratically legitimate fundamental rules, provides a sturdy foundation for PAS.

With the current magnitude of global challenges, no PAS can act alone. Articulations that bring PAS together to express shared directionality are necessary to establish such fundamental shared bedrocks. There is clearly no other way forward if we are to meet the unified call to action from 14,000 climate scientists to "come together as a global community with a shared sense of urgency, cooperation, and equity" (Ripple et al., 2021, p. 897). The UN Agenda 2030 has the potential to serve as such an underlying ethical framework for sustainable development for a eudaimonic PAS encapsulated in the principles of "people, planet, prosperity, peace, and partnership" and the principle of "leaving no one behind." To make this a reality, any eudaimonic PAS must emphasize interdependence, in essence: "to save the planet and live meaningful lives, we need to start seeing one another not as competitors but as collaborators working towards shared interests" (Bednar, 2023). To deepen our reconnection to such global bedrocks, we should let General Secretary Guterres' insightful words lead us: "The SDGs aren't just a list of goals. They carry the hopes, dreams, rights and expectations of people everywhere" (UN Secretary-General's address to the UN General Assembly, September 2023).

To illustrate the role of system intent and directionality in a eudaimonic PAS, let's revisit our analogy of the group heading to the camping site by the beach. Only after establishing the broader purpose of their journey can the group discern what they aim to achieve. For some, experiencing the destination may be paramount, while others may value the journey itself. The shared intent and directionality represent a collective aspiration, likely focusing on strengthening bonds among members for the long term. Establishing shared ambitions is crucial for group cohesion, yet it also involves reconciling diverse needs and desires within the group. Negotiating and managing this process effectively become integral to eudaimonic well-being within PAS.

3.1.1 Choosing the Right Metrics

Ideally, these agreed-upon higher-order needs and wants should manifest in core metrics for gauging progress. However, achieving consensus on such metrics can be challenging, often leading to different challenges such as abstract formulations to accommodate diverse perspectives or a significant legacy inertia where established metrics are kept in operation even if the evidence against them being helpful for the necessary progress is significant. The reliance on GDP as a primary measure exemplifies this challenge.

Despite decades of critique, GDP remains prioritized in political economies. A famous example is Robert Kennedy's remarks about GDP at the University of Kansas already in 1968. In his speech he starts by addressing all the harmful things that count as progress in GDP and he concludes, "Yet the gross national product does not allow for the health of our children, the quality of their education or the joy of their play. It does not include the beauty of our poetry or the strength of our marriages, the intelligence of our public debate or the integrity of our public officials. It measures neither our wit nor our courage, neither our wisdom nor our learning, neither our compassion nor our devotion to our country, it measures everything in short, except that which makes life worthwhile. And it can tell us everything about America except why we are proud that we are Americans" (Robert F. Kennedy's remarks at the University of Kansas, 2024).

It is clear that GDP is nowhere near to capturing the intended directionality of a eudaimonic PAS. We clearly can and urgently need to do better. Many leading economic thinkers and international organizations, like the OECD, have made significant efforts to "go beyond GDP," but more than half a decade since Kennedy's speech, GDP still takes precedence. Since Kennedy, criticism of GDP as a sole measure of progress has persisted, despite landmark reports (e.g., Stieglitz, Sen & Fitoussi, 2009). On the other hand, it should be clear to all that economic resources do matter for people, communities, and societies. This clear fact brings about a more nuanced discussion: "What we've got to do is really scrutinize: When is GDP useful? When is it not? In what circumstances does it have a place? But also recognize the huge number of perverse incentives […] bound up in GDP" (Kathrine Trebeck in Plough, 2020, p. 86). To be able to do this, it is helpful to look at the original intent of this measure when it was developed.

The GDP was developed by Simon Kuznets as a measure for economic activity to respond to the need to understand if America's economy had regained speed after the economic crisis in the late 1920s. As much as GDP could be recognized for its usefulness in gauging economic activity at the time, we should take Kuznets clear warning against its potential misuse for broader progress assessments seriously.

Relying solely on GDP is akin to a group measuring their journey solely by step count, neglecting the direction toward their desired destination and whether they destroy everything valuable on their way to get there. In neoliberal economics, GDP-centric views risk divorcing any activity from any higher-order goals as any growth is considered a success, even if it does not lead to any progress and can even be destructive for people or the planet. Mick Cooper (2023) describes this as "rogue goals," where short-term gains detract from long-term objectives, akin to straying from the metaphorical beautiful lake destination.

So, why is it so hard to dethrone GDP? Beinhocker (2020) makes an interesting case for this stickiness of GDP. He states that it could be explained by its consistent layered "ontological stack" that builds solid coherence. This coherence unifies ideas about how people function (as self-serving utility-maximizing), how efficiency is defined (by trying to externalize any challenges to natural systems, to other parts of the globe, etc.) deeply rooted in neoliberal politics based in utilitarian philosophy. It is obvious that we must find other and more appropriate metrics. Such metrics must be better equipped to indicate real progress toward a defined higher-order end and it needs to portray economic progress freed from perverse incentives that hollow out the opportunity to achieve what we really aspire to.

To establish relevant metrics able to dethrone GDP, we need to develop credible alternatives, built on a similarly coherent ontological stack but firmly grounded in eudaimonic principles (Beinhocker, 2020). Such a stack needs to be framed by eudaimonic well-being principles of solidarity, dignity, agency, capabilities, and sustainability. At the core is a view of people as "other-regarding," multi-directed, and focused on meaning. Importantly, that eudaimonic alternative is aligned with recent developments in cognitive, social, and natural sciences. Establishing this kind of measure would be a significant step forward since it would help frame a political economy within a fundamental political and ethical stance toward future generations and the survival and flourishing of people and the planet.

In conclusion, activity can only lead to progress toward higher-order needs and wants when we can navigate and move actively on the path to where we want, respecting the do's and don'ts in the context along our path. We need metrics to know that we are heading fast enough in the right direction, and they need to be developed with the higher-order needs and wants and the conditions for good navigation in mind. We will not be able to get there if we rely on measures of progress that hold us back by incentivizing doing harm to what we want to achieve. However, while new metrics are necessary, creating conducive conditions for this endeavor is equally essential. Let's now shift our attention to governance, organization, and later to the skills necessary for a eudaimonic PAS, equipped for hard, dedicated, intentional work to deliver on the policy direction we want.

3.2 Governance and Organization of a Eudaimonic PAS

"We are at an unprecedented moment in the history of humankind and in the history of our planet. Warning lights—for our societies and the planet—are flashing red. They have been doing so for some time, as we well know" (UNDP, 2020). The situation is critical, but the path forward is both complex and uncertain.

The internal dynamics of PAS play a crucial role in fulfilling their eudaimonic directionality. Therefore, it's imperative to closely examine how PAS should be structured to embody eudaimonia under current conditions. Responsibly stewarding a eudaimonic PAS involves navigating toward fulfilling the system's highest-order needs, necessitating flexibility, adaptability, and innovation. A PAS capable of stewardship and innovation is essential for ensuring equitable outcomes in the areas of greatest value for the flourishing of people and the planet over generations. Historically, efficiency in public administration systems has been equated with automation and standardization of work processes. This approach prioritized task differentiation and separation, akin to a well-oiled machine following set protocols. However, such a model, while efficient in specific domains, risks suboptimization and hindering progress in other areas. It mirrors Kahneman's System 1 of fast, automatic thinking, prone to cognitive biases and myopic views. In light of the need for this magnitude of transformative change, incremental approaches to change just won't do.

Embracing uncertainty and complexity demands continuous learning, adaptation, and innovation at all levels of the system. This requires that governance and organization within PAS undergo a radical paradigm shift toward becoming dynamic, interdependent systems rather than rigid, mechanistic entities. In essence, transitioning from a mechanistic to a dynamic system perspective transforms how a PAS functions. It shifts the focus from mere efficiency to effectiveness in promoting collective stewardship and innovation for maximally flourishing outcomes for all people and the whole planet.

Returning to our group of friends on their journey, the living, dynamic system approach prioritizes the group's higher-order needs and wants over mere efficiency. If each individual in the group optimised for the most efficient route to the lakeside retreat, they might choose to drive there in their cars as fast as possible. But when their primary intent is to strengthen group bonds, shared meaningful experiences are core. This is achieved much better by making the journey together as it helps to create shared memories, laughter, and opportunities for reciprocal exchanges based on every individual's strengths and capabilities. Navigating their challenges collectively will foster a sense of responsibility for each other and provide opportunities for mutual support. Under conditions of complexity and uncertainty, it is likely that groups who are able to recognize interdependence and build upon the diversity of capabilities among the members are most likely to succeed.

As eudaimonia revolves fundamentally around human-centric principles, eudaimonic PAS inherently prioritize people at their core. Let us now turn our attention to the people in an eudaimonic PAS.

3.3 *Fundamental Skills Necessary for Public Servants in the Eudaimonic PAS*

PAS are first and foremost built by people, namely those individuals who serve in various roles and work with other people inside and outside of the PAS. The OECD outlines four primary areas of activity for public servants: policy advice and analysis, service delivery and citizen engagement, commissioning and contracting, and managing networks. Assuming that equity, stewardship, and innovation are central to eudaimonic PAS, our focus in this section pertains to the skills required by individuals to effectively embody these values at this moment to build the future we want. The massive challenges for sustainable development and the transformations driven by AI may likely challenge what is considered to be the core capabilities of public servants in all these four areas of activity.

Table 2, from the OECD report, summarizes key attributes of public servants and senior leaders, along with the cultural conditions necessary for proficient, strategic, and innovative task execution in these domains (OECD, 2017a).

It is clear that current and future skill demands for all decision-makers, team members, leaders, partners in multi-sector collaborations, and contributors to policy development and directionality will have to be reimagined. It is likely that the need to act professionally will remain while at the same time the ability to act strategically and innovatively will increase exponentially. With support from emerging AI, our ability to be values-driven and ethical will just become more and more essential.

Table 2 Toward a professional, strategic, and innovative civil service (OECD, 2017a)

	Professional	Strategic	Innovative
Needs civil servants who are:	Qualified Independent Values driven Ethical	Outcomes driven Evidence based Future oriented Proactive Networked	Iterative Data literate Citizen centered Curious Storytellers Insurgent
In a civil service that is:	Merit based Capable of integrating soft skills, ethics, talent management (future potential vs. past performance) Able to structure the right balance of generalist and specialist professions and career paths	Agile Attractive to skilled job seekers Planned and managed to ensure the right skills and competencies are effectively allocated to areas of current and emerging need Future oriented and responsive	Open and collaborative cultures, leadership and management Engaged Autonomous (e.g., work design) Mobile Diverse Learning oriented
Led by senior civil servants who are:	Trusted policy advisors and effective transactional managers	Transformational leaders, change managers	Collaborative leaders and adaptive managers

The significant reliance on technical expertise in the machine metaphor of the PAS will have to be expanded significantly to include strategic and innovative capabilities.

The primary challenge for leaders today is to foster adaptability within organizations and among their members in the face of increasingly dynamic and demanding environments. Adaptive leadership (introduced by Heifetz, 1994) is defined as the practice of mobilizing people to address tough challenges and thrive. Leadership is considered separate from formal managerial positions and concerns how to confront problems for which there are no simple, painless solutions. Today, adaptive leadership must be recognized as a fundamental set of skills to drive positive transformation as we so frequently encounter complex problems that compel us to develop radically new and uncertain approaches to challenge the status quo.

Complexity leadership (Uhl-Bien & Arena, 2018) is a later theoretical development, much aligned with Heifetz's theories. Here, the argument is that in complexity, leaders have to facilitate their organizations to develop adaptability. This asks organizations to cultivate ambidexterity, integrating both approaches to existing production and new innovation. The role of leadership is to create an adaptive environment that enables everyone to navigate this complex process effectively. Both scholars of complexity and adaptive leadership propose that we need distributed leadership instead of seeking leaders who are solitary saviors. Organizations need to develop a collaborative approach that leverages the capacity of everyone across, inside and outside traditional boundaries of the organization. Therefore, everyone in the eudaimonic PAS needs to be an agent for transformation towards sustainable development. But what is required to empower this individual and collective agency?

One significant aspect addressed by sustainability researchers is the notion of the inner-outer transformation pathway. According to Ives et al. (2023), to successfully engage in transformation, it is imperative to nurture values, beliefs, and worldviews that support transformative relationships with oneself, others, nature, and future generations. This happens through building cognitive, emotional, and relational qualities. A recent narrative review of this pathway introduces the acronym IMAGINE to capture core principles to strengthen this inner-outer pathway:

- **I**nterdependence of inner/outer and individual/collective/system phenomena
- **M**ultiple potential that is latent within each of us to enable transformative change
- **A**ctivation of inner dimensions across individual, collective, and system levels
- **G**eneration of inner transformative capacities through intentional practices
- **IN**clusion of diverse perspectives
- **E**xpanding knowledge systems for sustainability

This focus on key personal skills takes us to the individual and inter-personal aspects of the eudaimonic PAS.

3.3.1 Inner Development Goals: A Framework of Transformational Skills for Sustainable Development

A eudaimonic PAS necessitates that practical wisdom be exercised by public servants and elected political leaders alike, in both everyday decisions and critical strategic processes. However, achieving this requires a greater focus on developing the inner-outer transformation pathways of every public servant and political leader, within adaptive spaces for transformation built within PAS.

Despite decades of research in this field employing a broad and diverse range of theoretical approaches and conceptualizations, leaders have often found themselves lacking clear guidance on how to navigate this crucial space. To address this gap, the global NGO Inner Development Goals has been established with the mission of bridging the gap between science and practice of critical transformational skills for sustainable development. Drawing on input from leaders and experts, the Inner Development Goals framework (see Fig. 1) aims to distil and communicate the complex scientific findings about inner development goals, i.e., the transformational skills for sustainable development, to a broader audience of change makers in all sectors across the world (Inner Development Goals, 2021).

Each of the framework's five dimensions holds a set of transformational skills, encompassing a spectrum ranging from fundamental human qualities to developed human capabilities and intentionally trained and learned skills. All of these skills can, to varying degrees, be cultivated through specific approaches or through experiences of leading and working with complex challenges at work, in the community and in personal life.

For our group heading to the lakeside camp, these five dimensions guide their individual and collective process to better be able to achieve their mission of

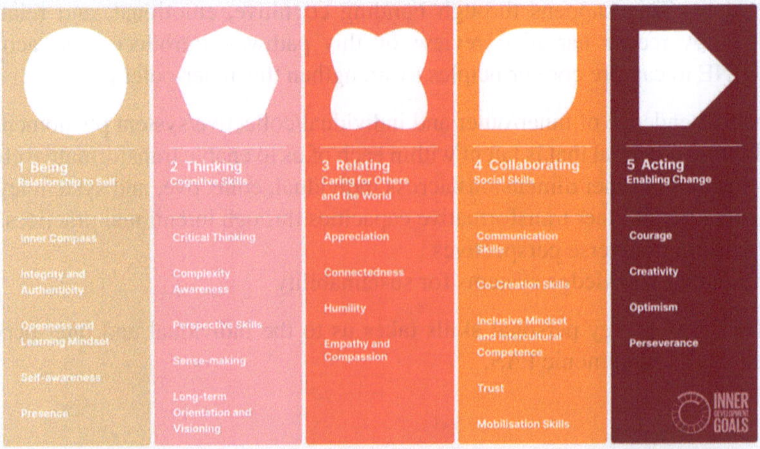

Fig. 1 Inner Development Goals framework

enhancing their collective well-being in a future together. By actively engaging with their surroundings, they recognize that achieving their goal of learning together requires them to resist the temptation to take shortcuts, such as driving individually to the camp or staying home when faced with heavy rain. Although these options may offer immediate comfort, they do not align with the group's core values and mission. Instead, embracing challenges as valuable learning opportunities and remaining open to growth must be prioritized. Leveraging their individual and collective cognitive abilities helps them make informed decisions on a foundation of interconnectedness with each other and the natural environment. Embracing the diversity within the group, in terms of varied forms of mobility and other personal strengths, necessitates a deep respect, understanding, and appreciation of each member's uniqueness. Effective collaboration enables them to strategically utilize each person's abilities to overcome obstacles along the journey. Remaining committed to their overarching mission empowers them to approach challenges with creativity and resilience, even when faced with adversity such as long, arduous journeys in unfavorable terrain and weather conditions.

4 Core Components of a Transformation Agenda for Eudaimonic PAS

After analyzing the three essential elements of a eudaimonic PAS, we arrive at the final dimension of the analytical framework: the necessary transformation to achieve a PAS fit for its purpose. At a time when citizens rightfully expect transformation, governments able to tackle twenty-first-century challenges are required, still we oftenind PAS based on twentieth-century paradigms. Now, more than ever, we urgently need successful transformations aligned with the intent of eudaimonic directionality in policy and core metrics. Now is the time to invest in developing eudaimonic PAS, which are able to enhance their systemic nature to better serve as stewards for transformation toward a juster, fairer, and more sustainable world. If we do not become better at balancing short-term efficiency gains with long-term environmental, social, and economic considerations, we risk providing unhelpful solutions from centuries back. At the 2023 United Nations General Assembly Meeting, a shared political declaration was taken to reaffirm the 2030 Agenda ambitions. This shared reaffirmation states that we need to establish a new normal where comprehensive, far-reaching, people-centered, indivisible, and interlinked goals and targets are universally accepted and acted on. However, that requires that we develop a new form of PAS, built on policy directionality, governance, and organization-based system design principle built on synergy and feedback loops, and investments in staff capability development. These are all essential and complementary in a eudaimonic PAS. So, what actions can further facilitate this emergence and what existing initiatives can illuminate the path forward?

4.1 *Exploring Synergy in Policy Direction for Human and Planetary Well-Being*

Synergy lies at the heart of the UN 2030 agenda, aiming to create coherence between policies and sectors to pursue sustainable development priorities. However, a review of progress up to 2021 suggests that while there's been a discursive shift in how sustainable development is understood and communicated, tangible impact remains limited. Nonetheless, conceptual innovation and shared political declarations reaffirming the 2030 Agenda's ambitions could hint at deeper reflections and early signs of systems change. Still we need to get to urgent action since, as American political scientist Jenna Bednar succinctly puts it, government has become a tool of the economy rather than a vehicle steered by citizens to achieve collective goals (Bednar, 2023).

To ensure meaningful progress, new measures must monitor progress toward equitable, cohesive, and synergetic development. Luckily, we already know a lot of what would be more helpful. Scholars following the eudaimonic line of ethical reasoning have proposed outcomes such as dignity, community, beauty, sustainability, prosociality, and biophilia and many more, in essence focusing on creating the conditions conducive to life (Bednar, 2023; Beinhocker, 2020; Raworth, 2017). While progress metrics have improved, they still prioritize GDP over such eudaimonic goals. Measures aligned with human and planetary higher-order needs and wants must be given priority to drive genuine progress.

Important breakthroughs can be seen in policy (some examples are the OECD's Better Life Index, the UN's 2030 Agenda, and the UNDP Human Development in the Anthropocene measure). While they all manifest steps in the right direction, they have also been critiqued for not going far enough. Progress metrics based on a full eudaimonic ontological stack would be necessary, but right now, however, a more pressing problem is that those measures are so far from being considered at par with legacy measures such as GDP.

Perhaps another solution could help overcome this status quo. Economists such as Erich Beinhocker at the Institute for New Economic Thinking at University of Oxford have taken on the core measures of growth and prosperity directly during the last two decades. They argued that "if the concept of growth is to have significance it should represent improvements in lived experience" (Beinhocker & Hahnauer, 2014, p. 5). They are developing much more direct measures of society's prosperity focusing on the availability of solutions to human problems for all. To replace GDP as a more and more irrelevant attempt to estimate material living conditions, they have developed measures much more directly on "the rate at which new solutions to human problems become available" (Beinhocker & Hahnauer, 2014, p. 5).

4.2 Governance for Eudaimonic PAS

The UN Secretary General asks us to see that the UN Agenda "is more than the sum of measurable goals, targets, and indicators. It is both a normative orientation and a guide for action for identifying and pursuing sustainable development priorities and creating coherence between policies and sectors, in all contexts – local, regional, national, transnational and global" (UN Secretary General in his remarks to the High-Level Political Forum on Sustainable Development, 2023). In line with this, a eudaimonic PAS excels in fostering institutional integration and policy coherence vital for sustainable development progress. However, a stocktake of the progress of Agenda 2030 suggests that so far the "impact has been largely discursive, affecting the way actors understand and communicate about sustainable development" (Biermann et al., 2022, p. 795). More structural changes in PAS have rarely happened due to significant inertia, bureaucratic hurdles, and short-term political agendas that hinder progress (ibid.). Flexible, adaptable structures, processes, and cultures are imperative to allow for adjustments based on evolving circumstances. Engagement of diverse stakeholders ensures comprehensive understanding and legitimacy of administrative actions. Monitoring, evaluation, and learning systems focused on outcome-based measures are crucial for assessing PAS impact on sustainability goals.

The missions approach to innovation and its more operational structural form in innovation ecosystems (Mazzucato, 2021; VINNOVA, 2022) offer promising steps forward. They are designed to address messy, interdependent, and complex challenges through better bringing the best ideas from different societal sectors and areas of expertise together. Missions approaches are tested in many different contexts, including in the EU Horizon Program for Science and Innovation, and the OECD has established a specific Mission Lab. While a promising breakthrough, these approaches may often not be able to go the whole way they intend. The OECD Mission Action Lab's critical examination of the practice of missions and mission-oriented policies conclude that missions easily encounter at least 13 major pitfalls. Most of these may not be driven by the mission approach itself, but the inability to provide the right systemic cultural, structural, and processual governance conditions for them to be able to be designed and deliver as intended (OECD 2024).

These lessons ask us to address deeper levels of PAS governance to find models that can address the structural and cultural problems of policymaking processes aligned with the eudaimonic PAS. "Humble Government" is a promising attempt by Demos Helsinki, a think tank connected to the Prime Minister's Office in Finland. Their model is based on the fundamental assumption that policy fit for the twenty-first century requires a shift to be able to better address complex adaptive problems. To align with the law of requisite complexity, each level of complexity requires solutions with a corresponding level of complexity. On this foundation, Humble Government builds necessary capacity to run four crucial approaches for policy-making under complexity: (a) focus on phenomenon as they actually function in the real world, (b) establish long-term perspectives, (c) establish processes that build trust, and (d) assure continuous learning throughout the process. Provision for ongoing revision under public oversight is fundamental for this loop to work (Annala et al., 2020).

Humble policymaking acknowledges complexity and unavoidable governmental fallibility. To manage this complexity, a continuous investigation of options tested in real-world contexts gives information to develop and revise policy goals only after enactment. To make it practical, four conditions to support a coherent process between a diversity of stakeholders can be proposed including (1) start with a "thin consensus" around overarching goals, (2) provide autonomy by devolving problem-solving to key stakeholders whom you ask for accountability, (3) create feedback loops, and (4) revise and make framework goals more specific through the learning that the process of experimentation has provided. That challenges the conventional model of policymaking, which is built on ideas of predictability and establishes a linear process of designing complete and definitive policies, enforced by comparing behavior to requirements to execute it as efficiently as possible.

A eudaimonic PAS is actively focused on supporting such adaptability to complexity to emerge. However, the conditions of success in such a process require significant motivation and capacity to build the critical dynamic capabilities for risk-taking, experimentation, learning, and failure tolerance (Piening, 2012, Castillo & Trinh, 2018). That will only be possible if the value of such capabilities is deeply integrated into each organization's culture, providing the core direction for its design and feedback (Schoemaker, Heaton & Teece (2018). However, both the pace and scale transforming governance structures to respond effectively to key challenges for people and the planet are currently not high enough. "More fundamental change is needed for the SDGs to become 'the bold and transformative steps … to shift the world on to a sustainable and resilient path' that the 2030 Agenda of the United Nations has promised" (Biermann et al., 2022, p. 798).

To establish the governance and organization for eudaimonic PAS, the most important step would be to see the PAS itself as a key problem. It is challenging but necessary (1) to be clearsighted enough to accept that current ideas about policy directionality, governance, and skill sets in PAS will not be able to deliver what is required and (2) to be humble enough to accept that we do not know how to solve the complex adaptive challenges and 3) D at the same time remaining firm in our commitment to start exploring and learning this new way of governing and organizing. Our best bet is to start actively looking for such "entry points" for cross-boundary collaboration where opportunities for focused action to progress toward collective goals through learning and experimenting together exist (Martínez Orbegozo et al., 2022).

4.3 Public Sector Staff Development: Building Capability

In cultivating eudaimonic public administration systems (PAS), the focus shifts fundamentally toward the individual actors and their relationships inside and outside of the system. Their capacity, both individually and collectively, becomes the linchpin for realizing the desired structural, cultural, and procedural shifts. Thus, the pivotal question in eudaimonic PAS pertains to how individuals across various roles can develop and apply practical wisdom, enabling them to realize their fullest potential within the system. Stewardship involves managing organizations individually and as an integrated part of multi-stakeholder ecosystems, while innovation entails introducing novel ideas or approaches. These values intersect in the concept of adaptability, which requires striking a balance between established norms and emerging ideas to respond effectively to social, ecological, and economical challenges.

Eudaimonic PAS are established through a distributed leadership mode, where the traditional model of "administration as a machine" is left in favor of more relevant models of administrating as organic flows of activity in networks built for requisite complexity (Uhl-Bien & Arena, 2018). Enabling leadership facilitates creating an adaptive space where systems designs can be rewired to create conditions for accelerated new responses (Allen & Malekpour, 2023).

It is understandable that the established model of predictability and order could hold PAS in an exceptionally strong grip compared to other kinds of organizations. The role of PAS in societies to prevent system breakdown may be experienced as posing more significant risks than for other kinds of organizations. Also, the impossibility to legitimately start and run a parallel PAS, as a firm and other organizations can do to establish a radically different version of themselves, raises a unique challenge for the PAS. But based on the eudaimonic principles of flourishing at the best of one's potential, this should rather ask of PAS to be more capable to drive transformation toward adaptability within itself. Thus (re-)establishing conditions for practical wisdom to excel in its own dynamic capabilities should be a core priority for the eudaimonic PAS.

In our contemporary context, theories of leadership for organizational adaptability in complexity offer insights into the skills required of public servants moving forward. The paradigm shift toward adaptability aligns closely with the core values of eudaimonic PAS, particularly stewardship and innovation. They require the right forms of leadership for organizational adaptability to empower individuals to navigate and thrive in dynamic environments.

Illustratively, initiatives like "Samverket" in Sweden (Samverket, 2023) and the Global Leadership for Sustainable Development Program (Inner Development Goals, 2023) offer promising avenues for building dynamic capabilities within PAS. Samverket, translating to "the agency for collaboration" in Swedish, explores how shared physical spaces and intentional processes can foster innovation and learning to find new collaborations within the PAS. During Samverket´s first test phase it was able to facilitate collaboration among public servants from diverse governmental agencies as well as regional and local authorities, leading to new innovative approaches that have been adopted nationally and internationally.

Similarly, the Global Leadership for Sustainable Development Program, spearheaded by the Inner Development Goals foundation, provides a structured learning platform for leaders across sectors to strengthen their impact on sustainable development. By integrating individual and organizational transformational skills, the program fosters a systemic approach to capacity building within PAS. Moreover, its emphasis on knowledge sharing within and across countries promises progress in institutionalizing transformational capacity within PAS and their interactions with other sectors.

In summary, nurturing the capabilities of individuals within PAS emerges as a pivotal strategy in advancing eudaimonic governance. Initiatives that facilitate collaboration, innovation, and transformative leadership hold promise in realizing this vision, contributing to the evolution of PAS fit for the challenges of the twenty-first century.

5 End Reflections

As an applied academic and PAS leader, writing this chapter on eudaimonia and PAS has renewed my hope for our collective journey. Reflecting on what is and what needs to be reveals both promising opportunities and the need for innovative approaches to make our PAS truly serve us and engage our citizenry.

It is clear that the journey will be long and hard, but we are many who are able and willing to engage in this together. Significant shifts are necessary on each of the three levels we have explored and they are deeply interdependent. As we see positive signs as, for example, in clear statements made in concert between country leaders in the United Nations General Assembly but at the same time limited progress in aligned cross-level transformative change one has reason to feel despair. However, when we consider these statements as invitations, clear calls to action giving us entitlement to bring our best forward, we must get together to seize this opportunity and unlock our joint full potential wherever we can act to allow the creative forces of experimentation and learning on how to make our PAS truly eudaimonic living systems fit for the journey we must make.

To make the stated intent real and drive change based on such shared directionality, comprehensive sustainability strategies that can guide system-wide long-term governance decisions should be a core focus. To assure horizontal coordination, these sustainability considerations should be integrated into all policies, and not be treated as isolated initiatives. They should be based on a holistic approach that considers social, economic, and environmental dimensions as interdependent. They should take equity seriously and assess how different segments of the population are affected in access to resources and opportunities. As transformation initiatives, especially those focusing on new synergies, require initial investments and longer time horizons for benefits to materialize, planning for the longer-term should be the core focus rather than an exclusive focus on immediate efficiency gains.

In uncertain, volatile environments, stagnation will always be a bigger risk than tests not delivering everything we hoped for. As Peter Drucker famously said, "the greatest danger in times of turbulence is not the turbulence – it is to act with yesterday's logic" (Peter Drucker in Castillo & Trinh, 2018). Our best bet is to be guided by equity and other core values and then work from the principles of active stewardship where we make sure to constantly equip ourselves with the right conditions to innovate and collectively learn. Such conditions need to be built both as core to the day-to-day tasks and in strategic longer time frames for a flourishing eudaimonic PAS. Then and only then can we assure that we focus on helping everyone to achieve their maximal potential for practical wisdom at the core of any eudaimonic PAS. Anything less poses a significant strategic risk to achieving what we value most—creating a sustainable future for ourselves and our planet together.

References

Acharya, K., & Plough, A. L. (2020). The imperative for well-being in an inequitable world. In A. L. Plough (Ed.), *Well-being: Expanding the definition of progress*. Oxford University Press.

Allen, C., & Malekpour, S. (2023). Unlocking and accelerating transformations to the SDGs: A review of existing knowledge. *Sustainability Science, 2023*(18), 1939–1960. https://doi.org/10.1007/s11625-023-01342-z

Annala, M., Leppenän, J., Mertsola, S., & Sabel, C. (2020). *Humble government: How to realize ambitious reforms prudently. Government's analysis, assessment and research activities*. Demos Helsinki.

Bednar, J. (2023). Governance for human social flourishing. *Dædalus – Journal of the American Academy of Arts & Science*, Winter 2023.

Beinhocker, E. (2020). Toward a new ontological framework for the economic good. *Global Perspectives, 1*(1). https://doi.org/10.1525/gp.2020.17578.

Beinhocker, E., & Hahnauer, N. (2014). Redefining capitalism. *McKinsey Quarterly, 3*(3).

Biermann, F., Hickman, T., Senit, C.-A., et al. (2022). Scientific evidence on the political impact of the Sustainable Development Goals. *Nature Sustainability, 5*, 795–800. https://www.nature.com/natsustain

Carayannis, E. G., Barth, T. D., & Campbell, D. F. (2012). The Quintuple Helix innovation model: Global warming as a challenge and driver for innovation. *Journal of Innovation and Entrepreneurship, 1*(1), 2. https://doi.org/10.1186/2192-5372-1-2

Castillo, E. A., & Trinh, M. P. (2018, July). Catalyzing capacity: Absorptive, adaptive, and generative leadership. *Journal in Organizational Change Management, 32*, 356–376.

Cooper, M. (2021, April). Directionality: Unifying psychological and social understandings of well-being and distress through an existential ontology. *Journal of Humanistic Counseling, 60*. https://doi.org/10.1002/johc.12148

Cooper, M. (2023). *Psychology at the Heart of Social Change. Developing a Progressive Vision for Society*.

Díaz, S., & Demissew, S. (2015). The IPBES conceptual framework – Connecting nature and people. *Current Opinion in Environmental Sustainability, 14*, 1–16. https://doi.org/10.1016/j.cosust.2014.11.002

Eudaimonia. (2024, March 28). *Wikipedia*. https://en.wikipedia.org/wiki/Eudaimonia

Heifetz, R. A. (1994). *Leadership without easy answers*. Harvard University Press. https://doi.org/10.2307/j.ctv1pncrt0

Independent Group of Scientists appointed by the Secretary-General. (2019). *Global Sustainable Development Report 2019: Future is now*. United Nations.

Independent Group of Scientists appointed by the Secretary-General. (2023). *Global Sustainable Development Report 2023: Times of crisis, times of change: Science for accelerating transformations to sustainable development*. United Nations.

Inner Development Goals. (2021). *Inner Development Goals: Background, method and the IDG framework*. Retrieved from www.innerdevelopmentgoals.org/about/resources

Inner Development Goals. (2023). *Global Leadership Program for Sustainable Development. Evaluation report*. IDG_GLSD+Summary+2023+Evaluation.pdf (squarespace.com)

IPCC. (2022). In H.-O. Pörtner, D. C. Roberts, M. Tignor, E. S. Poloczanska, K. Mintenbeck, A. Alegría, M. Craig, S. Langsdorf, S. Löschke, V. Möller, A. Okem, & B. Rama (Eds.), *Climate Change 2022: Impacts, Adaptation, and Vulnerability. Contribution of Working Group II to the Sixth Assessment Report of the Intergovernmental Panel on Climate Change* (p. 3056). Cambridge University Press. Cambridge University Press, Cambridge. https://doi.org/10.1017/9781009325844

Ives, C. D., Schäpke, N., Woiwode, C., & Wamsler, C. (2023). IMAGINE sustainability: integrated inner-outer transformation in research, education and practice. *Sustainability Science, 18*, 2777–2786. https://doi.org/10.1007/s11625-023-01368-3

Kahneman, D. (2011). *Thinking, fast and slow*. Penguin Books.

Lavecchia, S. (2025). The present of an ancient notion: Eudaimonic well-being as a concept for sustainable entrepreneurs? In M. Mühlböck (Ed.), *Sustainable transformation and well-being. The role of business for individual and collective flourishing* (pp. 15–27). Springer.

Martínez Orbegozo, E. F., de Jong, J., Bowles, H. R., Edmondson, A., Nahhal, A., & Cox, L. (2022). Entry Points: Gaining Momentum in Early-Stage Cross-Boundary Collaborations. *The Journal of Applied Behavioral Science, 58*(4), 595–645. https://doi.org/10.1177/00218863221118418

Mazzucato, M. (2021). *Mission economy: A moonshot guide to changing capitalism*. Penguin UK.

Meadows, D. (1999). *Leverage points: Places to intervene in a system*. The Sustainability Institute.

OECD. (2017a). *Skills for a high performing civil service* (OECD Public Governance Reviews). OECD Publishing. https://doi.org/10.1787/9789264280724-en

OECD. (2017b). *Systems approaches to public sector challenges: Working with change*. OECD Publishing. https://doi.org/10.1787/9789264279865-en

OECD. (2024). *Designing Effective Governance to Enable Mission Success*, OECD Science, Technology and Industry Policy Papers, No. 168, OECD Publishing, Paris.

Piening, E. P. (2012). *Dynamic Capabilities in Public Organizations: A literature review and research agenda. Public Management Review, 15*(2), 209–245. https://doi.org/10.1080/14719037.2012.708358

Plough, A. L. (Ed.). (2020). *Well-being: Expanding the definition of progress*. Oxford University Press.

Raworth, K. (2017). *Doughnut Economics: Seven Ways to Think Like a 21st-Century Economist*. Random House.

Robert, F. Kennedy's remarks at the University of Kansas (2024, March 28). Wikipedia. https://en.wikipedia.org/wiki/Robert_F._Kennedy%27s_remarks_at_the_University_of_Kansas

Ripple, W. J., Wolf, C., Newsome, T. M., Gregg, J. W., Lenton, T. M., Palomo, I., Eikelbloom, J. A., Law, B. E., Huq, S., Duffy, P. B., & Rockström, J. (2021). World Scientists' warning of a climate emergency 2021. *BioScience, 71*(9), 894–898.

Samverket. (2023). *SAMVERKET – A philosophical, but practical, guide to co-operation and innovative power in the public sector*. Downloaded from website Samverket in English – Samverket, March 25 2024.

Schoemaker, PJH., Heaton, S. & Teece, D. (2018) Innovation, Dynamic Capabilities, and Leadership. *California Management Review Volume 61, Issue 1*.

Stieglitz, J, Sen, A. & Fitoussi, J-P. (2009) Report by the Commission on the Measurement of Economic Performance and Social Progress. *Commission on the Measurement of Economic Performance and Social Progress*.

Uhl-Bien, M., & Arena, M. (2018). Leadership for organizational adaptability: A theoretical synthesis and integrative framework. *The Leadership Quarterly, 29*(1), 89–104.

UNDP. (2020). *The next frontier Human development and the anthropocene. Human development report 2020*. United Nations Development Program.

UNDP. (2022). *Uncertain times, unsettled lives shaping our future in a transforming world. Human development report 2021/22*. United Nations Development Program.
United Nations Secretary General. (2023, September 18). *Secretary general's remarks to the high-level political forum on sustainable development*. UN.
VINNOVA. (2022). *Designing missions Mission-oriented innovation in Sweden— Mission-oriented innovation in Sweden— A practice guide by Vinnova*. Retrieved from. https://www.vinnova.se/contentassets/1c94a5c2f72c41cb9e651827f29edc14/designing-missions.pdf?cb=20220311094952

Fredrik Lindencrona is a father of three and an international expert with significant practical experience in leading policy and systems governance processes toward sustainable well-being impact. His work during the last 20 years has focused on supporting such processes at city and regional levels. He is often engaged with international organizations including OECD, as an expert and to facilitate international exchange among senior executives and political leaders. Dr. Lindencrona is one of the initiators of the Inner Development Goals, a global movement to strengthen leadership for multisectoral collaborative processes to achieve sustainable development where he currently leads research co-creation.

The Philanthropic Sector Creating Momentum for a Sustainable Transformation and the Well-Being of Changemakers

Tomas Björkman, Claudia Garuti, Susanne Hillebrand, and Marisa Mühlböck

Abstract Serving the common good has been at the heart of philanthropy for hundreds of years. Institutions that have been created to organize philanthropic actions such as foundations are characterized by stability and continuity. However, in an ever-changing world of growing complexity, the sector is challenged to transform in a way that being of service for society can still be realized. This includes not only tackling new topics and fields of impact but also questioning current frameworks, structures, and processes inherent to philanthropic institutions. In this light, the promotion of sustainable development as a path to secure and increase the collective well-being and the flourishing of the planet should be seen as paramount. This is closely tied to the well-being of individuals, in particular of those agents of change who drive such sustainable transformation. Based on interviews with philanthropists who have already acknowledged the need for transforming the philanthropic sector in this sense, this chapter strives to question the role of modern philanthropy in this context. Drawing on practical experience, recommendations are presented on how philanthropy could create a stronger momentum for sustainable transformation and well-being in the future.

Keywords Philanthropy and sustainable development · Transforming philanthropic institutions · Sustainable transformation · Well-being of changemakers · Philanthropic sector challenges · Agents of change · Modern

T. Björkman
Ekskäret Foundation, Stockholm, Sweden

C. Garuti
Better Way Foundation, Vaduz, Liechtenstein

S. Hillebrand
Hil-Foundation, Vienna, Austria

M. Mühlböck (✉)
Independent Researcher, Vienna, Austria
e-mail: research@marisamuehlboeck.com

M. Mühlböck (ed.), *Sustainable Transformation and Well-being*, CSR, Sustainability, Ethics & Governance,
https://doi.org/10.1007/978-3-031-75566-8_15

philanthropy · Impact of philanthropy on well-being · Philanthropic frameworks and structures · Practical recommendations for philanthropy

1 Introduction

The history of philanthropy can be traced back to Ancient Greece. Not only the word itself stems from the Greek word *philanthrôpía*, which means "love of humankind" (Grant, 2012) or the "gods' love for humans" (Muukkonen, 2010). It was Plato himself who facilitated the establishment of his famous Academy through stipulating in his will that the revenues of the family farm be used for its foundation (Strachwitz, 1996). In Europe, charitable foundations and trusts were established as early as in the Middle Ages, with those for nonreligious purposes particularly flourishing in Protestant areas such as Germany, Scandinavia, and Britain. While endowments of institutions such as hospitals or universities were the main targets of philanthropic endeavors, grant-making foundations that indirectly supported beneficiaries through funding intermediaries only emerged in the nineteenth century on a broader scale (Grant, 2012).

In general, philanthropic activities have historically been focused on funding initiatives that are more short-term and easy to measure—all prerequisites that do not genuinely align with a topic such as cultivating the development of individuals, even though research has shown that this might lead to an increase in well-being for the individual and the collective (e.g., Pöllänen et al., 2019). Yet, some philanthropists have already recognized that in a world of increasing complexity, positive social change is dependent on healthy change leaders. However, dealing with this complexity requires skills and capacities that haven't really been the focus of management and leadership theories in the past. Consequently, mental health problems among changemakers[1] are on the rise and not few of them are even at risk of burning out. In this book, several contributors underline this fact: Lux (2025) openly shares a very personal story, whereas Vandor and Meyer (2025) deliver study results and refer to social entrepreneurs' tendencies such as vulnerability to illusionary goals, over-engagement, disappointment, and exhaustion that might ultimately lead to a burnout.[2] They also point out ecosystem actors' responsibility in this context. Thus, what are the implications for the philanthropic sector in this regard? Facing those challenges, what is the role of modern philanthropy? How does it have to change and which barriers need to be overcome in this context?

[1] According to Ashoka, which is an international network that builds and cultivates a community of changemakers, "[a] changemaker is someone who is taking creative action to solve a social problem" (https://www.ashoka.org/de-at/collection/changemaker-skills).

[2] At the same time, an increase of mental health issues becomes evident not only in the changemaker community but also in business on a broader scale. According to survey results from the Business Development Bank of Canada, almost half of Canadian entrepreneurs are experiencing mental health challenge (Business Development Bank of Canada, 2023).

Based on interviews with philanthropists who have already acknowledged the need for transforming the philanthropic sector, this article strives to answer those questions and make recommendations on how philanthropy could create a stronger momentum for sustainable (business) transformation and well-being in the future.

2 Recognizing the Importance of the Topic: Personal Drivers

Three personalities with different backgrounds and experience have provided their expertise as the interview partners and co-authors of this article. What unites them is their recognition that investments in the well-being of grantees are valuable strategies for effective philanthropy. They share the unanimous realization that successful philanthropic work must take into account changed framework conditions, where well-being and human flourishing play a central role. To start off this topic, this section presents, in their own words, these personalities' personal motivations behind the acknowledgment of the importance of the topic and their views on key skills for modern sustainable leaders.

Claudia Garuti: *I have personally been dealing with the question of how we become a more flourishing version of ourselves for a long time. Eventually, I, in my role as a consultant and facilitator, decided to create a physical space that allows such investigation together with other people. There, groups sit in circles and a dialog on a level playing field takes place, even on delicate issues such as climate change or women's empowerment. Even though this is a small intervention, it is effective, since it creates a space where people connect better. Multiple times, I have witnessed people in such a circle open up in a vulnerable way, but also in a way that activates them as agents of change. Thus, it is important to me to bring in all those learnings in the work of the Better Way Foundation,*[3] *our family foundation, in which I have been actively involved for about three years. The foundation particularly tries to serve as a platform for social entrepreneurs and for individuals who have great ideas or solutions which might contribute to sustainably transforming society and our planet. From my personal experience as well as from the experience in our projects, I strongly believe that inner work that creates a mindset shift from our own narrow self to the common good—what is generally referred to as "from ego to eco"—is key in this matter. Consequently, we have started to think of how to integrate this element of inner well-being and inner development into our philanthropic work.*

Tomas Björkman: *I used to be a businessman and a serial entrepreneur for many years. About 20 years ago, I had the opportunity to sell my banking business, with the commitment to remain as the chairman of the banking group in Sweden and*

[3] https://www.betterwayfound.org

on the board of directors of the Swiss bank for some years. In my role as a chairman, I had the opportunity to instigate a leadership development program and when I myself participated in it, I saw the deep effect it had on me and the management team. Even back then, business complexity had reached a high level and through the program we were able to manage this challenge better. Today, the business environment is even more complex and rapidly evolving. Furthermore, I noticed the positive effect in my private life. And slowly I started to look beyond my private and the company sphere to what our impact on society and the planet actually was. This experience made me ponder, why, if we in business know the importance of supporting our individual flourishing and its effect on the business environment, do we not talk about this in the general public. In 2008, after my contract with the bank ended, I set up the Ekskäret[4] *Foundation in Stockholm, where we have been focusing on exploring the connection between personal inner growth and development and societal change, both from a theoretical but also a practical perspective. The experience from more than ten years finally led to the idea of the Inner Development Goals (IDGs),*[5] *an initiative that was launched three years ago. I co-founded other initiatives in this small ecosystem such as 29 k,*[6] *a non-profit open source digital platform for personal growth, the London-based think tank Perspectiva*[7] *or the Emerge project,*[8] *which highlights individuals and organizations that are emerging in this complex world and seeks to respond to the most urgent challenges of our time. Interestingly, we can also find valuable concepts in this matter when we look into our own history books. Some of these can be found in a book that I co-authored which is titled* The Nordic Secret. *Our intention was to explore how Denmark, Norway and Sweden went from poor feudal agricultural societies to rich industrialized democracies thanks to the German educational concept of "Bildung". The book also investigates the close relationship between "Bildung" and contemporary developmental psychology,* i.e., *the concepts of "ego-development" and "transformative learning", and it concludes with a discussion about what we can learn from this positive transformation and how to apply it in the current global crises as co-creators of a modern society* (Andersen & Björkman, 2017).

Susanne Hillebrand: *12 years ago, I got the chance to set up our family's philanthropic organization. One of the Hil Foundation's*[9] *focus areas is the education of and awareness-raising among kids and young adults on topics such as diversity, intersectionality and gender sensibility. Being a lesbian myself, with decades of involvement in LGBTQIA+ activism, I have been able to bring in an authentic perspective on the opportunities and challenges in this field. I believe that dealing with*

[4] Ekskäret is the Swedish word for Oak Island. https://www.ekskaretfoundation.com

[5] https://www.innerdevelopmentgoals.org/

[6] https://29k.org/

[7] Perspectiva is a community of expert generalists working on an urgent 100-year project to improve the relationships between systems, souls, and society in theory and practice. https://systems-souls-society.com

[8] https://www.whatisemerging.com

[9] https://www.hil-foundation.org

that topic might also have triggered my interest in inner development, well-being and human flourishing, which is another pillar in our philanthropic work. In this matter, we particularly address the social entrepreneur and changemaker community, since we perceive this as an effective approach to contributing to sustainable transformation. One aspect I believe to be especially crucial in this context, is the ability of leaders to show their vulnerability. How can they become aware of their authentic selves and where do they find a safe space to develop and cultivate the corresponding mindset for this? Where can they open up about their insecurities, anxieties or fear? This is what our foundation tries to tackle. We mainly do this in co-creation with project partners such as The Wellbeing Project, an organization that strives to catalyze a culture where changemakers are well and their work thrives,[10] *or the Impact Hub,*[11] *a network which aims at inspiring, connecting, and enabling impact makers.*

3 Is the Topic Really Hot? Practical Learnings and Challenges Ahead

As the interviewees are still pioneering the philanthropic sector with their work, it is legitimate to ask how their endeavors have been perceived in the general public, among their philanthropic peers as well as among their grantees so far, what the first real-life learnings in this matter have been, and which challenges lie ahead of them. Their feedback on this is summarized in this section.

Learning: *People and organizations in different sectors show interest in the topics of sustainability transformation, inner development, and well-being and how they are all connected.*

Drawing from the Inner Development Goal project's first 3 years, a significant public interest can already be noticed. The initiative was launched with the intention to kick off a conversation about the fact that outer transformation cannot be achieved by just focusing on the outer aspects of life but that it is related to inner change. Those ideas became popular at a surprisingly high speed. The experience is that there is a need to talk about these issues, and the IDG project seems to deliver a framework for and legitimate the conversation about them. Without very much effort from the central organization, more than 500 local IDG hubs all over the world were established, mainly in Europe and North America, but also in South America, Africa, and Asia. That can be interpreted as a very promising sign that more and more people are recognizing the importance of also looking on the inside of individuals and of organizations to create well-being, flourishing, and societal change that we so urgently need.

[10] https://wellbeing-project.org

[11] https://impacthub.net

Challenge: *Finding a common language that at the same time takes into account and reduces the complexity that comes with this topic.*

Human beings are complex and so is their inner development. Acknowledging this complexity also requires overcoming the very simplistic view of our mind as a rational machine and with this the idea of homo economicus. The image of humankind of just making decisions with our brain like a computer, with complete knowledge of what is good for ourselves, that feelings don't matter, and the inner world is of no importance is simply an illusion. The challenge ahead will be to find a common language to explain this extremely complex and abstract matter without getting too simplistic. Only when we succeed in conveying to the larger part of the population why inner development and well-being is important and what that means exactly, that inner work is a lifelong project, and that this is important not just for the individual, but for the development of society, will we be able to transform entire sectors. An important part of this message is that this is not just about a single individual in their development. Inner does not only mean individual, but the organizational and the societal cultures represent very important elements in this context, too.

Learning: *Changemakers highly appreciate philanthropic support for more well-being.*

All three philanthropists have been providing formats and tools that aim to increase the well-being of individuals and their organizations. All of them have been receiving positive feedback for their endeavors, which, for instance, comprise approaches such as coaching by a professional, providing spaces for peer-to-peer exchange or safe spaces and programs for inner work. Their experience aligns with findings of The Wellbeing Project that evaluated input from 85 grantmakers and concluded that the creation of space for open sharing is a key factor of nourishing well-being in the changemaker community (Shams-Lau & Wilberding, 2020).

Furthermore, the support to fund collective moments, where teams and partners come together in a physical place (something that can be costly in particular for organizations that operate internationally), has been identified as a relevant factor in the well-being context. It seems that it can be catalytic in a sense that coming together enables collective learning, speaking with one voice, and a transformation in the way people want to effectuate and affect change.

What the philanthropists realize is that through such measures conversations emerge with a new quality of openness, authenticity, and vulnerability, where a culture of "we together" instead of "me against you" is nurtured. Yet, exploration and experimentation on the most effective ways to support changemakers in their well-being is still ongoing. Nevertheless, there is a unanimous conviction among the three philanthropists that such offers should be commonly integrated into funding practices.

Challenge: *Finding more people in the philanthropic community who are aware of this need.*

Although recognition of the need for well-being support seems to be gaining momentum, the funding in this matter lags behind the need. In general, it still seems to be hard to find allies in the sector. One reason might be that such endeavors challenge philanthropists themselves to do their own inner work and to show themselves in an authentic way. Some interviewees share the experience that other philanthropists shy away from that. Other explanations might be found in the fact mentioned above that a clear, common language has not yet been fully developed in a manner that this need is fully understood. This underlines that the search for such a language should also take into consideration what elements a narrative for the philanthropic sector requires. That narrative could, for instance, contribute to guiding philanthropist through challenging situations such as conflicting interests and inner tensions: e.g., how do I prioritize, if my ambition to have all projects milestones fulfilled and reported on time clashes with a decrease of well-being for the project team?

Challenge: *Resolving the requirement to measure well-being support.*

It goes without saying that the use of earmarked philanthropic funds must not be arbitrary. Rightly, it must be comprehensible and traceable where the money is spent. This principle is tightly connected to the question of how the impact of the funding is measured. The demonstration and documentation of individual well-being effects on a measurable scale, and in particular their impact on the related projects, are still a challenging task and can lead to a board's refusal to implement well-being initiatives. However, one should keep in mind that measuring often brings a short time perspective, whereas inner development is certainly a process over many years. That area of conflict should therefore be taken into account.

4 A Modern Role of Philanthropy

The philanthropic world as we know it today comes from an age where the world was much more stable. Philanthropic work was relatively simple, building hospitals and schools or handing out alms to poor people. Today, many of those things are supplied by the state and, consequently, philanthropists have to find their new role in society. While identifying blind spots where the public sector is failing today, it soon becomes obvious that the territory to enter is much more complex than it used to be. Therefore, not only new fields but also new approaches in impact creation have to be found. Yet, the sector seems to be stuck in the old patterns in many ways. This section sketches, again mainly drawing from the philanthropists' practical expertise, elements and recommendations for an adequate new role of philanthropy in modern times. In this endeavor it becomes clear that sustainable transformation and well-being are not just new topics on philanthropists' agenda, but issues that influence the sector's self-image and even funding strategies, practices, and organizational structures.

From the provision of basic goods to the facilitation of real innovation: Philanthropy could transform into a place where real innovation is happening. As the world is evolving, new possibilities arise, but also new problems that need new solutions. It makes absolute sense that the philanthropic sector provides grants to those who offer potential solutions and thus support the implementation of those.

However, what has so far been rather neglected is funding the work of ideation and idea development, since that is often seen as a sort of overhead. Most philanthropic organizations assume that that has to happen before someone files an application where a clear idea with its theory of change is already presented. But if we really want better ideas and better projects, we also need to fund the process of ideation. This is also true when it comes to the fields of sustainability transformation, inner development, and well-being, where many questions are waiting for truly innovative answers.

From scattered and selective action to transforming sectors: Philanthropy can put topics that are not yet seen as a priority in the public or the business sector at the top of the agenda. Many of the pressing issues of our time are directly or indirectly connected to a system change or a sustainability transformation and well-being (e.g., agroecology or regenerative agriculture in the field of ecology, neurobiology, or relational aspects of our minds in the social science sector, to name just a few). The philanthropic sector could see its role in transforming or creating new sectors that will be better for people and the planet. That involves spearheading and facilitating transformation in one sector so that other sectors can then take up what has already been tested (e.g., new technologies, new policies). Furthermore, that requires agile, flexible, and innovative thinking, by which the provision of risk capital is not seen as a waste but a necessary step in the process of innovation. That might be a frustrating role from time to time, but in this way, philanthropy can really create momentum for a better world. The following metaphorical quote sums it up very well: "We are end-of-life companions of an old culture and midwives of a new culture, not knowing exactly what that is.[12]".

From funding only a few small but "secure" projects to a braver portfolio approach: It seems that most philanthropic organizations have not gotten used to the principles of an uncertain world yet. In this regard, the sector might benefit from learnings in the business world, in particular from the venture capital sector, which has learned to navigate in an uncertain environment. Drawing from the portfolio approach, one accepts from the beginning on that a large part of the supported projects will fail. That is part of the strategy, because through such a radical approach, your chances increase that the others will be a tremendous success.[13] First attempts in this direction are represented in the venture philanthropy approach, which is strongly oriented toward entrepreneurial principles (OECD netFWD, 2014).

From a mere project view to a holistic view where entire organizations are transformed: Whether one works with partners in the business, public, or NGO sector, it is very important that projects are not being captured by the "old" system. Philanthropy should rather fill the gap that is left by the public sector, which, due to political systems' mechanisms, focuses too much on short-term solutions. The

[12] One of the philanthropists remembered this metaphor from a workshop she was attending.

[13] The philanthropists underlined that success for them is not only defined in financial terms. It is measured against their aims. In this sense, success could also be if a project delivers more impact than one hoped for or if the organizations they support are able to continue their work and expand their reach.

system of philanthropy allows for a long-term systemic view. Unfortunately, until now only few philanthropists have managed to incorporate such a view, but are stuck in the same 1- to 3-year project cycle as the public sector.

At the same time, more and more organizations have already realized that the technological and business environment evolution is now so fast that one cannot operate like one used to do in a large organization (i.e., to gather all information and forward it to the management board and then some very clever person tries to make sense of the world and designs a masterplan). They understand that they need to be much more agile and operate in some sort of self-organizing way. What might happen in such an environment is that many people in those organizations start to feel empowered, which might lead to more trust overall.

From hierarchic to participatory philanthropy: Philanthropists should walk the walk and talk the talk, which includes doing their own inner work. This involves how the relationships with partners and grantees are managed. Many philanthropic structures impede a healthy culture in the sense that they create top-down relationships that primarily empower those who have the money. Thus, the rules of the relationships are solely made by the philanthropic organization, and the partner has to follow those. However, if we would like to nurture a culture of trust that leads to healthy and successful long-term relationships, we should ask the question of how to create a more horizontal culture that then supports this concept of well-being and flourishing on both the funder's and the grantee's side. One approach in this context could be to give more power to the grantees who have the knowledge of their specific field and group of beneficiaries. A good example for efforts to shift power relationships in the philanthropic sector is Guerrilla Foundation, which has established an Activist Council and a Founders Circle with the intention of evolving to a more diverse and heterogeneous organization.[14] Similarly, Hil Foundation hands over 100% of the decision-making authority to their girl's and boy's advisory boards who decide how the dedicated project funds are spent.[15]

From neglecting unhealthy patterns to actively nurturing well-being: Talking the talk also comprises a conscious observation of how innate patterns in the giver-receiver relationship influence the project outcome. One aspect is the management of expectations. Giving money comes with certain requirements, which in many cases adds pressure to grant receivers. Oftentimes, their own aspirations in regard to impact creation are very high. Changemakers are generally impact-driven and thus, as mentioned above, not seldom at risk of burnout. Philanthropists should not promote or ignore a "never stop working for the good cause" attitude, but rather start to actively open a conversation about the partner's well-being and ask them what they need and how they could support them. They should signal that it is alright to slow down if this prevents burnout, since this is the best way to create the impact one is trying to achieve with the grants given. What one could additionally do in this matter is encouraging change leaders to find their focus and optimize the structure of their organization. Many of them tend to work at all angles in their projects and act like nothing will be good enough without their input. At the same time, many

[14] https://guerrillafoundation.org/about/managing-power-and-conflict/

[15] https://www.maedchenbeirat.at/ and https://www.bubenbeirat.at/

philanthropists are disposed to consciously or unconsciously supporting this tendency, since they often believe in this one often very charismatic person in the lead.

Another realization in this context is that a significant number of social entrepreneurs and changemakers are not aware of the underlying personal driving forces behind their commitment to creating a better world. Not everybody is motivated by a healthy source; sometimes it is rather negative emotions such as anger, revenge, disappointment, or guilt that come into play (see, e.g., Rodgers, 2010; Weetjens 2025). If such feelings are not brought into consciousness and professionally processed, there is a much higher risk for mental health problems or, turning it upside down, research has shown that making a meaningful connection with their inner selves has helped changemakers to increase their well-being (Severns Guntzel & Murphy Johnson, 2020). However, many social entrepreneurs work in such a resource-poor environment that this important inner work cannot take place. Thus, if the awareness among philanthropists about those interdependencies were increased, the whole sector could benefit from it. Both the support of projects to include support for the change agents and the development of special well-being programs are meaningful ways in which philanthropists can contribute in this matter.

Again, neglecting or encouraging such patterns might backfire when mental health problems come into play. By bringing in different perspectives and openly communicating about unhealthy behavior, one is also taking into consideration and honoring the complexity of the people involved in such a relationship.

From stewards of the money to conscious transformation leaders: The philanthropic sector is part of this society and also has to transform. People in that sector should work on their own inner transformation in order to consciously recognize what is really needed in the world. The perspective should shift from protecting the assets to perceiving their role as authentic and conscious transformation leaders.

From pure funding to the provision of different resources: The assumption that it is only money that contributes to the change is not valid anymore. As much as this was true in the past, today, in many cases, in the Western world, but also in the global South, it is not just about handing out blankets or building schools, but it's also about other resources such as knowledge transfer. Money has to come with knowledge and experience. That is a combination that aims at empowering the beneficiaries to create what is needed instead of providing some final product. Thus, the goal is to shift the power in a way that ultimately the philanthropic support will not be necessary anymore. What all that requires is a mindset shift that decreases the focus on the effect of money and sheds light onto the effect of many other things that social entrepreneurs bring to the world through their projects.

5 Barriers for a Transformation of the Philanthropic Sector

While someone who intends to establish a new philanthropic institution in the here and now might just focus on the end points of the transformative paths presented above, the implementation of those might be much harder for existing organizations. We see two types of barriers in this matter, structural ones and some sort of an "old" mindset that prohibits the realization of such a transformation.

5.1 Structural Barriers

Many foundations came into being during a time when topics such as sustainability, well-being, and human flourishing were definitely not on the radar. As such, their statutes and by-laws were not designed to consider them, either. Some are so restrictively stipulated that an expansion of the foundation's purpose in a way that includes such zeitgeisty issues becomes very challenging. This is of course first and foremost dependent on the country-specific laws and regulations, but many of them are quite rigorous in this context and require a lot of effort and overcoming of bureaucracy to change the foundation's purpose or beneficiaries accordingly (Philea, 2022). Thus, to overcome such structural barriers, adaptions in the regulative contexts are required.

5.2 The "Old" Mindset

Philanthropic organizations that were established in the second-half of the last century were very much influenced by a modernistic neoliberal managerial mindset. The need to control the project regarding every little detail and with it all financial resources were underlying principles in that era. This included the requirement to implement tools and strategies for measurement such as the definition of KPIs. Even if this is legitimate in the sense that such approaches help to avoid a waste of money in the short run, the problem is that in order to achieve systemic effects such as the transformation of entire sectors, measurement in the short term becomes very difficult. Therefore, such old principles could actually hinder important work today. To shift to a new and more flexible mindset that balances the needs of today with a sensible use of money through appropriate measuring tools and procedures, the entire governance structure, in particular the managing director, but also the foundation's board must be aligned in this matter.

The concrete advice to other philanthropic organizations would therefore be to have a clear review of their own governance structures and to make sure that the people on the board as well as the trustees do really understand these issues of systemic transformation, the interrelations of inner transformation for outer change, and of individual as well as collective well-being, and the complexity that comes with of all these issues.

Further advice in the context of the use of financial means is to set aside a small part of the budget (approximately 10–20%) for long-shot projects where potential failure is consciously taken into account in order to test new ideas or, metaphorically speaking, to build bridges into the fog. This uncertainty or "fog" shouldn't stop us from funding projects that we believe could have a chance of a breakthrough. If a significant number of philanthropists adopt such a mindset, then at least some of those projects will be successful and lead to valuable insights within the philanthropic community. This will certainly require a greater openness and cooperativeness within the philanthropic sector, which seems to have already started to develop. At the same time, a lot of protectionism in the philanthropic world can still be noticed, where, for instance, NGOs compete for philanthropic funding and see each other more as competitors than cooperators. We have to overcome such a mindset so that transformation on a broader scale can be realized. One way of doing that could be by earmarking certain funding opportunities exclusively for projects that result from collaborations between two or three grant-seeking organizations.

6 Conclusion

The world is changing more rapidly than ever. It seems that even a large part of decision-makers has not anticipated and consequently not been prepared for that. Thus, a shift to a more agile and flexible mindset seems to be the call of the hour, also for philanthropists. At this point, one can speculate that deep societal transformation is unavoidable with two potential scenarios as an outcome: a breakthrough that might lead us to a new way of relating to ourselves and to each other, society, and the planet or a breakdown that would entail a fragmentation of the world order with our political structures collapsing even before the planet does. With many forces in the political and the business sector being part of the problem, everyone, and in particular the philanthropic organizations as agents of a more independent sector, is called upon to facilitate the path into the first rather than into the second scenario. If they do not succeed in that, they will become irrelevant as philanthropists in the current turbulence and ongoing transformation. Summing up, there is a huge opportunity for philanthropists to be of tremendous relevance in the current world. Yet, they need to take on a new role as described above. They need to be careful that they do not just replicate what the market or the political system does and they should start thinking outside of the box. They should become enablers of people who do bold things, of bold conversations and of bold projects. As such, philanthropists need to become bolder themselves.

References

Andersen, L. R., & Björkman, T. (2017). *The Nordic secret. A European story of beauty and freedom* (1st ed.). Fri Tanke.

Business Development Bank of Canada. (2023). *Survey on SME owner's mental health and support*. Research report.

Grant, P. (2012). A short history of philanthropy. In *The business of giving*. Cass Business Press. Palgrave Macmillan. https://doi.org/10.1057/9780230355033_3

Lux, H. (2025). Lessons from a wasp sting: A plea for inner development in order to change the world for the better. In M. Mühlböck (Ed.), *Sustainable transformation and well-being. The role of business for individual and collective flourishing* (pp. 311–324). Springer.
Muukkonen, M. (2010). Philanthropy, ancient. In H. K. Anheier & S. Toepler (Eds.), *International encyclopedia of civil society*. Springer. https://doi.org/10.1007/978-0-387-93996-4_139
OECD netFWD. (2014). *Venture philanthropy in development: Dynamics, challenges and lessons in the search for greater impact*. OECD Development Centre.
Philanthropy Europe Association (Philea). (2022). *Comparative highlights of foundation laws. The operating environment for foundations in Europe*. Philea.
Pöllänen, E., Osika, W., Bojner Horwitz, E., & Wamsler, C. (2019). Education for sustainability: Understanding processes of change across individual, collective, and system levels. *Challenges* 2023, *14*(1), 5. https://doi.org/10.3390/challe14010005
Rodgers, K. (2010). Anger is why we're all here: Mobilizing and managing emotions in a professional activist organization. *Social Movement Studies, 9*(3), 273–291. https://doi.org/10.1080/14742837.2010.49366
Severns Guntzel, J., & Murphy Johnson, N. (2020). *Wellbeing inspires welldoing: How changemakers' inner wellbeing influences their work*. Report.
Shams-Lau, J., & Wilberding, L. (2020). How funders can support individual well-being. *Stanford Social Innovation Review*. https://doi.org/10.48558/n117-8k47
Strachwitz, R. (1996). Stiftungen in Deutschland: Geschichte und Gegenwart. Die Idee des Guten – eines der obersten Ziele Platons. In: *Stiftungen*. Grünenthal GmbH, Aachen.
Vandor, P., & Meyer, M. (2025). Impactful ventures, burned-out founders? Insights from two studies of social entrepreneurs. In M. Mühlböck (Ed.), *Sustainable transformation and well-being. The role of business for individual and collective flourishing* (pp. 159–173). Springer.
Weetjens, B. (2025). The skull in our suitcase: What we can learn from Zen in challenging moments. In M. Mühlböck (Ed.), *Sustainable transformation and well-being. The role of business for individual and collective flourishing* (pp. 281–289). Springer.

Tomas Björkman, after many years in business as an entrepreneur and investment banker, is now a social entrepreneur and the founder of Ekskäret Foundation in Stockholm. He is also the co-founder of the research institute Perspectiva in London, of the Co-creation Loft and the media platform Emerge in Berlin, the 29k.org digital personal development platform, and the Inner Development Goals (IDGs) framework. He is a member of the Club of Rome and a fellow of the Royal Swedish Academy of Engineering Science. He is the author of three books: *The Market Myth* (2016), *The Nordic Secret* (together with Lene Rachel Andersen, 2017), and *The World We Create* (2019). Tomas divides his time between London, Stockholm, and Berlin.

Claudia Garuti is the Co-director of the Better Way Foundation, a family foundation enabling social entrepreneurs to develop climate adaptation and mitigation solutions with a climate-justice and gender-inclusive lens. In addition, Claudia is developing Circles, a new initiative to promote awareness and dialogue as well as to increase support for climate action. Claudia worked in the philanthropy sector for over 10 years, delivering international projects while working at Ashoka and acting as a strategic advisor to civic participation initiatives. Prior to that, Claudia worked in London in the arts sector, where she managed Prix Pictet, the leading photography prize dedicated to sustainability. She holds a master's degree in Business Administration from ESADE, Barcelona.

Susanne Hillebrand is the founder and director of the Hil Foundation, which promotes equal opportunities, particularly for children and young adults, all genders, sexual minorities, as well as people that are mentally strained or ill and strives to raise awareness for the crucial role of resilience, well-being, and self-care among changemakers within the social entrepreneurship ecosystem. She is a board member of the Sinnbildungsstiftung, an association of foundations to promote pioneering projects in the education sector.

Marisa Mühlböck holds a doctorate in social and economic science. Over the course of her career, she has gained extensive insight into different fields of work, which has fostered her understanding of diverse organizational structures, thinking patterns, and social context. As a former social entrepreneur, she also knows the entrepreneurial life from practice and incorporates this experience into her research. As a Research Country Co-lead, she is involved in the Inner Development Goals (IDGs) Initiative, which is dedicated to researching and promoting inner development to achieve the UN Sustainable Development Goals. She also served as a founding board member of Verband für gemeinnütziges Stiften, Austria's first association representing and advocating for charitable foundations. Her current research interests are based on her theoretical background as well as on her practical experience and focus on sustainability in different areas, sustainable entrepreneurship, and well-being.

Part IV
Transitional Chapter II

Teaching and Implementing Ethics for Well-Being in Business: A Field Report

Mathias Johannes Christof Schüz

Abstract The author of this paper looks back on more than three decades of experience in teaching ethics at business schools and universities as well as in teaching and implementing them in business enterprises. Even in his academic career as a physicist, philosopher, and educator, he worked out the implications for ethics in a chapter of his doctoral thesis on the philosophical consequences of quantum physics. During his entry into the business world at IBM, he gained his first experience regarding the importance of ethical behavior for business success. He seized the opportunity to take an in-depth look at the possibilities and limits of implementing ethics and thus also at the individual and social well-being in a large company after his employment with the Gerling Insurance Group, at that time the largest industrial insurer in Germany with an international orientation. Together with the third-generation owner, he initiated an "Academy for Risk Research" there in order to implement an ethics-based risk culture in the company. For 16 years, he successfully helped to shape the corporate culture, also with the involvement of a corporate university for whose contents he was responsible. After the company was sold, he returned to university teaching and research. The rare combination of practical experience and transdisciplinary knowledge enabled him to develop and successfully implement a new didactic approach to teaching applied business ethics. Both collective and individual well-being prove to be a consequence of ethical behavior.

1 Introduction

Already during my studies of physics, philosophy, and education, I realized that science is not without presuppositions and has effects on our living world. The ecological problems with which humanity, and indeed all life forms on our planet, are

M. J. C. Schüz (✉)
Gachnang, Switzerland
e-mail: mathias.schuez@zhaw.ch

M. Mühlböck (ed.), *Sustainable Transformation and Well-being*, CSR, Sustainability, Ethics & Governance,
https://doi.org/10.1007/978-3-031-75566-8_16

struggling are the effects of a way of thinking and behaving that is methodologically rooted in the modern Enlightenment. René Descartes, for example, postulated how nature as an object can be reconstructed according to our own ideas and needs by means of analysis, quantification, and synthesis. Man, as the only thinking and thus also the only feeling being on this planet, can and should fathom the mechanism of nature and, like a watchmaker, reshape it to his liking. Consequently, he regarded animals and plants merely as unfeeling machines (Cf. Descartes, 1997, 90 f).

In this spirit, humanity, with the help of science, technology, and economics, has actually changed our planet in such a way that 75% of its natural surface has been destroyed by human impact so far (UNEP, 2023). Today's ecological polycrisis, to which climate change and species extinction are only two of many contributing factors (cf. Schüz, 2025), now threatens the survival of mankind despite its scientific and technological achievements.

The destructive forces of one-sided science and technology would not have been possible without the economic drive for efficiency and profit maximization. At the latest with the beginning of industrialization in the nineteenth century, the age of mass production began. The efficient exploitation of nature also included the instrumentalization of human beings, who themselves operated machines controlled by their bosses, first the principal, then the executing superiors, the agents. As executing organs, they had little insight into the exploitation of themselves and others. The key to the irreversible downfall of our present civilization, but also to its salvation, lies in understanding how the economy was run then and how it is run now. Ethics plays a central role.

Ethics, as an "art of getting along well with each other," (Schüz, 2023, p. 538) aims to maintain the well-being of all those affected by an action and sustains the viability of all living beings. Those who destroy nature by their actions in business enterprises or exploit people under slave-like conditions have the opposite effect. This provokes resistance, which in the long run comes back to haunt the perpetrator. Therefore, the knowledge of quantum physics, i.e., the knowledge that everything is connected with everything and has an effect on everything, also has ethical implications and concrete effects on well-being (cf. Schüz, 1986, 263 ff). Small children, at least, can often experience the interconnectedness of life firsthand. When I was 4 years old, I personally witnessed an angler cutting up a freshly caught fish with a knife. I felt as if the knife had penetrated me, and for weeks the memory of it disturbed my well-being.

Let us remember: Humans are part of nature and are now confronted with its catastrophic effects. Living well with all forms of life therefore requires a mindful approach to them. However, our finite consciousness easily loses sight of the far-reaching consequences of our actions. Therefore, for self-protection alone, the first step is to recognize the consequences and the risks associated with them. The second step is to assess whether they can be avoided; the third step is to mitigate them; and the fourth step is to compensate for the damage caused.

I developed and published some of these insights in my Ph.D. thesis (Schüz, 1986). With these ideas I started my career in business, first at IBM, and then for 16 years at the Gerling Insurance Group. There I had the opportunity to test a possible implementation together with the owner Rolf Gerling. What succeeded, what failed, and why?

Today's polycrisis (cf. Schüz, 2025) prompts me to reflect on my experiences in order to offer some suggestions on how to deal with the complex modernity that threatens the future existence of humanity. In this article, I trace my reflections on the theory and practice of business ethics and explore the possibilities and limits of applied business ethics. To this end, I will first describe my experiences in business enterprises and then analyze research and teaching in the university sector, especially in continuing education (MBA). The aim is to show the reader the different dimensions as well as possible pitfalls in teaching and implementing ethics in organizations and thus to point out ways to better implement them. Depending on the success of the implementation, this also has an impact on the well-being of the entire company and its employees. We are all invited to rethink how to move forward instead of reproducing old solutions that are part of the problem.

2 Experience in Commercial Enterprises

2.1 At IBM

I started my career in business at the beginning of January 1986 as a trainee and then as a key account manager at IBM. Their business philosophy is based on "three basic beliefs": (1) "respect for the individual," (2) "superlative customer service," and (3) "pursuit of excellence in all tasks" (IBM, 2023). Originally, the company founder Thomas Watson Sen. formulated: (1) "give full consideration to the individual employee," (2) "spend a lot of time making customers happy," and (3) "go the last mile to do things right; seek superiority in all we undertake" (Collins & Porras, 1997, p. 89). This makes it clear that the founder was concerned with the well-being of his customers *and* employees. A prerequisite for this is "Think" in all you do, as Watson repeatedly emphasized (cf. Olwell, 2021). Without thinking, there is no innovation.

2.1.1 Visionary Culture

The prevailing culture at IBM was described as "visionary" by Jim Collins and Jerry Porras in their famous study *Built to Last.* Visionary companies are characterized by the fact that they are not focused on "maximizing profitability" (1997, p. 149). Nevertheless, they are much more successful in terms of their stock market value than their more profit-oriented peers (ibid, 4 f). Their success is based on the driving force of their respective core values. According to the founder's son Thomas Watson Jr., the core values ("basic beliefs") were never questioned despite all the changes encountered at his company (ibid, 74 f).

In fact, throughout its decades-long history, IBM has always been able to successfully overcome any crisis and the associated strategic realignment thanks to the upheld core values (ibid, 244; cf. IBM, 2023). Since the DNA of IBM's culture is based on a strong customer and employee orientation and innovation in solving customer problems, rather than on specific products, it has always been capable of reinventing itself in terms of contents.

2.1.2 Well-Being and Customer Orientation

I myself can confirm that I felt well received and respected at IBM right from the start. Two-thirds of my 18-month trainee period consisted of full-week courses, most of which took place at the well-equipped training center in Herrenberg near Stuttgart and were often concluded with exams. The comfortable accommodation and the range of meals, including vegetarian food, contributed to the well-being of the participants. In addition, a good group spirit quickly developed in the classes of trainees. There was no obligation to compensate for the expensive training with years of commitment to IBM.

The trainee program was equivalent to a full course of study for me. After graduating, I was offered the position of key account manager at Hoechst (now Aventis). I was Mr. IBM there, so to speak, who had to satisfy all kinds of inquiries about IT on site.

With regard to my understanding of ethics in companies at that time, I learned how the "basic beliefs" were actually also lived, i.e., that corporate principles do not have to be paper tigers. However, it sometimes takes reminding to overcome bureaucratic inertia. For example, at his request, I once provided the head of research at Hoechst with the latest PC from IBM for testing purposes. However, he was missing the announced innovative mouse as an input device. He regretted that very much, since he particularly wanted to test the performance in conjunction with it. So I started a long search for a mouse, which nobody in Germany could or wanted to provide at that time. Also, my last inquiry at the IBM head office failed at first. Only my reference to the IBM slogan "Everything and everyone for the customer" activated the ambition of the employee I asked. It took less than 2 h before she proudly informed me that she had been able to find a mouse for the customer in Paris. Two days later, I received the desired input device via internal mail.

In the new position, I was taught even more principles in dealing with customers: We should only bring about a customer purchase decision through the quality of the problem solution and avoid any other form of influence. This also included never speaking negatively about competitor products or their manufacturers. We should even value the competition, because without it, we would become sluggish and lazy in our thinking. Our innovative strength lives precisely from competition. Without it, I realized years later that there can be no fair economy.

On one occasion, I arranged for the head of IT of the insurance department at Hoechst to use complex software from a third-party provider that ran on IBM hardware. It was supposed to display all insurance solutions and the associated premium and claims volumes from primary insurance and reinsurance. For days, I sat with the customer at the vendor's site to test out the product's capabilities. In the process, I gained my first insight into the complexity of managing risks in companies through insurance services. At the time, I had no idea how useful this knowledge would become. A short time later, I received a personal offer from the owner of the large industrial insurer Gerling for a new position. At the job interview with him, I was able to present myself as no longer completely clueless, at least in terms of insurance terminology.

2.2 *Phase of Transition from IBM to Gerling*

2.2.1 An Idea Was Born

How did the invitation come about? Günter Neske, the publisher of my dissertation, mentioned in passing during a meeting that he knew the owner of the Gerling Insurance Group very well. His son and heir had just published a dissertation on risk consulting in industrial insurance. With this keyword, the idea of a Gerling Academy suddenly appeared in my mind's eye, which would convince both employees and clients of the need for more ethical and ecological responsibility in business. No one would dismiss the recommendations of an established insurer as "green" nonsense, but would recognize in them an economic self-interest: After all, they would have to pay in the event of a claim. Neske was immediately enthusiastic about the idea and asked me to write a concept. That very night I wrote four pages on the idea. He forwarded them to Hans Gerling. It wasn't until weeks later, after several preliminary meetings with top managers, that I was invited to Zurich for an interview.

2.2.2 Doubts About the Change

The preliminary talks made a good impression on me at first. However, I had my doubts after reading Günter Wallraff's massive criticism of the Group. In 1973, Wallraff worked undercover under a false name as a porter and messenger at the headquarters in Cologne. Together with Bernt Engelmann, he published the abuses he uncovered, especially in the treatment of employees on the lower levels of the hierarchy—long working hours, low pay, intimidation by superiors—he published under the title *You up there—us down there* (original: *Ihr da oben—Wir da unten*) (1978, pp. 207–258). Using concrete examples, Wallraff showed how the working atmosphere was experienced as very stressful for those "down there." Above all, the paradoxical behavior on the part of superiors and including Hans Gerling caused insecurity and fear. People often did not know what was expected of them at certain moments and how their own behavior would be judged from above. In the lower ranks, one could hardly speak of well-being.

Obviously, conditions in many German companies were hardly any better in the 1960s and 1970s. This was confirmed in numerous other investigative reports by Günter Wallraff (1970, 1972). Strictly hierarchical, efficiency-oriented management prevailed in the German economy, albeit under the guise of a "new management style," as proclaimed by the *Academy for Managers* ("Akademie für Führungskräfte") in Bad Harzburg. Hundreds of thousands of managers from more than 2000 German companies have been trained there since its founding in 1956 (Chapoutot, 2021, p. 88 and 116). The founder of the academy was Reinhard Höhn—a qualified lawyer and professor with a relevant Nazi past (ibid, p. 88). He promoted a new "elastic warfare" in business (ibid, p. 103), namely, a "management by delegation of responsibility" (ibid, p. 109). His so-called Harzburg Model meant for subordinates "freedom ... in the choice of means, never in the choice of ends." Operating procedures were no longer to be authoritarian, but still hierarchical. The boss was not to give detailed instructions on how to achieve the set goals, but was to monitor, control, and evaluate the results. According to Höhn, the "command hierarchy" should be transformed into a "responsibility hierarchy" (ibid, p. 116).

But the consequences were dire. Employees who chose the wrong means could easily be disqualified by those who set the goals. Those "up there" did not feel bound by their own targets—according to the Roman saying "qoud licet iovi non licet bovi" (What is allowed to Jupiter is not allowed to the ox). For example, Wallraff described how at Gerling a newly hired doorman was quickly fired because he took the prescribed access control too seriously and demanded identification from one of the top managers (1978, p. 219).

It was clear to me that this kind of arbitrariness, which the higher-ups repeatedly demonstrated to those below them, would create a culture of fear in the company. I would not feel comfortable in such a company. It would be very different from what I had experienced at IBM.

2.2.3 A New Challenge

On the other hand, in 1987, the described working climate dated back more than 14 years. Much had changed since then. As a result of Wallraff's publication, Hans Gerling had initiated the correction of some grievances (cf. Engelmann and Wallraff, 286 ff). By the way, the Harzburg leadership model had been exposed as "misleading" by the end of the 1970s at the latest. For it led the promised freedom "into a certain alienation." When goals were not achieved, it was no longer the superiors who were responsible, but the managers. They had chosen the wrong means and thus failed. They then had to serve as scapegoats for their superiors. "States of anxiety and exhaustion, 'burnout' as well as the 'bore-out syndrome,' these 'psychosocial symptoms' were a form of 'internal dismissal'" (Chapoutot, 2021, p. 118, cf. 138).

In addition, Hans Gerling had in the meantime lost a majority stake in his Group in 1974. The private bank Herstatt, in which he held an 85% stake, had gone bankrupt as a result of speculative currency transactions—the first bank crash in German postwar history. Gerling was forced to sell 51% of his company to satisfy the claims of creditors and the lost savings of bank customers. Legally, he would not have had to do so. However, he explicitly acknowledged his "social responsibility" (Gerling, 1974, p. 34), which left him no other choice. In 1986, he succeeded in buying back all the shares in his Group. As a result, he was prominently featured in a cover story in the May issue of *Manager Magazin* as "the man who never gives up" (Hoffmann & Poweleit, 1987, pp. 228–243). Gerling's portrait on the cover looked out at me for weeks at my workplace at IBM. It was during this time that I had my first preliminary discussions with Gerling's top managers. Uwe Volker-Bilitza, the new Personnel Director and member of the Holding Board, offered me the prospect of contributing to the "mental framework" and cultural transformation of the Gerling Group.

The announced meeting with Hans Gerling himself finally took place in July 1987. Hans Gerling was very open to my idea of a Gerling Academy. After an hour, his son Rolf joined us. Although his main focus at the time was on his second-degree in depth psychology at the C.G. Jung Institute, he recognized the need for his father's insurance group to deal with ecological risks. The future owner's ideas about holistic risk consulting overlapped with my vision of an academy. There was a resonance, both personally and in terms of content. In addition to economic responsibility, the group should also assume ethical and ecological responsibility. As a new father of a daughter, he also felt a personal responsibility to leave a better world for the next generation. This also coincided with my own aspirations. The conditions for a fruitful collaboration seemed to be in place.

Despite the promising job offer, it was not easy for me to say goodbye to IBM, especially since I had to call my superiors up to the highest level in the region to persuade me to stay. When I handed in my resignation, they assured me of their goodwill because they had lost an employee but gained a potential customer.

2.3 At Gerling Group

2.3.1 First Experience

I started work at the company's head office in Cologne on October 1, 1987. Initially, I developed the "Conferences and Publications" department independently in the newly founded Gerling Risk Consulting Group. Only later, I was promised, would it be possible to spin it off into a Gerling Academy. Even in the first few weeks, I noticed major cultural differences between IBM and Gerling. Hierarchies were much less important there than here. There, business success was what counted, whereas here, closeness to the owner was important.

The associated hierarchy was reflected outwardly in the architecture, which was a defining feature of the Cologne headquarters for more than 3500 employees. The buildings from different eras radiated powerfully across an entire city district. Visitors were impressed when they entered the various lobbies for the first time and were greeted by liveried porters. Reactions ranged from intimidation to resentment and admiration to awe. The size and furnishings of the offices were carefully balanced according to the employee's position. The higher up in the hierarchy, the larger the room, the desk, the visitors' corner, and the more valuable the furnishings, such as original paintings from Hans Gerling's wife's gallery. Hans Gerling's office was legendary. It was 7 m high and covered more than 200 m^2. In the center stood a noble desk upholstered in white leather. Günter Wallraff had once disgraced it when, at the end of his research, he sat down on it and played with the golden Gerling globe like Charlie Chaplin in his movie "The Great Dictator."

The loyalty of the entrepreneur to his employees was generally very high, as was the loyalty of the employees to the entrepreneur, especially since Hans Gerling became more actively involved in their welfare after the Wallraff affair and the Herstatt crisis. He gave the personnel director free rein to create a better and fairer working atmosphere.

Changing old patterns was not easy. Hierarchies were still very important and omnipresent, as I was soon to discover. Since my employment had been initiated by the owner, my rank could not be too low, and so I was given the title of "Prokurist" for a probationary period until the promised "department head" a year later. This included the privilege of access to the canteen for authorized officers. There was the Century Hall for clerks and the Garden Casino for directors and board members. Members of the holding company's board of directors had their own private rooms with personal service. The long-serving sommelier, master of tens of thousands of bottles of wine, recommended the right choice for the appropriate menu.

Because of my obvious closeness to the owner and his family, I was often invited into these "sacred" rooms. In general, people were careful with me because no one knew exactly how close I really was to the Gerlings. My special position gave me easy access to all levels of the hierarchy. This made it easier to analyze their behavior. In any case, I no longer encountered the arrogant, self-important behavior that Wallraff had described. A beneficial generational change had already taken place.

Many were pleased with the spirit of optimism and supported the idea of a new kind of risk consultancy and the planned activities of a future academy. These included a conference on environmental liability (Gückelhorn & Steger, 1988) and the publication of a two-volume anthology of interdisciplinary risk research, Risk and Venture ("Risiko und Wagnis," Schüz, 1990), a selection of which was published in English years later (Obermeier & Schuez, 1996). However, there were also skeptics, even resisters, who feared innovations because they might cause them to lose their privileges. Accordingly, they tried to create a negative atmosphere against our work.

It was not least because of such ambivalent influences on my work that Rolf Gerling decided to pursue the founding of the Academy in Zurich, where he had his private residence. A Swiss work permit for me had to be obtained with good reasons and an influential lawyer. In Otto-Peter Obermeier, we found another qualified person who was enthusiastic about the idea of the Academy. Together we founded the Gerling Academy for Risk Research as a Swiss corporation in 1990, after the anthology had been published. Three years later, we founded the Gerling Academy Press (now Murmann Verlag).

2.3.2 Change of Generation

It was fortunate for Rolf Gerling that the Academy was not founded later, as his father passed away just 1 year thereafter. This forced him to assume management responsibility for the entire Group. For him, the Academy was now a think tank with which he could consult independently of corporate interests. Unlike his father, he did not want to run the company strategically or even operationally. He left that to the holding company under the direction of a hired CEO. However, as the new chairman of the board, he reserved the right to shape a new corporate culture and the freedom to choose top executives.

He understood that this was a fundamental paradigm shift for the company. After all, the existing culture was tailored to the person of Hans Gerling. Without this figurehead, to whom every employee could relate directly or indirectly, employees lacked orientation in the face of all the uncertainties surrounding concrete decisions. Rolf therefore suggested formulating a written vision and mission that would define the company's future. A corresponding set of values would provide guidance on how to implement them in day-to-day operations. In other words, he wanted to put the desired culture on paper.

2.3.3 Steps for Cultural Change

Together with Obermeier and me, he took on this task. First, we examined what the latest management research had to say about the importance of corporate culture for business success, which had only been discovered in the 1980s (cf. Ouchi, 1981; Bate, 1994; Schein, 2004). It was clear to us that cultural change is a long process that cannot be prescribed. Behavioral patterns have been ingrained in an organization for too long to simply be unhinged by a piece of paper. The desired vision, mission, and values may be proposed by the entrepreneur, but they should be supported—at least by middle and senior managers. Therefore, a first draft was presented to a selection of them, discussed, and suggestions for improvement solicited. Many controversial discussions led to a final version, which was then made available to all employees in printed form. At the same time, we invited managers at all levels to 1- or 2-day workshops where the owner and the other two authors presented and discussed the desired culture.

These were the first steps in a lengthy process that would give the entire company a new direction. The key to implementation, however, was the owner's commitment to provide the resources to make it happen. Above all, it was about a corporate responsibility that was not only economic, but also social and ecological—expressed through the three core values of "performance, behavior and creativity," meaning "commitment to performance and success," "respect for people and nature," and "joy of beauty and creativity," each of which was specified by additional values (see Fig. 1).

These values formed the basis for how the mission was to be carried out: "Our mission is: to provide the best possible comprehensive risk management and

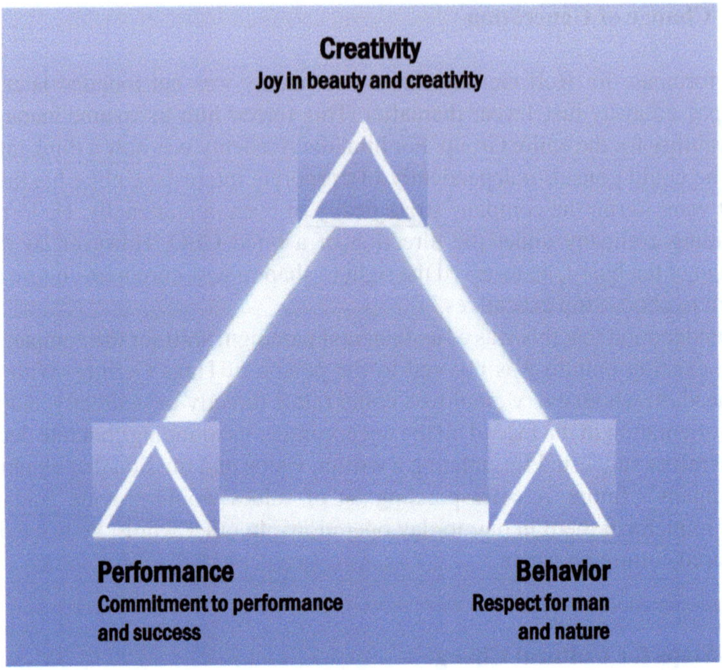

Fig. 1 The three main responsibilities demanded by Gerling Corporate Culture

insurance services in dialogue with our clients, to shape an exceptional and responsible company, and to work in harmony with creation, for the good of man and nature." (Gerling 1995, 2000).

Neither the mission nor the corporate values were chosen arbitrarily. Ultimately, they also reflect the entrepreneurial intentions of the late Hans Gerling. The aim was to preserve the continuity of the Group's history while at the same time accepting the challenges of the new era. Even then, the focus was clearly also on responsibility for future generations, i.e., "sustainability"—a new concept at that time. It was first introduced to the general public in the UN's Agenda 21 and presented and discussed at the first environmental conference in Rio de Janeiro in 1992.

After almost 2 years of preparation and coordination and numerous workshops with employees, the first version of the culture brochure was published in 1995—a second revised version was published in 2000—and made available to all employees. The workshops gave all employees the opportunity and resources to contribute their own ideas. The Human Resources Department promoted cultural projects as a top priority.

For example, the department that manages the canteen at the company's headquarters in Cologne, which feeds 3500 employees every day, implemented an unusual idea to provide regional and organic food. They contracted 40 local organic farmers to deliver their seasonal produce daily. The question was what to do with the leftover food. They resold them to the people of Cologne at cost price in their own organic food store.

Many other ideas for sustainable responsibility (e.g., in the design of insurance products) were implemented internally at Gerling. The external impact was so great that it had a positive effect on recruiting. Younger employees were enthusiastic about the opportunity to contribute to the preservation of the social and ecological environment through their own initiatives. In sustainability ratings such as those from OEKOM or SAM (now RobecoSAM), the Gerling Group was assessed as "best practice" among insurers (No. 2: Swiss Re). SAM evaluated it in regard with economic, social and ecological sustainability.

2.3.4 Corporate University

Overall, the Corporate University ("Marienburger Seminare") for middle and senior managers, which was founded in 1990 and for which I was primarily responsible, also contributed to a change in culture. The private estate of Rolf's parents was converted into an educational facility in Cologne specifically for this purpose. Hundreds of managers had to choose at least two events from the multi-day seminar program. In the most recent edition in 2003, 100 events were offered on 54 different seminar topics (with 61 internal and external lecturers) ranging from insurance, value, general management, behavior, leadership, and communication seminars to needs-based trainings (Gerling, 2003). Self-leadership and a corporate management game were also important and very popular topics for more than a decade. From the beginning in 1990, I also offered seminars on business ethics. In the process, I gradually developed the first approaches to risk ethics, which were published as a monograph under the title *Werte – Risiko – Verantwortung. Dimensionen des Value Managements* (1999) published by Gerling Academy Press.

2.3.5 Hinderances for Realizing Ethical Culture

Over the years, I have discovered why the noble goals of the desired corporate culture have not been unconditionally accepted by every employee or manager. Here are some of the reasons:[1]

1. Managers are people of different characters. Not all of them are motivated by ideals, but by greed for power, prestige, and money. Such personalities, when in powerful positions, can make decisions with devastating consequences for a company. In 1997, for example, Rolf Gerling was persuaded by a member of the holding management board to acquire a reinsurer in the United States. His seductive due diligence report, as it later turned out, emphasized the opportuni-

[1] Cf. my *Applied Business Ethics*, where I listed 18 other reasons why employees in organizations can lose their ethics competence (2019, 172 ff).

ties and downplayed the risks. Whatever his hidden motive was, the billion-dollar purchase was the beginning of the end for the privately held group a few years later. The company had a large package of undisclosed risks (notably asbestos) in its portfolio. Combined with other insured losses, such as the attacks on the World Trade Center in New York and the stock market crash in the following years, this led to losses running into billions, forcing Rolf Gerling to sell the company in 2005 at a substantial devaluation.

2. Not all of them are mature personalities who can deal constructively with their shadow sides. Some live out their narcissism, Machiavellianism, or psychopathy, known in social psychology as the "dark triad," at the expense of their employees. In some departments, there has been bullying from the top, but also from the bottom—a challenge for the HR department to contain such excesses.[2]
3. Especially among the older generation of leaders, there were some who secretly opposed the new culture. They were uncomfortable with the idea of giving up their old privileges in their positions. Such personalities were rather toxic for the implementation of ethical leadership. Officially, the kind of behavior Wallraff had denounced was no longer acceptable. As a result, the privilege of hierarchical canteens was abolished immediately after Rolf Gerling took over the Group. The desired flattening of hierarchies still had its opponents.
4. On the other hand, hierarchies are also built from the bottom up. I myself once experienced how we were walking out of the main entrance with Rolf Gerling when the porter ran ahead of us and cursed an employee because he had cleaned the street with water. He shouted that was unacceptable when the boss walked by. Rolf had to calm him down and tell him that it was not necessary to interrupt someone's work because of him. Either the porter was still stuck in the old culture of fear, or he wanted to exercise his small, momentary power over someone else in this way.

2.3.6 Main Learnings

Looking back, 16 years of work at the Gerling Academy taught me essential lessons about the importance of ethics as an "*art of getting along well*" (Schüz, 2023, p. 538) in an organization. An artist should consider the possible consequences as well as the commitment to accepted rules and values. However, depending on the situation and the abilites of the actors, the outcome can be very different.

In particular, the prevailing group dynamics, leadership style, and organizational form[3] help determine whether the culture evolves as more "adaptive" to stakeholder concerns or as "non-adaptive".[4] Overall, the desired ethical culture was implemented

[2] Several of the experiences I have had with the associated behaviors have been described in our Lügen in der Chefetage (Schüz et al., 2007).

[3] Such experiences provided me with my LOGIC scheme as a structured consideration of corporate values or vices; see Schüz (2021, p. 229; 2019, 209 f; cf. 1999, p. 186).

[4] This is the distinction made by Kotter and Heskett (1992, 44 ff).

very well. Internal anonymous surveys attested to the company's well-being and the commitment of its employees—notwithstanding the few pockets of resistance.

However, they were so dangerous for the development of the company because they pretended to be loyal on the outside world, but cultivated their selfishness on the inside. Here, the *Performance Culture Matrix* of Jack Welch, the former CEO of General Electric (GE), should have been taken into account: Ideally, you should promote employees who are "high performers" with a "high cultural fit." Low performers with a low cultural fit, on the other hand, should be dismissed. But many companies, like Gerling, keep the high performers with low cultural fit. GE became strong in the 1980s and 1990s by firing those individuals.[5] This approach would have saved Gerling a lot, perhaps even Rolf Gerling's personal defeat. Instead, it took away the well-being of thousands of employees over the course of their working lives.

3 Academic Research and Teaching

Without almost 20 years of experience in companies, I would not have been able to develop my more than 20 years of teaching at universities in its current form. The practice taught me to better bridge the gap between textbook knowledge referring to hundreds of relevant publications in the field of business ethics and the culture really lived in companies. The challenge was to develop a didactic approach that would engage young business students and, more importantly, employees who were pursuing their MBA through continuing education. For me, this meant speaking their language and connecting with their way of thinking while at the same time also conveying the theoretical and philosophical foundations of ethics.

Interestingly, when I started working at the Zurich University of Applied Sciences (ZHAW) in 2006, I was primarily employed as a senior lecturer in *strategic management*. Here, I had to familiarize myself with the relevant literature and convey the content in a memorable way, especially to the bachelor students. I realized that ethics is an important part of strategic management decisions. If I do not take *stakeholder reactions* into account, I provoke their resistance and the transaction costs increase considerably, as the example of Gerling showed after Wallraff's research.

3.1 Implementing Ethics in a Business School

The first opportunity to teach business ethics as a stand-alone subject was in the form of an elective module, i.e., students could voluntarily choose it as a major before graduation. The fact that ethics was suddenly integrated as a crosscutting theme in most of the programs was due to the efforts of our dean at the time to

[5] CEO Today (2020).

obtain accreditation from the *Association to Advance Collegiate Schools of Business* (AACSB). This association first asked what the *School of Management and Law*'s attitude to ethics was. The answer initially left the impression that ethics was more a secondary issue for a business school. When the organization expressed its great skepticism, the dean remembered my elective. After that, my topic was in great demand and had to be implemented accordingly. This was the beginning of my personal triumphal procession, during which I have been able to familiarize and even inspire thousands of students with the ideas of ethics in business over the course of more than 18 years. However, this was only possible with a healthy mix of theory, practice and application, as well as lively, authentic teaching.

3.2 A New Model for Ethical Responsibility

For this purpose, a simple model had to be developed that established the link between responsibility and ethics (cf. Schüz, 1999, 183; 2012). I finally succeeded in doing so by defining the triangular relationship of responsibility (as a so-called three-point relationship) between the responsible subject (individual or company), who has to answer for the manifold consequences of its activities, and the affected stakeholders (see Fig. 2). Around the three points of the responsibility triangle, the

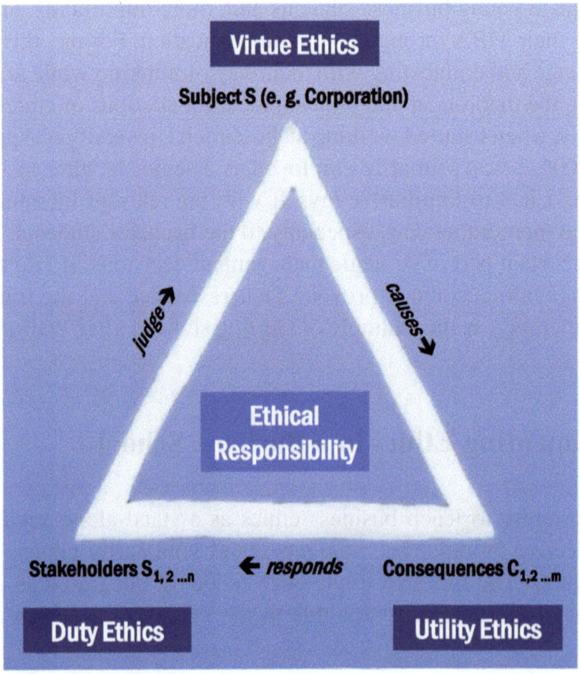

Fig. 2 Ethical responsibility as a three-point relation

three most important ethical approaches of philosophy can be plausibly assigned: utilitarian ethics examines the consequences, duty ethics establishes the rules, and virtue ethics tries to influence the character of the individual or the culture of the company for the better. In this way, all three approaches complement and control each other -like a separation of powers in ethics - and, in a sense, form the tools for the ethical evaluation of an action.

3.3 Economics of Fair Exchange

Of course, the didactic tools also include Socratic dialog, which is used to awaken the participants' awareness of the problem. I usually start my seminars by asking about the nature of economic activity. The most common answer is profit maximization, followed by the question of where the profit comes from. Rarely does anyone immediately come up with the idea that it is achieved through an exchange on the market. On closer inspection, however, exchange is only good if all exchange partners make a profit, i.e., when there is a balance of give and take. Profit maximization would consistently mean that I take everything and give nothing.

This brings the problem into focus: None of the participants would like to be ripped off. But are they willing to do the other person no wrong? This is where the golden rule of ethics comes into play: "Treat others as you would like to be treated." While most agree with this, they must admit that the desired middle ground of a fair exchange between robbery and gift is difficult to achieve when dealing with long supply chains. They quickly realize that the so-called value chain of a smartphone, for example, is actually made up of robberies along its individual links and is thus actually a chain of value destruction for many exchange participants. This actually starts with the fact that nature, which, for example, provides rare earths in the Congo, gets nothing back except its own destruction. Likewise, the countless child slaves are treated unfairly. As end consumers, we hardly care about this and thus become part of the robbery ourselves (see Fig. 3).

Years of experience in teaching business ethics had enabled me to create a new foundation for how to gradually replace *"robber economics,"* as I call it, with the idea of *"economics as the art of fair exchange"*, conceived by Aristotle (cf. 2009, 1132b10-20; Schüz 2023a, 525f; 2023b, 3451). It leads finally to a *circular economy* - nowadays, in demand worldwide. Making people aware of the risks of robbing nature and humanity is the first step, the next is *to avoid* them, and if that doesn't work, *to reduce* and *compensate* them. If everyone does his or her best in this regard, perhaps an even better world could be created in which nature and its countless forms of life are treated fairly as partners in exchange.

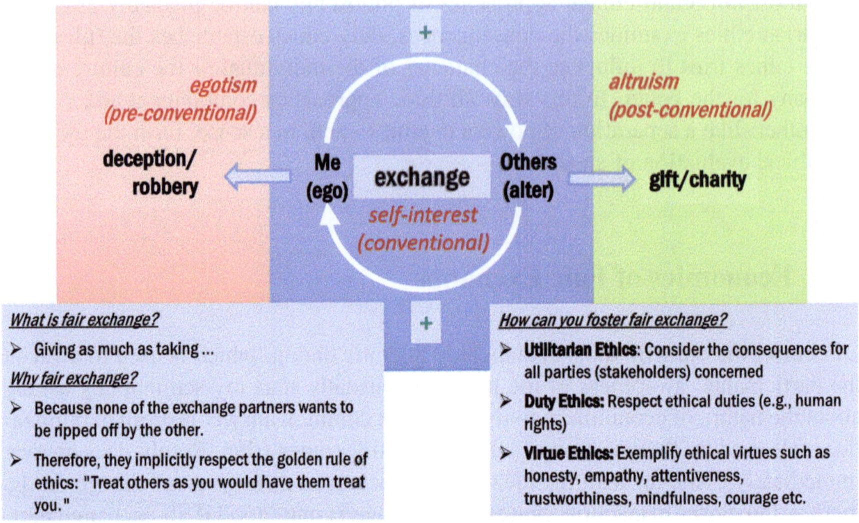

Fig. 3 Fair exchange (economics) as the middle ground between robbery and charity

4 Conclusion

My decades of experience in teaching and implementing ethics in the corporate world have shown me that the ethical commitment on the part of participants in my seminars depends on how they themselves feel about being affected by unethical actions. The VW emissions scandal showed this clearly. The actors who installed cheating software in the emissions control system were driven to do so by a culture of fear that prevailed in the organization. Those who are afraid of their superiors have their well-being massively disturbed. In order to restore mental equilibrium, one does everything possible to eliminate the fear-inducing moment, in other words: to do everything that the superior demands. Failure must be avoided. The psychological pressure exerted by their superiors to find a solution to the stringent emission tests in the United States robbed them of the ethical competence to even consider the side effects, repercussions, and long-range effects of their fraudulent actions on the numerous stakeholders (Schüz, 2021, 271 ff). They became victims of the ubiquitous tunnel vision triggered by a one-sided shareholder value thinking and an authoritarian management culture (cf. Ewing, 2017). Those who do not perceive the social and environmental impacts cannot take precautions.

The implication of fear as the main driver of unethical behavior shows something else. Those who are free of fear may not be as greedy for profit maximization. After all, greed is rooted in a kind of fear, namely, not getting enough, e.g., in comparison to others. Consequently, freedom from fear means being at peace with oneself and others, i.e., getting along well with one's social environment. This *well-being* is reciprocal. Only when others feel well in an environment I can feel equally well. In a sense, well-being is contagious. Conversely, a healthy, natural environment reinforces well-being in the individual. Seen in this way, well-being is an indicator of a systemic balance between my inner and outer worlds.

It is not the place here to elaborate the complementarity of the three ethical approaches mentioned above and their applications under real conditions. The following diagram summarizes this and provides three *test questions* as a first therapeutic indication. They can be used to test whether a proposed decision can be considered ethically acceptable. Each of these questions also tests the well-being of the individual. Those who (a) trade fairly do not have a guilty conscience; those who (b) are publicly pilloried do not feel well; likewise, those who (c) feel compassion for an injured party by putting themselves into their shoes cannot really feel well because they are sharing their pain (see Fig. 4).

I also used such test questions in several management workshops at the Swiss Post. The company had to deal with a scandal in its Postbus division after it was discovered that it had received unjustified subsidies from the state for more than 10 years. Those involved were happy to take the profits, suppressing their guilty conscience in the process with the help of the "diffusion of responsibility" (Schüz, 2019, p. 175): The more involved an individual is, the less they feel called to intervene. Well-being through unjust profit was thus deceptive. Because after it became known, the perpetrators were put under massive pressure, had to accept dismissals, or even face criminal charges (Baur, 2018; Schüz, 2021, p. 262). The well-being of all involved was destroyed once and for all by the loss of reputation. Incidentally, the entire Swiss postal service subsequently adopted the meaning of the test

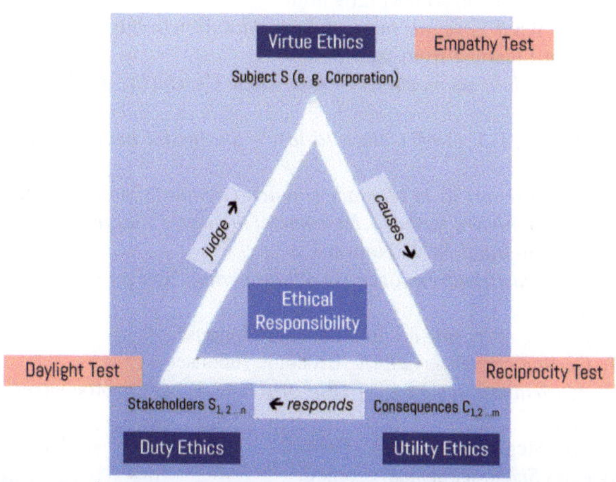

Utilitarian Ethics / Teleological Ethics	Reciprocity Test	Does the decision guarantee fair exchange conditions along the entire value chain?
Duty Ethics / Deontological Ethics	Daylight Test	Can the decision be made public (among friends, in the media, etc.)?
Virtue Ethics / Aretological Ethics	Empathy Test	How would I feel about being affected by the decision?

Fig. 4 Ethics approaches with test questions

questions in its "code of conduct" (Swiss Post, 2019). They also serve as a search pattern for the so-called skeletons in the closet. Detailed derivations, applications, and justifications for the three ethical approaches are presented in my textbook (Schüz, 2021).

My decades of experience in both business and academia have taught me that by appropriately communicating and justifying ethics in practice, many managers, as well as ordinary employees, are willing to align their economic actions more responsibly for the benefit of society and nature. Time and again, I receive feedback to this effect from former seminar participants. Many of them were even able to convince their bosses to adopt ethical business practices and implement them in their companies. I also hear the resigned words: "As a small individual, what can I do?" I reply that someone who walks slowly like a snail is still infinitely faster than someone who doesn't move at all. I am convinced of this myself: If everyone takes on just a small part of the responsibility for the whole, it could lead to the salvation of the whole.

References

Aristotle, (2009). Nicomachean Ethics. Oxford University Press.

Bate, P. (1994). *Strategies for cultural change*. Butterworth-Heinemann.

Baur, T. (2018). Wir brauchen eine neue Firmen-Kultur. Interview, September 19, 2018. *Blick*. https://www.blick.ch/wirtschaft/wir-brauchen-eine-neue-firmen-kultur-postauto-chef-baur-sagt-seinen-chauffeuren-sorry-id8877269.html

CEO Today. (2020). *What have we learned from Jack Welch*. https://www.ceotodaymagazine.com/2020/03/what-i-learned-from-jack-welch/

Chapoutot, J. (2021). *Gehorsam macht frei – Eine kurze Geschichte des Managements von Hitler bis heute*. Ullstein.

Collins, J. C., & Porras, J. I. (1997). *Built to last – Successful habits of visionary companies*. Harper Busines.

Descartes, R. (1997). *Discours de la Méthode (French-German)* (2nd ed.). Meiner.

Ewing, J. (2017). *Faster, higher, farther – How one of the world's largest automakers committed a massive and stunning fraud*. W.W. Norton.

Gerling, H. (1974). *Mein Privatvermögen ist relativ gering*. Spiegel Interview (Der Spiegel No. 32, pp. 31–34).

Gerling. (1995). *Gerling corporate culture*. Brochure Inhouse.

Gerling. (2000). *Gerling corporate culture*. Brochure Inhouse.

Gerling (2003). *Marienburg seminars – Our program*. Program brochure (English-German), Inhouse, Cologne.

Gückelhorn, H., & Steger, U. (1988). *Umwelt – Unternehmen – Haftung. Strategien zur Bewältigung eines Schlüsselthemas*. ecomed.

Hoffmann, K., & Poweleit, M. (1987, May). *Hans Gerling greift wieder an: Der Mann, der nicht aufgibt*. Manager Magazin No. 5 (pp. 228–243).

IBM. (2023). *A business and its beliefs*. Harper Business. https://www.ibm.com/ibm/history/ibm100/us/en/icons/bizbeliefs/

Kotter, J., & Heskett, J. (1992). *Corporate culture and performance*. The Free Press.

Obermeier, O. P., & Schuez, M. (1996). *Risk – Paradise lost or gained? A holistic approach to the environment*. Gardner Press.

Olwell. (2021). *The original story of "Think."* https://www.discerningreaders.com/watson-sr-original-story-think.html

Ouchi, W. G. (1981). *Theory Z – How American business can meet the Japanese challenge* (Vol. 24, p. 82). Avon.
Schein, E. H. (2004). *Organizational culture and leadership*. Wiley.
Schüz, M. (1986). *Die Einheit des Wirklichen – Carl Friedrich v. Weizsäckers Denkweg*. Neske.
Schüz, M. (1990). *Risiko und Wagnis – Die Herausforderung der industriellen Welt* (Vol. 1 and 2). Neske.
Schüz, M. (1999). *Werte – Risiko – Verantwortung. Dimensionen des value managements*. Gerling Akademie Verlag.
Schüz, M. (2012). *Sustainable corporate responsibility – The foundation of successful business in the new millenium*. Central Europe business review, no. 2, Prague 2012 (pp. 7–15). https://cebr.vse.cz/pdfs/cbr/2012/02/02.pdf
Schüz, M. (2019). *Applied business ethics – Foundations for study and practice*. World Scientific Publishing.
Schüz, M. (2021). *Angewandte Unternehmensethik – Grundlagen für Studium und Praxis* (2nd actualized ed.). Pearson.
Schüz, M. (2023a). Developing sustainably responsible strategies in business. In P. Molthan-Hill (Ed.), *Sustainable management – A complete guide for faculty and students* (3rd ed., pp. 525–558). Routledge.
Schüz, M. (2023b): 3. «Sustainable Corporate Responsibility (SCR)», in: Idowu, S. O., Schmidpeter, R. (eds-in-chief.): Encyclopedia of Sustainable Management, Vol. 4, Cham: Springer Nature, pp. 3450–3469.
Schüz, M. (2025). Transdisciplinary management of polycrises – A philosophical approach. In M. Schüz (Ed.), *Transdisciplinary thinking and acting – Fundamentals and perspectives*. Springer.
Schüz, M., Wirth, S., & Bode, A. (2007). *Lügen in der Chefetage. Gesammelte Unwahrheiten aus dem Management*. Wiley.
Swiss Post. (2019). *Our code of conduct – A clear stance*. https://www.post.ch/-/media/post/ueber-uns/dokumente/verhaltenskodex.pdf?la=en
UNEP. (2023). *What you need to know about nature crisis*. https://www.unep.org/facts-about-nature-crisis
Wallraff, G. (1970). *Industriereportagen – Als Arbeiter in deutschen Großbetrieben*. Rowohlt.
Wallraff, G. (1972). *13 Unerwünschte Reportagen*. Kiepenheuer & Witsch.
Wallraff, G., & Engelmann, B. (1978). *Ihr da oben – wir da unten*. Rowohlt Hamburg.

Mathias Johannes Christof Schüz is Professor Emeritus for International Business and Responsible Leadership with exploratory focus on corporate responsibility and business ethics at the at ZHAW (*Zurich University of Applied Sciences*), Switzerland. He studied physics, philosophy, and pedagogics and did his doctorate on philosophical consequences of quantum physics. He started his career in business as a trainee and key account manager at IBM. Together with the insurance entrepreneur Dr. Rolf Gerling, he co-initiated the Gerling Academy for Risk Research in Zurich and ran it for 14 years before becoming a senior lecturer at various universities and finally professor at ZHAW. He has published numerous articles and books, such as one outlining a new philosophy of economy in German: *Werte – Risiko – Verantwortung. Dimensionen des Value Managements* [Values, Risk, Responsibility: Dimensions of the Value Management] (Munich, Germany: Gerling Akademie Verlag, 1999); another, published with two co-authors, deals with lies in management: *Lügen in der Chefetage* [Lies on the Executive Floor] (Weinheim, Germany: Wiley, 2007). His textbook *Applied Business Ethics – Foundations for Study and Practice* was published in 2019 (Singapore/London/New Jersey: World Scientific Publishing), while the German original appeared in a second, expanded edition at Pearson in 2021 under the title *Angewandte Unternehmensethik – Grundlagen für Studium und Praxis*. As an emeritus professor, he is still a visiting faculty member at the ZHAW and gives guest lectures and sessions at numerous universities internationally. Being a transdisciplinary researcher, he is co-editor of a new book series published by Springer-Nature titled *Transdisciplinary Management of Social and Ecological Crises* (also editor of the first volume to be published in 2025).

Part V
Real-Life Stories: Individual Experience and Subjective Perception in the Context of Sustainable Business Transformation and Well-Being

The Skull in Our Suitcase: What We Can Learn from Zen in Challenging Moments

Bart Weetjens

Abstract It is relatively easy to take good care of oneself in times when things are going well. But what about managing your wellbeing in challenging moments? The author of this chapter is a Zen monk and serial social entrepreneur. As one of the co-leads of The Wellbeing Project, an initiative that aims at catalyzing a culture of inner wellbeing for changemakers, he has been investigating the relevance of inner work for personal wellbeing in the light of his rich experience. In this chapter he tackles the potential threats behind the motivation to become a social entrepreneur, outlines why the emergence of The Wellbeing Project has been relevant for the changemaker sector, and shares a personal story of stress, anxiety, and despair to illustrate how Zen practice qualifies to overcome even the most challenging moments in our lives.

> To study the Way is to study the self.
> To study the self is to forget the self.
> To forget the self is to be confirmed by all things.
> To be confirmed by all things is to remove the barriers between one's self and others.
> – Dogen Kigen (Tanahashi, 2013)

1 Introduction

Imagine driving to the airport to fly home. When you get there, you walk in, and just 5 min later you are being arrested, handcuffed, and told that you're facing 30 years in prison. Do you feel the shock?

Now let me tell you: I don't need to imagine that, because it happened to me for real. Let me take you on a journey that led me to deep insights about how to apply Zen practice in real life and how those learnings can be applied to enhance the level of wellbeing both individually and collectively.

B. Weetjens (✉)
Antwerpen, Belgium
e-mail: hello@bartweetjens.be

M. Mühlböck (ed.), *Sustainable Transformation and Well-being*, CSR, Sustainability, Ethics & Governance,
https://doi.org/10.1007/978-3-031-75566-8_17

2 When a Nightmare Becomes Reality

It was 2015, the evening of 9 of February. I was about to travel from Tanzania, where I had lived and worked for 15 years, back to my home country of Belgium. Since 2000, Tanzania had been the home base of my initiative APOPO, a project that trains giant African pouched rats to save human lives by detecting landmines. What started as an unlikely research project evolved into a sizeable global humanitarian operation, recognized and awarded by the main social entrepreneurial networks. I was labeled a "social change leader" and my innovation was applauded publicly. Though this success served as an inspiration to many, on a personal level, it left me depleted. When I was about to move back to Belgium that evening in 2015, I was determined to let go of my executive role and move on to explore the intersection between personal wellbeing and sustainable entrepreneurial action.

It was dusty, noisy, hot, and humid, as usual in Dar Es Salaam. When my driver dropped me off shortly before dusk, I took time to take in the magnificent sunset in paramount colors. I would surely miss that back in Belgium. I strolled to the entry and pushed my suitcase through the scanner. It contained some personal belongings that I didn't want to leave in Tanzania. Among them was a skull, the human skull that had decorated my desk since I was a student. When I put it into the suitcase, I thought for a moment that I might possibly have to explain the origin of that skull at customs. But my attachment to the object, which served as a reminder of the impermanence of life to me, was too big to simply let go of. I argued to myself that my Swahili was fluent enough to explain the skull's origins, even if it was poorly documented. Moreover, I was a familiar face at Dar Es Salaam airport, since I traveled almost weekly on international flights those days, and my humanitarian work received a lot of attention. Usually, my luggage was never checked. Well, this time my suitcase was opened, and there was no room for explanations. In no time I was arrested, handcuffed, my passport confiscated, and I was moved to the airport's catacombs, where I learned that I was accused of possession and trafficking of human body parts, facing a 30-year sentence in prison. Prisons in Tanzania are not particularly a wellness resort. In just a few moments, my savoring experience of the last tropical sunset had suddenly turned into my worst nightmare.

3 My Life as a Social Entrepreneur

This wasn't the first time that my inner balance was disturbed by external factors. In fact, my 20 years of entrepreneurial rollercoaster had been a constant juggling of conflicting personal needs and expectations from the outside world. In that process I had been putting enormous pressure on myself and on my team, wanting to succeed in developing an appropriate detection technology, while it seemed that most systems in place were opposed to making that happen.

Even if we had the most peaceful intentions, that entrepreneurial journey often felt like a fight. We worked under huge operational pressure, under permanent funding scarcity, and we faced a big burden from staff turnover, lack of engagement, and burnout among our staff. The social change space is all about innovation, strategy, impact, and scale and requires an almost supernatural 24/7 performance from the social change leaders. On top of the pressure regarding performance and scale, it is taken for granted that we are "obsessed" with our idea.

But isn't obsession a psychiatric syndrome? Are we trying to solve the world's problems in the right way? Would it be possible that we unconsciously carry our anger, our fear, and our dark complexities into our real-world actions and thereby risk propelling what we are desperately trying to defeat? In other words: what is the relationship between inner development and the quality and effectiveness of our entrepreneurial actions in society?

At that time, a number of us in the field of social change were already noticing that there was a serious threat that needed to be addressed. A fellow social changemaker approached Ashoka, one of the key network organizations in social change, in this matter to begin to explore this. This initiative then led to a research project where 50 social change leaders around the world were interviewed. The results showed that the great majority of them had never made time for their own personal inner work. They had worked their fingers to the bone, while ignoring not only their personal needs, but also their family needs, and in some cases, they even ignored their organizational needs. Most of them over-identified with their work. And quite often, there were narratives of heroism, sacrifice, and martyrdom swirling around those leaders. Their reason for existence had become their work, and they would sacrifice their personal wellbeing to fill a hero role on stage. Furthermore, in many cases personal traumata had ignited their entrepreneurial journey. And while the general public was applauding the heroism, the innovation, and the societal impact, on a personal level, those leaders didn't learn healthy ways of working with trauma or let alone opening up to their personal vulnerability.

4 The Emergence of the Wellbeing Project

Although these insights revealed important trends, the data gathered was not sufficient to serve as a basis for decision-making that would consequently lead to more wellbeing and a healthier culture in the social entrepreneurship sector. However, these first learnings eventually triggered a keen interest in the sector to look deeper into the matter, which resulted in key spearheading organizations in social innovation, such as Ashoka, the Impact Hub, Synergos, the Skoll Foundation, Fetzer Institute, and Esalen Institute, to join forces and co-create an initiative called The Wellbeing Project.[1] I myself got involved in the initiative and joined the senior

[1] https://wellbeing-project.org

leadership team. This happened in 2015. From the beginning on, The Wellbeing Project has been aiming at:

- Shifting the culture in the field of social change to a more caring and compassionate one
- Responding to the huge need for personal support among social change leaders
- Researching the relationship between inner work and social change
- Exploring a required infrastructure of support within the sector

By doing so, The Wellbeing Project brought to life a different way of looking at social change, one that works from the inside outward and definitely starts by addressing the inside. Eighty seasoned social entrepreneurs, with ages ranging from 35 to 65, were selected from 45 countries worldwide and invited to take part in an 18-month high-value offer, featuring 3 retreats with outstanding facilitators, a customized personal support program, wisdom teachers, and peer learning. They went through this 1.5-year program in 4 cohorts of 20 people, in which they prioritized their inner journeys. In each cohort, researchers participated and collected data for a developmental evaluation study that explores the connection between personal development and the quality and effectiveness of social change.

The results delivered compelling evidence that there is indeed a strong relationship. The shifts seen in the participants' personal and work environments were extraordinary (Severns Guntzel & Murphy Johnson, 2020):

- In general, they became more balanced and more aware of their personal needs, which made them happier and healthier people.
- They integrated this sensibility and approach into their relationship to their work, allowing for sufficient family and recovery time.
- In their work environment, they learned to share leadership in more horizontal ways throughout the organizations and built more trust in their colleagues.
- They were able to be more present in the situation and approachable and learned to listen closely.
- They embraced emergent learning, collaboration, and holocratic models in their organizational culture.

Furthermore, that change at the leaders' personal level eventually led to more dedication among their employees, happier staff, and less staff turnover, which all contributed to the sustainability of their impact.

Those insights corroborated the hypothesis that lasting change begins within each and every single one of us and that the path toward inner wellbeing might be the most effective approach to solving the world's pressing social and environmental issues—a theory that I had intuitively felt to be valid much earlier on. It became particularly obvious to me in my darkest hours while being deprived from liberty in Tanzania. There is a quote by Sufi poet Rumi that underlines this: "Yesterday I was clever, so I wanted to change the world. Today I am wise, so I want to change myself" (Helminski, 2000).

5 The Wellbeing Benefit of Zen Practice in Challenging Situations

So back to that unforgettable night in Dar Es Salaam. I was held in custody at the airport police station, where I was sitting on a piece of cardboard on the concrete floor. I felt broken. I was stressed, anxious, and in despair, as well as being filled with anger. I was innocent, but would the Tanzanian justice system acknowledge that? The odds of spending the best part of my life in jail were extremely daunting to me. It felt like every possible negative emotional state occupied my mind simultaneously. Self-judgment was a big part of that. How could I have been so mindless, so ignorant? How stupid was I? For how long would I be isolated? How long before I would see my family again? Was there any chance for me to escape from this absurd situation?

All these thoughts, judgments, and projections made me completely NOT present, and even more stressed. By dramatizing my own situation, I descended into a universe of grief, sadness, and self-pity. I tried to control my tears, but I failed. While sitting in tears on my piece of cardboard, a wandering cockroach passed by. It stopped, turned in my direction, and stayed there immobile while moving its antenna up and down, left and right, and in all directions. I imagined it was sensing my misery. A few seconds later, it wandered off again into the dark on the other side. But at just that moment, face to face with the most unlikely of all creatures, my attention was drawn back into the present moment. Thank you, cockroach!

Where was my relaxed, content mind, the free and fluid mind that I had experienced during long Zen retreats? Where had that deeply embedded feeling of harmony, of interconnectedness gone? In my misery, I remembered my teacher's guidance: "Always stay connected to the body posture and the breathing," which is all about PRESENCE. So I tried to do that. And what did I feel? It felt as if I was being strangled by a snake. I noticed my heart was beating strongly and quickly in my chest, and I noticed my breath was shallow. There was nowhere to go, and it seemed there was nothing sensible left for me to do but surrender to that reality. So I stopped fighting it. I stayed in the reality of the present moment. I welcomed my fear and faced it. And that was key: just sticking with noticing how the fear was impacting my body, how my shoulders felt tense, how my chest felt cramped, how the curved posture blocked my diaphragm, and how all these tensions kept my breathing shallow, and my mind overexcited.

In Zen training, I had learned to re-center the mind by adopting a grounded, balanced, natural tonus in the posture, by consciously relaxing unneeded tensions. I had learned to feel the support of the earth's gravity, that incredible force that supported me in any given moment. I had learned to naturally extend the body posture by straightening my back, tucking in my chin slightly, and by letting the top of my head lean toward the sky. By relaxing my shoulders, chest, and belly. By inclining my pelvis slightly forward so that the diaphragm can relax, which frees the way for breathing to descend into the lower belly. Visualizing the downward movement of my exhalation gradually helped me to slow the rhythm of my breathing and lower my gravity point back to where it belonged, in the hara, this boundless center of life energy, just below the belly button.

And what did this focus on posture and breathing do to my mind? I can tell you: It was kind of magic. For a moment, I got a glimpse of harmony and a sense of wellbeing, even in these circumstances.

Of course, that good feeling didn't persist. My dualistic "thinking" mind immediately took over again, and again, and again. Neuroscientists would say this is normal, as we are wired to do so through evolution. But every time I noticed a mental construct again, I brought my focus back to a graceful posture, and I tuned into the breathing, in particular to the exhalation. My mind in crisis slowly made room for stillness. What happened there? This kind of "landing" in the stability of a graceful posture transformed my fear into trust. A regained trust that resulted in calmness and clarity. There was an opening to new possibilities.

What came up in me while sitting in stillness was a haiku[2] by the renowned Japanese poet Ueda Chosu: "Broken, broken again, the moon on the surface of the sea, it so easily mends."

So how could I mend my brokenness in this unfortunate situation? I remembered the profound Zen teachings I received just before I moved to Tanzania. My teacher advised me to pursue three foundational practices to lead a harmonious life in any given circumstance:

- The first practice is stillness. Generally, it consists in stopping doing anything, stopping acting in the outside world, stopping speaking, in my situation an inner dialogue, and stopping thinking, by changing from doing to feeling and from thinking to being. Being in stillness means to open up to the direct experience of reality, without the filters of our conditioning. It means to become aware of the living reality of the present moment, which is eternal. We are always here in our bodies, and the present moment is always now. The practice of stillness is needed to build presence, the continuation of embodied awareness.
- Traditionally, stillness is practiced by sitting quietly in full awareness, but it is not limited to the sitting posture. We can all practice stillness in our daily lives—standing up, walking, traveling in a vehicle, waiting in line at the supermarket, or being held in custody, like I was. It only requires intimately taking refuge in and remaining aware of the body while continuously adapting one's posture in relation to circumstances and personal needs.
- While presence directly results from the first practice, the practice of stillness, the second practice, which is the practice of reperception, directly results from tuning into and staying with the breathing.

So, what is reperception? By reperception, I mean the ability to re-assess reality with the enhanced awareness that we develop in the intimacy of the stillness. When we embrace stillness, when we focus our attention to the embodied experience of the present moment, something magical happens. We notice our breathing, we feel

[2] A haiku is a short unrhymed poetic form that originated in Japan.

our heartbeat, we become aware of the tensions accumulated in our muscles. We get a better sense of our internal processes, of the state of our body.

When our mind is overexcited, stressed, anxious, or angry, we can use breathing, especially breathing out, to naturally, automatically calm down. Simply focusing our mind on breathing out completely, until the gravity point of our body lands back in the hara, automatically stops the excitement while reconnecting with our original life force and creating space for something new: a new inhalation, a new insight, a new emotion, a new action. When we breathe out completely, there is non-thinking, and when we breathe in again, there is thinking. By consciously staying tuned in the present moment to this process of breathing out and in and out again, our disturbing thoughts related to an uncertain future or a regretful past gradually fade away while making space for a fluid mind, a mind that embraces reality as it is. A mind that is open to possibility, the possibility of fresh, natural, harmonious, loving, and caring thoughts.

The practices of stillness and reperception may lead us to all kinds of positive personal outcomes, like more calmness, more clarity, self-determination, confidence, courage, and so on. But without the third practice, the outcomes remain limited to the personal sphere, while true meaning is born in a caring relationship to the world.

- To become truly universal and boundless, the third practice is key. And that is to practice without any personal goal, without aiming for any personal benefit, without any expectations. Not pursuing anything in particular, nor rejecting anything in particular. Just embracing reality. That is how we can develop a free mind, a mind that isn't attached to anything, that lives an undivided life, wholeheartedly. A life embedded in a deep sense of belonging. And then it becomes natural, almost automatic to embrace reality generously, with love and care. For me, sitting on the cardboard in the dark, that meant in the first place to accept my complicated situation, which was far from being easy, but it worked well enough to survive the night in the prison cell.

Therefore, as I myself have practiced it the hard way, I am convinced that these three practices have the potential to lift you up in any given situation, no matter how challenging they might be:

1. Welcome to the body.
2. Breath consciously.
3. Generously embrace reality.

With this advice, I could end this text, but I am sure that all the readers who have kept on reading until now would like to know how the story with the skull ended.

6 What's the Skull in YOUR Suitcase?

Early in the morning, after a sleepless night, I heard a vehicle driving into the courtyard. A representative from the President's office came by. He wore a black suit, golden glasses, a golden watch, two golden rings, and his blinding smile revealed a golden tooth. In a friendly manner, he apologized for what had happened to me the night before, and he encouraged me to be patient. Since my case had already made it to the public prosecutor's desk, there was no way he could intervene anymore. I needed to sit out the investigation. He gave me his number and I was set free under the condition that I'd report daily. When I walked out on the sandy street, there was a newspaper stand with flocks of men standing around it, discussing the headlines. My skull story was the front-page story, luckily only in Tanzania, though. It felt odd to buy these newspapers.

It took 6 weeks for the forensic DNA testing to be concluded in South Africa, and my passport to be return. After that, it took another 7½ months and five court hearings before my innocence was acknowledged and I was finally set free. I felt relieved and deeply grateful for the insights I had gained over the course of this adventure.

Whenever I now have moments of despair, I remember those hours on the floor and I recall what I did to mend easily like the broken moon on the surface of the sea. And this is also what I want to give to you today: The certainty that, in any given circumstances, you always carry a superpower within you that can be activated at any time.

Now, to conclude, go ahead and ask yourself: What is the skull in YOUR suitcase? Let me invite you to open up your luggage, and explore the hidden stuff inside, the stuff that we are ignorant of. And whenever you become aware that you are shocked, anxious, or depressed, whenever the outer challenges become too urgent, too complex, too pressing, take it as an invitation to intimately reconnect with yourself: May I find stillness, may I reperceive reality, and may my motivation to do so be one of care and service to all. Because inner wellbeing inspires sustainable welldoing.

References

Helminski, K. (2000). The Rumi Collection: An anthology of translations of Mevlana Jalaluddin.

Severns Guntzel, J., & Murphy Johnson, N. (2020). *Wellbeing inspires welldoing: How changemakers' inner wellbeing influences their work* [Report].

Tanahashi, K. (2013). Treasury of the True Dharma Eye: Zen Master Dogen's Shobo Genzo.

Bart Weetjens is a Zen Buddhist monk and social entrepreneur. With a vision to empower the most vulnerable communities of subsistence farmers in sub-Saharan Africa, he took the initiative to train giant African rats as sustainable detectors in response to the global landmine problem. His social enterprise APOPO was recognized and awarded by networks like Ashoka, the Skoll Awards, and the World Economic Forum. When Mozambique officially declared itself to be free of landmines in 2015, he transitioned from his executive role to focus on the practice of Zen. In the same year, he got involved in *The Wellbeing Project*, which aims to catalyze a culture of inner wellbeing for all changemakers. While exploring the relationship between personal growth and social innovation, he co-founded the initiative *innerpreneurs.org*, which is a community of entrepreneurial leaders in pursuit of their noble life purpose.

From Paragliding in the Himalayas to Finding Purpose in Unleashing Women's Full Potential: A Personal Report

Veroniek Vermeulen

Abstract This chapter portrays the real-life case of Veroniek Vermeulen and her transformation from a ratio-driven successful businesswoman in an international consumer goods company to a passionate entrepreneur who has found meaning and purpose by supporting companies to create workplace environments that help women strive. Veroniek takes the reader on a journey that started in Nepal when she trusted her inner feeling and opted for paragliding in the Himalayan mountains instead of taking a meditation course. She openly shares her doubts and troubles as an entrepreneur and as a human being and provides insight into her private understanding of well-being and her learnings from entrepreneurship. Furthermore, this chapter reveals factors that affect women's well-being at the workplace and offers solutions to overcome such challenges.

1 A Life-Changing Trip to the Himalayas

The very first time that I got in contact with meditation was when I signed up for a retreat in Nepal. That was in 2012, when I took time off from my corporate career. At that time, you could have said that I had made it: I had a good education, a successful corporate career, a top salary, I was married, travelled lots, had lived in many places, had a beautiful home and dressed in designer clothes. I had ticked practically all the boxes, at least the ones that society presents as a 'success'. Nevertheless, I looked at my life and realised that I wasn't happy. I felt unfulfilled and wondered if that was all there was to life. Hadn't I done everything to be successful? And only then I realised that they weren't MY boxes. Everything I lived for was according to what my parents valued, or my friends were doing, the things that were considered to represent status and success, but everything I had achieved, everything I had

V. Vermeulen (✉)
Silatha, Amsterdam, the Netherlands
e-mail: veroniek@silatha.com

M. Mühlböck (ed.), *Sustainable Transformation and Well-being*, CSR, Sustainability, Ethics & Governance,
https://doi.org/10.1007/978-3-031-75566-8_18

pursued weren't MY dreams. So the question arose, what did I actually want? And who was I?

A good friend had encouraged me to try out meditation, but, honestly speaking, I was far from being convinced of its benefits. Still, I trusted her and signed up for a short meditation course in Pokhara, Nepal. The venue was located in a village not too far from the base of the Himalayas. Soon after my arrival, I spotted paragliders floating effortlessly up in the sky. That sparked an idea: Being so close to the Himalayas, how cool would it be to fly over the top of the world! I quickly found out, though, that it wasn't so simple to find someone who would be willing to take me to the high altitudes. I was told later on that I had become the first woman to take on this challenge. When my guide sketched out the plan for the trip, which included several days of climbing up to 5500 m, where we would unpack the parachute, I realised that I would have to choose between this adventure and the meditation course. But I had already put my mind to it and clearly desired to go for this once-in-a-lifetime experience. So I did it. The paragliding was amazing. However, it was not what ultimately changed my life.

On my way up, I encountered a Buddhist man who was interested in learning about my motives for coming to Nepal all by myself. During our conversation he introduced me to Buddhist teachings and shared the following wisdom with me that touched me deeply: 'One can live without food for months and manage to survive without water for weeks, but it only takes a few minutes without air in one's lungs to bring about death. So what I do with meditation is to honour my breath for the life it gives to me'.

As a Dutchwoman, the altitude scared me and I decided to practice his teachings while walking up, which was anyway in silence, since the porter who accompanied me didn't speak much English at all, and my Nepali is rather basic. And while walking up in silence, his words kept resonating within me and I focused on my breathing. Never before had I thought of how special it actually is to breathe and the live it gives you, and I started to open myself up to the idea that there might be more behind the concept of meditation than I had assumed. The breathing exercises and calm around me really helped me to get to the top. After the paragliding, I was a bit annoyed when I learned that I had to stay a day at a specific altitude to give my body time to acclimate to the lung-crushing conditions. I was a very impatient person, to be honest, and I still am to a certain extent. Yet, that involuntary stop became one of the most beautiful days in my life. While I was resting in the calm, I suddenly entered into a state of being in which I simply existed without doing anything in particular—me on my own in that magnificent nature surrounded by the majestic mountains. For the first time that I could remember, I felt pure happiness that came from the inside. Until then, I had just experienced the pleasure from, for instance, achieving a particular goal. At that very moment, I realised what happiness truly was, when it did not depend on external factors. Real happiness emerges when you are just in the moment. And I was just in the moment looking at that breathtaking landscape. Also, for the first time, I felt happy on my own. Before, I wasn't so comfortable just being by myself. That day I discovered the wholesome effect of being alone. To this very day, it is something I truly appreciate, at least from time to time.

Looking back, I believe that this adventure turned out to be the best meditation course I could have ever taken. It showed me the connection with myself and my heart, as well as the powerful effect of being in the present moment with its positive effect on my well-being. In contrast, our modern lifestyles are busy and preoccupied. We spend most of our time thinking about the past or future, rather than simply living in the moment.

2 A Purpose-Revealing Journey

After that initial experience, I deep-dived into a 10-year journey on which I learned about different meditation techniques, took in more wisdom from Eastern traditions, and discovered how I could benefit from them in the long term. I must say that change of direction resulted in some questions from people close to me. Had I become part of a sect or some spiritual hippie? What had happened to the ratio-driven Veroniek they had known? Some were afraid I was lost, whereas I, actually for the first time, started to feel ground under my feet, starting to get a bit of a sense of what I was doing.

I kept attending meditation courses, Vipassana and Zen Buddhist retreats, and stayed in several monasteries. I discovered that even monks are still very human, that most of them struggle like anybody else, that most of them are not released from pain and that they also have their individual sorrows, attachments and egos. However, I have been able to meet some head monks who have attained a level where they seem to radiate pure love and energy. It was a powerful revelation to see that it's possible to enter a stage where you really feel connected and in a constant flow of happiness. To make this last, it takes years and years of work, but it was so powerful to see those head monks, full of love and how inspiring they were. That one can effectuate change, through love.

Nonetheless, my focus was eventually on how to integrate the valuable wisdom that I had collected on my journey into my daily life. I understood how meditation enables consciousness and peace of mind, and that motivated me to transform the way I live my life. In our modern Western lives, there is always more to desire and that is, by nature, deeply unfulfilling. Meditation offers an alternative, to connect inwards. It allows us to nurture our own inner quality and strengths for long-term pleasure and true fulfilment.

Yet, it was still difficult for me to keep meditating on a regular basis and I discovered that it was not only me who struggled with that, but many other people as well. Thus, I kept pondering how I could change this. At that time, a lot of meditation apps were launched. However, the retention rate of those apps is only 5% on average, which means that 95% of their users stop using the app within a month. In particular in the Western world, meditation is hardly something that we learn at a young age. Knowing the value of meditation and familiar with the difficulties to make it a lasting habit, I kept pondering how to support people to show up to meditate regularly.

I started to research the subject extensively and found the following three key players in creating habits: B.J. Fogg (2020), Charles Duhigg (2013), and Nir Eyal (2014). Based on their findings, I understood that having a tangible object would be very helpful to establish a new habit. Starting small, very small, and integrating it into what you already do in life was another principle I reflected on. That led me to create a few simple meditations and an 'anchor', a beautiful piece of jewellery with a valuable gem of one's choice to reflect the good intention and serve as the tangible reminder to meditate.

3 Silatha: The Start of My Entrepreneurial Journey

Originally, I never had the ambition to be an entrepreneur. I really liked corporate life, which was very comfortable and luxurious for me. We had assistants to take care of many things, and it came with good money and power. I had the impression that it provided an environment for me that helped me grow. The decision to extend what started as a hobby and establish my own company stemmed from the need that was evident to me and the opportunity that I saw within—because meditation had had such a huge impact on me and had really changed my life. I saw it as the most precious gift, which I wished to share with everyone else in the world. I had also shared my experience and my insights with people in my company, but I felt that I needed to take on another role outside of the organisation to create a true impact. And I would do it together with a friend, so I thought, and that made it even easier to leave the safe haven of corporate life. After 3 months, my friend who had agreed to launch the project with me was pregnant with her second child, and I had to decide if I would still stay on track and set up the company all by myself. Emotionally, I felt as if I had been abandoned at that moment, but I am not a person that likes to quit after I have made a decision. And meditation had taught me that things happen for a reason, so I decided to go for it and established Silatha.[1] Looking back, I am not sure if I would ever do such a thing on my own again. It is a rollercoaster and having a partner by your side is really valuable. Though, on the other hand, decision-making is much quicker and easier when done alone.

At the beginning of my entrepreneurial journey, my mission was basically to share the benefits of meditation with women and support them through various stages of life with specific meditations that can help during their respective stage. Meditation and mindfulness had not only changed my habits but also encouraged me to reflect on myself and see my true identity. I had always thought that I was a very ratio-driven personality, as that was what I had seen as a kid, what I had perceived as 'the right character traits'. And thus I had given myself that identity. I realised that I am actually a hypersensitive human being and, for all those years, I had suppressed my emotions. That realisation made me aware of the benefits of

[1] 'Sila' stands for *nobel life*, whereas 'Atha' means *now* in Sanskrit.

being true to myself and showing who I really am. It helped me to open up and being honest about my own life, as well as to listen to other people's stories in much more detail. All of a sudden, people started to tell me stories I had never heard before. They dared to share their struggles with me, and that was how I learned in particular about women's health challenges, for instance, about miscarriages, fertility journeys or post-natal depression—very personal stories you do not talk about very often. At that point, I realised that meditation had helped me to become a different, more authentic person, but that a lot of people around me, specifically my female friends, rarely felt safe enough to stand up for themselves and their needs. That was when my mission shifted: I wanted to help create an environment for everyone to flourish, in particular for women, because I was deeply touched by their openness about the challenges they face in everyday life.

In my private sphere, I had already started to connect friends with similar stories and worries and I could see the valuable effect of such 'safe spaces' where they could discuss delicate and personal issues. One of my friends has just recently told me that she only has kids because I connected her with another friend of mine who had gone through the IVF journey and given her the confidence to do it as well. That made me realise that promoting meditation alone would not be enough to fulfil my mission and that providing other elements like coaching and peer-to-peer support would be necessary too.

At the beginning, Silatha's target group were private customers, but soon we realised that our impact would multiply in the business space, where more and more women are leaving the workplace because of the fact that they do not feel valued. The programme we offer today is thus directed towards employers. It has developed from the learnings and realisations over time and will continue to do so. We support the women and also educate the organisation, so that every employee gets a better understanding of what these taboo topics regarding women's health mean. We focus on taboo topics that women go through, such as endometriosis and polycystic ovary syndrome (PCOS), miscarriages and post-natal depression or menopause with all its different phases and characteristics, and in our conversations with our customers and their employees, we discover and integrate more taboo topics along the way.

4 Women's Struggle for Well-Being at the Workplace and Silatha's Approach to Transforming Business Culture

Today's workplace is generally still a place created by men for men. Ninety-five per cent of CEOs are male (Milhomem, 2020). As female role models are rare, many women do not even imagine themselves in such senior positions. Consequently, women do not live up to their full potential, which is a huge loss for the organisations they work for. Enterprises would clearly benefit if women had more direct influence. Many studies (Gallup 2022; Women Count Report, 2021; Virgin Puls

State of the Industry survey report, 2017) show the huge value that comes from a diverse workplace where women can flourish. Here are just a few:

- Increased productivity and performance
- Reduced risk of discrimination and harassment
- Increased creativity, innovation and problem-solving
- Expanded customer base
- Enhanced company image that attracts more women
- Reduced employee fluctuation rates and costs, because when employees feel valued and supported, they are less likely to leave their jobs

Yet, if companies do not provide conditions for women to flourish, that huge potential remains untapped. The 2022 McKinsey Women in the Workplace report reveals that, for every one woman that gets promoted, there are two women leaders that leave the company.

From our experience, I can share that many of them feel invisible, unvalued and not secure enough to stand up for themselves when they struggle with their well-being. That is proven by facts and figs. A Russell Reynolds study (2023) confirms that the number one reason for such women leaders to leave is because they don't feel valued. In Deloitte's 2023 Women @ Work report, among the 15% of women who report experiencing health challenges related to menstruation (which means most women do not even dare to report them), over 40% say they work through the pain and discomfort. Nearly one in five (19%) say they have taken time off for symptoms related to menstruation, but did not disclose the real reason why—and a small number (7%) report that they disclosed the reason, but believe that doing so adversely impacted their career (Deloitte, 2023).

As part of creating a thriving environment, companies need to address taboo topics such as hormonal realities, for instance, as well as menopause, post-menstrual syndrome (PMS), endometriosis, PCOS, fertility endeavours or a miscarriage and post-natal depression, in addition to the multiple caregiving roles and management biases.

The following are only a couple of examples that illustrate in more detail why that is relevant:

- Women arrive in senior positions in their 40 s and 50 s. On average, a CEO is in their mid-50 s. Peri-menopause normally starts around 45 and lasts until roughly 55. Around 80% of women who go through peri-menopause suffer from symptoms that vary from physical discomfort like hot flushes and heavy bleeding to mental struggles like anxiety, low confidence, depression and sleeping difficulties. There are about 34 symptoms that come and go over the years. Ninety per cent indicate that their menopausal symptoms have a negative impact on their work, 30% are considering leaving the company and 10% actually do quit. Nine out of ten women say they feel unable to talk to managers at work about their menopause symptoms. They do NOT feel valued.
- Another juncture where we lose many women is at the mid-senior management level, when women are in their mid-30 s. They have ambitions for a successful

career, but other things are also going on in their lives. Starting a family, for instance, can be accompanied by fertility struggles, miscarriages and post-natal depression. My personal experience supports those facts. Like me, most of my female friends are engineers, too, and they started off with great careers, but took a step back when they were starting their families. For 80% of my university friends, their husbands are now the main breadwinners. Interestingly, the only ones who earned more than their partner still have an equally strong career.
- Eighty-four per cent of women on a fertility journey mention that it affects their work and ability to concentrate, 68% of employees would consider changing their job to receive fertility benefits and 75% report no supportive workplace policies, resulting in significantly higher levels of distress.
- Just a few words on PMS: 89% of women report that their energy levels are negatively impacted and 77% struggle to concentrate. Forty-five per cent take sick leave due to symptoms, but most of them do not feel free to talk about their menstrual cycle at work.

Summing up, it is key for an employer to value their employees by caring about what is important and impactful in their lives. We lose women who could be valuable leaders and role models when they are not supported regarding issues such as those stated above. Hormonal realities are a totally overlooked factor when women leave their workplaces.

That is where Silatha comes in. Our mission is to empower individuals through building awareness, skills and open dialogues to establish a psychologically safe workplace that attracts and retains women and their allies. Silatha is an expert focused on supporting companies in creating a psychologically safe workplace, by educating and supporting on unspoken topics and biases, through a digital app. Our goal is to really transform a company's culture in which psychologically safety is prioritised. Many enterprises think they are on that track, but when one inquires as to how they deal with taboo topics like menopause, it becomes obvious that these issues have not been tackled yet. Thus, Silatha has grown into a company that offers the only holistic approach focused on female life cycles and the taboos around women's mental well-being in those phases of life. We help organisations to create environments where those realities are understood and offer specific tools to help manage them, such as the following measures:

- We facilitate awareness creation. As many employees know very little about these topics, we teach the whole organisation. During that process, we strive to normalise the conversations within the company on taboo topics, inviting people of all genders and all ages to understand what happens during this natural process, so people can learn how to be more supportive of and understanding towards each other.
- We create peer-to-peer support groups that focus on specific taboo topics so that participants can share experiences and do not feel alone. We support them with coaching and access to the right knowledge and experts so that they are informed, feel better, are less worried and isolated and are more efficient in doing their job.

- We reset expectations regarding work culture and create a psychologically safe work environment where women feel able to communicate with their manager about their needs, feelings and experiences.

By means of this product and service portfolio, we strive to support enterprises that believe they can be more successful by creating an empowering and inclusive environment for women so that everyone can flourish. We facilitate a mindset shift and system change within the company and, as such, support the sustainable business transformation of our clients.

5 What Well-Being Is for Me and How I Integrate It into My Daily Life

For me, well-being is pretty much every element in life that gives you joy, support and relaxation. If you do not take good care of your own well-being, you cannot really tend to someone else's well-being in the long run. If you run out of energy yourself, how can you give your positive energy to others? Thus, you need to make sure that you feel well and healthy to a certain extent. By really making my well-being a priority, I discovered I can have all the stress in the world, for instance, with my business, where so many things do not turn out as planned, but thanks to managing my well-being, I do not really feel stressed or burned out.

I believe that when you take care of some major elements, you really start to feel stronger and stronger in life and even when challenging events happen, they only throw you off for a moment, but you get back on track pretty fast.

Here are some of my favourites to foster my well-being:

- **Take care of and live your life according to who you are:** The ability to connect with your inner self repeatedly during your day gives you mental rest and stops your crazy mind from worrying and from thinking about all the things that might go wrong. That also includes finding time within your day to do what is important to you. We all have things that we don't like to do, but making sure you take time on a daily basis to do what you really like to do, even if it is only for half an hour, can make a huge difference. I, for instance, started taking a bath a few times a week, which helps me reduce the tension in my body. It is a great way to relax, especially when I do it before going to bed. It is always about getting to know yourself and your authentic needs better and better and discovering what is truly good for you.
- **Move your body:** Exercise and sports are extremely important elements in my life. Without them, I do not have the energy to perform and do not feel well. If you are not a sporty person, go for a walk and conduct your phone calls while in motion or take the bike instead of the car. For me, a long bike ride, run or boxing class resets my energy levels and boost my endorphins.

- **Get enough sleep:** I am convinced that sleep is the most important element that we often forget about. Having a good night's rest is a key factor to perform and has become my number one priority.
- **Eat healthy food:** You are what you eat. That saying is very true in my opinion. Even if it is not my strongest element, I try to pay attention to my nutrition and to drinking a sufficient amount of water.
- **Have the right people around you:** A key element is surrounding yourself with people who spread good vibes, whom you trust and who support you and cheer you up. My very best friends take me as I am. They would be honest to me about things that they question and even challenge me, which I believe is really important. At the same time, none of them would talk negatively behind my back, because they know they could share it with me, and the same applies for me. Positive and caring people help you grow and become a better person. On the other hand, getting rid of negative people might be hard, but equally essential for your well-being.
- **Open up about things that are troublesome:** When I was growing up, I definitely never learned to talk about emotions and feelings. Being rational felt like the way to 'fit-in', to be good. So I adopted and projected the image of a rational person. Consequently, I always suppressed my sensitivity and my emotions. I would never show tears, even when I felt tearful, since it did not align with how I identified with myself. I had to learn along the way that it would eventually prevent me from leading a happy life. I am in awe of those people that dare to show who they are and what they feel. That establishes new bonds and creates deep connections. I am proud of the extent to which I feel strong enough to show my true self nowadays, though there are still moments, or people, that cause me to struggle to share my deepest self. I learned that as soon as I do open up and talk about negative feelings, for instance, they actually disappear. They are no longer there.
- **And of course, meditate:** Meditation supports you to clear your mind, be in the moment and connect with yourself. I still have a rational mind in the sense that I struggle when I don't understand, when there is no logic. Meditation helps me identify what is bothering me and eventually let go of it. I have really learned that whenever I can influence something I would like to change, I will do everything I can to get the desired result. But if it is out of my hands, I rather ask myself what I could learn from the situation. How can I do it differently next time so that it doesn't hurt me or even happen at all. It is not that I just snap my fingers and it's gone. It doesn't stay with me for a long time anymore, though.

6 Learnings from My Entrepreneurial Life that Also Affect My Well-Being

- **Cultivating a team that you can trust is key:** Your team is your foundation and key success factor. Having a team around you that you can trust will also ensure that you can take care of your well-being and switch off from time to time.
- **Doing something that gives you purpose does not feel like work:** People sometimes ask me how I can work so many hours. I then tell them: 'You have a hobby. I have my business'. Of course, there are a lot of things that aren't fun. But feeling that the work you do is purposeful does make up for a lot. I worked in innovation at a consumer goods company for a long time. As I love creating, it really fulfilled me, for instance, to launch a new coffee that would make those few minutes in the morning more enjoyable or meaningful for millions of people. I really enjoyed the thought that I might impact their lives in a positive way. That they would start the day a tiny bit better than they normally would, and that gave me a sort of a purpose. However, what I do now is on a different level of purpose, because it has the power to really change people's lives in a positive way. And that is deeply fulfilling and motivates me to keep going.
- **You grow a lot as an entrepreneur, but you mostly don't realise it:** The funny thing is, as an entrepreneur you don't very often realise how much you learn and how much you start to master certain things. I only got an idea of my personal growth when I began to advise and support other founders that were just about to start their journeys. In corporate life, you get extensive feedback at least once a year and that gives you a clear picture of your achievements and potential. As an entrepreneur, you just see new problems popping up again and again, so you therefore never really feel that you have mastered them. You always feel like you are trying to survive. Thus, there are only a few moments when you look back or reflect with others on where you started. I am pretty sure that I have learned a thousand times more on my entrepreneurial journey than I would have learned as an employee.
- **Having the right advisors on board makes life as an entrepreneur a lot easier:** I recommend to everyone who is starting a business to look for capable people who are willing to advise and support your journey, e.g. to help you shape your strategy and vision again and again. I know that many people appreciate the high level of autonomy that the entrepreneurial role offers. I sometimes wish that I had someone next to me, but luckily I have excellent advisors, ask them for long- and short-term strategies and follow their advice a lot.
- **Be sure to always keep an eye on your own energy levels:** If you crash, your business crashes with you. So always ensure you check in with yourself and if you're too tired or don't feel the necessary energy to continue, take some time for your self-care. Whether that means exercise, a warm bath, a day off in nature or a weekend escape. Whatever is good for you when it comes to charging your batteries. When that is not in balance, you risk burnout.

- **Trust your inner voice:** I do believe, deep down in ourselves, we know everything that is good for us. One of the most precious insights I have gathered along the way is to listen more and more to and trust my own intuition and inner voice. Sometimes, that may feel a bit scary, especially when it means turning down a potential business opportunity. But so far, the story of my life has proved me right.

7 Conclusion: Creating More Value and Purpose in Other People's Lives Makes Me Love My Life

To summarise, I can confirm that, without a doubt, my trip to the Himalayas transformed my life fundamentally. Without that experience, I would probably not have discovered the gifts of meditation and the wisdom to reflect on life with a more helicopter view, which really helps to put things in a different perspective and to shake off worries and fears. That adventure sent me on a journey of self-discovery and deeply shifted the way I look at life. I can definitely say that my life is more fulfilled today and that I experience a higher level of joy, even if life itself has certainly not stopped confronting me with its inevitable twists and turns. At the same time, those learnings have enabled me to positively impact other people's lives to an extent I had never imagined possible. And that adds to this feeling of happiness and contentedness.

My entrepreneurial journey with Silatha definitely gives me purpose. Does it improve my well-being overall? Well, I still need to be careful in regard to my work-life balance. It is hard for me to take time off, because I'm so involved with and dedicated to what we do. Yet, I believe, if you talked to people on their death bed, they would tell you about how they regretted the things they had **not** done. And they would mention the moments of connection with others as what counted for them in the end. For me, I have either loved or learned from what has happened in my life, and people make the difference. I think, if we talk about well-being, a big part of it is doing good and feeling good about who you are. Thus, having a purpose in life makes me love my life. So, in the whole sense of well-being, it definitely brings me a lot.

References

Deloitte. (2023). *Women @ Work 2023. A global outlook*. Deloitte Touche Tohmatsu Limited.
Duhigg, C. (2013). *The power of habit*. Random House Books.
Eyal, N., & Hoover, R. (2014). *Hooked: how to build habit-forming products*. Portfolio Penguin.
Fogg, B. J. (2020). *Tiny habits*. Houghton Mifflin Harcourt.
Gallup. (2022). Advancing DEI initiatives: A guide for organizational leaders.
McKinsey. (2022). *Women in the workplace*. McKinsey & Company.
Milhomem, C. (2020). *Women on boards. 2020 Progress report*. MSCI.

Russel Reynolds. (2023). *What's motivating your women leaders*. Russell Reynolds Associates. https://www.russellreynolds.com/en/insights/articles/whats-motivating-your-women-leaders Accessed 6 June 2023.
The Pipeline. (2021). *Women count 2021. Role, value, and number of female executives in the FTSE 350*. Pipeline Consultancy LTD.
Virgin Pulse. (2017). *State of the industry survey report 2017*. Virgin Pulse.

Veroniek Vermeulen, founder of Silatha, is on a mission to create an equitable workplace where women and their allies feel accepted, trusted and valued. Silatha is an expert focused on supporting companies in creating a psychologically safe workplace, by educating and supporting on unspoken topics like PMS, fertility or menopause and unconscious biases, through a digital app. Veroniek is an engineer by education and spent 20+ years in global marketing. She has received many awards during her career among that the Influential Leader Award by BrainzMagazine, next to stars like Jessica Alba, Jay Shetty, Oprah Winfrey and Whitney Wolfe Herd.

If I Can Really Make an Impact, Then My Life Has Been a Success: A Personal Report on the Transformation of a Company and an Individual

Thomas Welser

Abstract The author of this article has been the CEO of a successful family business for 6 years, the roots of which go back to the seventeenth century. However, his personal path has been inextricably linked to the industrial company since his birth. In this article, he describes how this led him through successful but also challenging moments and how a change of perspective helped him to overcome a personal low point and start an entrepreneurial transformation process aimed at rediscovering the company's purpose and bringing it back to life. He also discusses selected situations, methods and projects from a practical perspective, as well as insights that have proven effective in his business environment for promoting a culture of wellbeing. He outlines why, for him, a successful life is always linked to the opportunity to achieve maximum impact and why sustainability and social responsibility are a logical consequence of this.

1 Introduction

Our family business has had a huge impact on my life to date. I was born at home and grew up more or less next to the production facilities, surrounded by the people who worked there and some of whom still do. Culturally, people at our company have always been on a first-name basis with one another, and that's still the case today. We are now a company with two and a half thousand employees, process around 400,000 tonnes of steel a year and have manufactured around 24,000 different products using our own process to date. As we are very broadly positioned and supply many industries, we have very good economic resilience and have been able to overcome all crises so far. I can say with a clear conscience that I manage a healthy and innovative Austrian family business.

T. Welser (✉)
Welser Profile GmbH, Gresten, Austria
e-mail: thomas@welser.com

M. Mühlböck (ed.), *Sustainable Transformation and Well-being*, CSR, Sustainability, Ethics & Governance,
https://doi.org/10.1007/978-3-031-75566-8_19

Our company is located in a rural area, in a village in Lower Austria's Mostviertel region. I also went to school there. Innovation and sustainability have always been essential values for us, but they also play a particularly important role in the transformation and further development of our company.

The history of our company can be traced back to the seventeenth century. My ancestors began with a simple pan forge. We are now the international market leader in the manufacturing of special steel profiles. I belong to the 11th generation. I joined the company 26 years ago and have been the CEO of the Group since January 1, 2019.

My personal history is closely interwoven with that of the company. This company history has challenged me to undergo an individual transformation that has made me the person I am today, and it has probably also had a significant influence on the further development of the company. It is difficult to say what triggered the individual and company transformation process—everything is simply too interrelated for that. As someone who is responsible for our organization, it is very important to me that we always treat each other as equals, both within the company and beyond. But one thing at a time.

2 Transformation at the Individual Level

There have been so many milestones and highlights in our long company history that I don't have enough space to list them all here. Of course, things have not always gone well for us either. The economic crisis around 2008 was also a challenging time for us, but we were ultimately able to overcome it successfully.

Now and again, there were also family tensions, which were very stressful for all family members. In 2012, when my father left the company for health reasons, a challenging phase began for me. The psychological pressure was clearly noticeable. As I know that I am not the only person in the entrepreneurial environment who has to deal with this kind of pressure, and because I realize that it is still a taboo that needs to be overcome in our society, I would like to be transparent at this point: I worked and lived at the limit of my capacity—and often beyond—for a period of several years.

Rather by chance, I ended up attending a management retreat at an institute for leadership development. One of the trainers, who still accompanies me today, managed to completely change my perspective within a few weeks. I realized that it was up to me to become capable of making decisions again. My perspective determines how I view the world around me and individual situations and how I subsequently act. From that point on, it was clear to me that it is always up to me, it is my own decision how I relate to the challenges—which perspective I choose.

A representative example will illustrate this: A few days after the seminar, I had a conversation with one of our key managers who had a massive influence on the company and whose decisions I rarely agreed with. Thanks to my change of perspective, I was able to approach the conversation with a completely new attitude. I was able to tell him my opinion in detail and openly, without personal and aggressive judgement. It was clear to me that something must have happened to the person that had made them into the person they are today, that had led them to this attitude and behavior.

I was deeply relaxed and looked at my conversation partner and his challenges. We had a real conversation as equals, for which my counterpart even thanked me afterwards. Those moments and experiences marked a turning point at which I regained my full strength. I began to grow with the challenges that had repeatedly made my life more difficult. The various health complaints that I regularly used to have are now all history. I haven't even had a toothache for over 10 years now. This change of perspective was the most healing thing I have ever experienced.

During this personal low point, there was a very special moment that made me very aware of the great human potential that lies dormant in our company. In 2014, the company celebrated 350 years of existence and the employees wanted to organize the main act of the celebration themselves. Customers, suppliers, political representatives and many people from our network were present at the celebration. Nobody, including us as a family, knew what to expect. The employees then performed our entire life's journey, the story from the forge to the present day, including a look into the future, on a 40-m-wide stage. With this performance, our people triggered one thing above all—an emotional tidal wave: all those present were enthusiastic and gave frenetic applause and standing ovations. At the end, everyone sang along to the song *We are Family*, which the employees had adapted to the company. In total, almost 300 people had worked on the show. At a time when I didn't see our family at its peak, when I didn't see much reason to be proud of ourselves and our part as role models and creators, the employees gave us a completely different perspective. And that was the day for me when I realized where I belonged to.

That cohesion has given me a very special strength. Because when the people in the company are so supportive of the company and ultimately make their very special contribution to its success and all this is not only visible but also tangible, then it is a turbo boost for your own motivation and energy. Thanks to that incredibly strong, unique signal from the workforce, I finally realized that it was about something much bigger than just my own tunnel vision. It was then that I decided that I would be there for this company and for the people in the organization and that I wanted to do my best to pass it on successfully to the next generation.

3 The Transformation Process in the Company

An initial management consulting process that lasted several years turned out in retrospect to be more harmful to our culture than beneficial to the company. The consultants had chosen a classic top-down approach. What we lacked was an examination of our rich history, our DNA and our self-image of daily interaction. It is extremely important to ask ourselves the questions "Where do we come from?", "Where do we stand?" and "Where do we want to go?" honestly and transparently.

In 2018, when my brother and I took over the management of the company together with a new generation of external managing directors, we were finally able to do just that. In contrast to all previous activities, this process was strongly driven by grassroots. Around 600 employees took part voluntarily. As part of this process, we finally succeeded in clearly defining our purpose and determining how we want to develop the company with a view to one day passing it on to the next generation.

One of the most impressive moments in this process for me was when we carried out a "purpose quest" as part of the transformation work—3 days in an alpine hut with 12 delegates from different areas of the company, from different locations and with different perspectives. These 12 people had in turn been selected from over 100 so-called multipliers or influencers that we had found within the company. I myself represented the perspective of the family and management. In preparation, we had used the "7 generations map" method. This involves systematically reflecting on where you come from and what has worked and what hasn't, what gives energy and what destroys energy. You also learn how to deal with what didn't work and to recognize the potential. In our case, we included three generations backwards, the current generation and three future generations. These "7 generations maps" were completed 62 times with over 600 employees. The result was more or less a whole book that all delegates had to read in advance. On site, we had to get up at 5 o'clock in the morning, then spend 2 h thinking about what it was all about, alone, sometimes in the snow. It was a very special and intensive task. It was about finding exactly one sentence out of those thousands of words and phrases that gets to the heart of our purpose. At the beginning, it was mainly overwhelming. But by the end, 5 of the 12 people present had even achieved a complete change of perspective for themselves. There were some incredible, intense and positive moments. And we actually managed to find our purpose and express it in one sentence: "Enthusiastic people create opportunities for generations".

Our four core values are still derived from this today: (1) We are pioneers out of enthusiasm—we are convinced that progress and innovation require courage and passion. That is why we also see mistakes as part of learning, encouraging us to try out new things. (2) We focus on joint further development—we are convinced that major challenges can only be mastered together. That is why we promote the development of the strengths and potential of individuals and teamwork. (3) We treat each other as human beings—we are convinced that every person is unique. That is why we promote diversity and encounters on a level playing field. (4) We take responsibility for generations—we are convinced that thinking in terms of generations is what makes a family business special. That is why the responsible, sustainable use of resources and regional roots are important to us.

The sentences we use to describe these values are still exactly as we defined them together back then. None of them had to be changed by communication professionals. They simply perfectly express what our backbone is all about. And this formed and continues to form the ideal basis and orientation for the transformation process we have been undergoing ever since. Since then, everything really does make much more sense. We can attribute everything that happens to this backbone, this orientation, and derive our decisions from it. I've hardly ever had any situations where this hasn't worked. It has its own logic and works like a compass.

Incidentally, our family history was also represented graphically as part of the process. The picture hangs in our training workshop today. We really worked through everything transparently and tried to draw our learnings from it. The picture traces the path of transformation from the small forge in the seventeenth century to the large industrial company of today. We have not shied away from pointing out the dark sides of history. At the same time, we can also show that we can be proud of who we are and what we have achieved.

4 Valuable Insights for a "Culture of Wellbeing"

Both my own transformation process and the one in the company, together with the people in the organization, have taught me a lot and provided me with valuable insights that I would not want to miss today. I would now like to outline the aspects that I believe are particularly important for a "culture of wellbeing":

- **Authenticity is important for successful development:** If we had not succeeded in dealing with our own history in a transparent, honest and authentic way, we would not be able to focus all our energy on the present and the future today. I therefore believe that sustainable transformation can never succeed without authenticity—whether at an individual or company level. What's more, in my opinion, the days when you could fool your employees are over anyway. Today, when someone asks me how I'm doing, I say exactly how I am, and that can be one way or another. When I enter into a conversation with our apprentices, which I always like to do, I don't prepare a story about our company's success or anything like that, but we introduce ourselves to each other, we make sure we get to know each other. They can ask any questions, there are no taboos, and I'm not afraid of the conversation, because with an open and authentic attitude, that's not necessary—and that makes a real difference.
- **Reflection is a pillar of our success concept:** Reflection has brought me a lot of positive things—for myself, but also for our organization. For us, the "7 generations map" is a method that has proven its worth in enabling reflection, structured as an outlook. It has helped us to initiate the necessary transformation and to take the first steps. I would even say that it has become a kind of success concept and is therefore of great value to us. We also encourage young people, such as the 15-year-old apprentices who join us, to reflect on themselves and their work in the company.
- **People want to be effective**: A key question for me is always where you can be most effective. Reflecting on this is crucial—it's always about fulfilling a purpose. For example, I've never met anyone who honestly says I don't work and yet I'm still happy. Most people who don't have a job or can't work are totally

unhappy and lose themselves in meaninglessness in the long run. I am convinced that you have to create something, design something, in order to be happy.

- **People need a "safe space" in order to be effective:** The many inputs in our transformation and purpose-finding process also showed that people want to be effective. However, it also became clear that they need to do so. In particular, they want to feel safe in being themselves. If you feel comfortable and enjoy going to work, you can develop your skills. This also applies to the internal error culture. If to feel safe employees are afraid of making mistakes, then the company will not develop, either. But if you can say, I made a mistake, make sure the same thing doesn't happen to you, then we can all learn something from it. The great thing is that we understood this together during our transformation process and defined it for ourselves.
- **Developing potential requires a supportive environment:** During my time as a manager, I have come to realize that it is not about motivating people, but about helping them to motivate themselves. We have therefore also stated in our vision that we want to create an environment in which the unique potential of our employees can unfold. The feeling of safety mentioned above is part of that. Today, however, I am also concerned with how we can create a "space of opportunities" that meets people individually exactly where they are. For some, it may be about giving them a push to leave their comfort zone for once. Others need very little to "learn to fly", figuratively speaking. In summary, it's all about attitudes and mindsets. In any case, I would like to make my contribution to creating this "space of opportunities" so that we can increasingly succeed in leading people to positive energy and thus contribute to greater wellbeing. You can't force anyone to do that and I don't want to, but I would like to create the opportunity. Incidentally, my favourite moments are always when people are successful, especially when they manage to change their perspective to a positive one. This is a concrete example: We take on 30 apprentices every year. And when the training manager approaches me with a smile and asks if we can take on a few more because we have so many good applications, that's incredible. Especially, when you realize how happy he is about it himself. That's something very special.

5 Sustainability, Social Responsibility and Creating Impact

I have already mentioned that sustainability and social responsibility have always played a key role in our company. For me, however, these topics are particularly closely linked to the topic of self-efficacy. Basically, it's always about creating a positive impact. Where can I be most effective? Where can I have the greatest impact? If I personally manage to answer these questions and implement the answers, if I can really make an impact, then I think that contributes massively to a successful life. At the top of Maslow's pyramid of needs is transcendence, a level that has unfortunately often gone unnoticed because Maslow only added it later (Maslow, 1969). By that, he means a need for meaning beyond individual development, a striving for the greater whole, which leads to a particularly strong feeling of fulfillment. And that is what it is actually about. So not only does it contribute to a successful life on an individual level, but it also automatically leads to more sustainability efforts and the assumption of responsibility for society, with a high degree of effectiveness.

On the subject of sustainability: Of course, we fulfill our duty first—we adapt to everything that the "Green Deal" entails: CO_2 balances, supply chain laws, taxonomies and so on. Not everything that comes along makes sense and is well thought through—but at least it is well-intentioned. Two years ago, we produced our first sustainability report and found that we are well positioned in this respect. We have built the third-largest PV plant in Austria and are currently investing in a wind power plant in Germany. However, our understanding of responsibility for generations goes far beyond this, for example, we organize sustainability days in the region together with schools, where we invite other companies to join in. The aim is to work together to reduce our carbon footprint.

One project that is particularly close to my heart is the beta campus, an association that was founded in 2019 to develop an ecosystem where people can learn from each other, spin innovative ideas and realize visionary technology, always with an eye on the next generation (Beta Campus, 2023). I am personally involved as the chairman. Part of the initiative is the publicly funded beta world project, where children and young people are given the opportunity to acquire knowledge about climate protection outside of school and apply it in practice. The strong involvement of the business community gives young people unique opportunities to work on practical projects and get to know regional companies. Conversely, this also gives companies the chance to get to know young talent and position themselves as an exciting future employer. The content is developed and worked on by the young people together with companies, municipalities and other partner organizations using a co-creation approach. All participants thus become active shapers of a positive future. Once again, it is important for us to anchor a different way of thinking and to give young people the feeling that they can make an impact. I am convinced that this makes a big difference—because if you can't make an impact, you are lost. And that's where you can make an impact—even if it initially involves smaller tasks on a smaller scale.

6 Outlook for the Future

One topic that is currently occupying us massively is the transfer of the company to the next generation. We have introduced a generation management system, both with regard to employees and, of course, the family. It is important for us to anticipate what the trends and future topics could be that we need to prepare for. How do we make the company fit for the future and capable of change so that it can always adapt well? In doing so, we want to further expand diversity, promote entrepreneurial thinking and action, anchor the willingness to innovate and change and continue to create business opportunities and development spaces. For the people in the company, but of course also for the young family members, this creates the opportunity to get involved wherever something new is developing. This brings us back full circle: "Enthusiastic people create opportunities for generations".

It is particularly important to me that we as a company continue on the transformation path we have embarked on. Everything we have done since then has proved successful, but a transformation in a large company is not a sprint, but a longer run. Along the way, it has also become very clear to me once again how closely the company's path is interwoven with my own transformation story. You can't deny your origins and you always take yourself with you. The good news is that this past always offers material to learn from—and by reflecting on it, everything can ultimately be attributed to something. After all, how am I supposed to know who I am if I've never been there?

There is a Chinese character that means opportunity and crisis at the same time. And that's the exciting thing. There are always two sides to a coin. You can't have one without the other. I'm excited to see what the future holds for our company and how my personal transformation will continue.

References

Beta-Campus – Verein zur Entwicklung eines Ökosystems für generationsübergreifende Innovation, Kollaboration und Fachkräftesicherung. (2023). https://beta-campus.at/

Maslow, A. H. (1969). The farther reaches of human nature. *Journal of Transpersonal Psychology, 1*, 1–9.

Thomas Welser is CEO of the "Welser Profile" group of companies—a manufacturer of custom metal sections with its headquarter in the beautiful Mostviertel region. The family business has always attached great importance to the joint development of the organization and its people—the constantly growing international customer base and the 2400 employees worldwide. Six years ago, Thomas Welser initiated an ongoing "cultural transformation" project. In addition, he sees his main tasks as building networks in the areas of innovation and technology, as well as the promotion, training and development of talents.

Lessons from a Wasp Sting: A Plea for Inner Development in Order to Change the World for the Better

Hannah Lux

Abstract Is it possible to make the world well without looking deep within, without making yourself well, so to speak? This chapter emphasizes the power of inner work for sustainable, outer transformation by sharing the author's incredibly personal story, as an award-winning social entrepreneur, inner development advocate, a driver of change, and as a woman committed to standing up for her truth. Her heartfelt account illuminates the impact of understanding how traumatic events shape our lives and responses, how cultivating awareness of personal and professional drivers helps pave a clearer path, and how healing the world within facilitates healing the world beyond. The author is convinced that sustainable entrepreneurship is not a career choice; it's a call to adventure and depth. And she invites those who are embarking on the journey of sustainable entrepreneurship to grow deeper in their impact while finding their way to wellbeing, through insightful prompts as part of a practical, developmental set of tools.

1 Breaking Open, Once Again

It is an afternoon in early April 2020. I find myself sitting on the kitchen floor in my tiny Viennese apartment. The spring sun is shining bright in the blue sky. It is so beautifully quiet outside—one positive side effect of this lockdown. But, inside me, my heart is racing like crazy. What in the world is going on?

Today marks the fourth week of the first COVID-19 lockdown affecting Europe. Like many people, we—my co-founders and I—find ourselves in a situation we could not have imagined 5 weeks prior. Our intergenerational coffee shops are closed and the elderly employees, who make up more than half of our workforce, are at home. Most of them are considered high-risk and require protection from the spreading virus, yet most of them urgently need their jobs to keep them out of

H. Lux (✉)
Vienna, Austria

M. Mühlböck (ed.), *Sustainable Transformation and Well-being*, CSR, Sustainability, Ethics & Governance,
https://doi.org/10.1007/978-3-031-75566-8_20

poverty. To our employees, their job provides their income and also their social safety net, often representing the only real sense of community and belonging they have. In fact, out of the near 80 people employed with us, 45 of them are bakers and hosts belonging to our group of "beneficiaries"—elderly people who are either living at or below the poverty line and/or are suffering from loneliness.

In the sixth year after founding our social business, a true heart project that we lovingly like to call Granny's Public Living Room, all the mixers in our coffee shops are still. Not a single granny cake is waiting to be eaten. Our guests are gone, our shops are closed, and the team sustaining the dream remains bound to their houses. Our coffee shops designed to create cozy, loving spaces for connection between generations have come to a complete standstill. The vibrant exchange between guests of all ages, from all corners of the world, whose support creates jobs for elderly people has ceased. The opportunity for our elderly employees to foster meaningful relationships within our multi-generational team as a way out of loneliness is seemingly gone. And with it, 6 years of giving everything and sometimes even more, the personal and financial risks undertaken to make our vision work, surpassing the scrutiny of almost every "expert" who advised us that this social enterprise would not and could not succeed. We, as founders, reflect on how far we have come to drive positive change while facing those times of uncertainty. But, how will we make it through this crazy pandemic?

With weeks of endless brainstorming between us and many co-founders, partners, and the team, we arrive at a solution for how to preserve and prolong this vision of ours, creating a world in which no one grows old alone, nor in poverty. After numerous days and nights of hard work together, we agree we must take out another loan. There is a risk of not being able to pay it back, since the fate of the world's economy is anyone's guess, but after close consideration, it is clearly the right and only way forward. That will enable us to pay the salaries of our team and cover our most important costs while compensating for our lack of incoming revenue. This will also give us time to pivot our social business model from offline-only to a digital income stream which, in addition, will keep us afloat. It makes total sense to sign this loan contract. I have signed many contracts before and have taken responsibility for bigger funding rounds and yet, today is different.

Something inside forbids me from holding myself together anymore. The contract papers and a pen to sign them lie in front of me on the kitchen floor, tears are streaming down my face. I can't stop crying. What feels like my body freezing and exploding at the same time "just" because I should sign the papers of a contract becomes a full-on panic attack and represents the first very loud and very apparent sign of an emotional burnout that, inevitably, leads me to turn my life around.

I text my fellow co-founder with a plea for help and he comes immediately. With his support and thanks to the results of inner work, including hours and hours of therapy and breathwork practice, I manage to calm myself down and sign the contract papers. In my calmer state I know this will help transform our organization into a better-prepared, more crisis-stable organization. (And now, almost 4 years later, our social business is doing well, and I can say with certainty the decisions made on that day were truly the right ones.)

But what had actually caused me to enter such an emotional state? From my early 20 s on, it was clear to me that I wanted to create a social enterprise just like this. It was clear to me that I wanted to do "more" than just earn money. I love everything about what we have achieved in our work. I love how my co-founders and I, together with an incredible and diverse team, have positively impacted the lives of so many. I love the people I work with, and I "get" so much in return from our collective efforts in terms of gratitude, creative freedom, recognition, and love. And yet, on this sunny afternoon in April 2020, I clearly feel I have hit a wall and something has to change.

That day, I had to sign or choose not to sign while something broke inside of me and I found it all too much. And by "it" I don't mean the extraordinary situation the whole world was facing as a result of the pandemic. The consequences of the pandemic, I believe, just sped up the process of arriving at the same conclusion. "It" was more the consequence of crossing my personal boundaries way too often over the years. The energy investment of forging different social ventures, the hours of working tirelessly to see them through to fruition, the constant hustle to maintain a productive workflow, and countless months of not being able to pay ourselves an adequate salary.

"It" was an accumulation of all these things, but, more so, "it" was a consequence of the extreme expectations I had placed on myself. Those expectations originated from a good yet probably unhealthy place within me, but they also existed as a detriment to my wellbeing. I had grand expectations that could never have been fully met, no matter how hard I worked or tried. I held myself to an impossible standard where there was always more that should and could be done.

2 From Within to the World

Returning to now, January 2024: I could have started my contribution to this book very differently.

I am a female social entrepreneur in her mid-30s and have co-founded different organizations and networks in the field of social and sustainable entrepreneurship. I have handed over the management of social businesses that I have co-founded and led, all of them still impacting the world and the lives of many in it. I sit on advisory and supervisory boards of different organizations close to the government, and I now consult at organizations of all sizes, coaching top-level executives and leaders of change in different industries. Awarded for my accomplishments, I have been named in the Forbes 30 under 30 list and have just been, together with my co-founders, voted EY Entrepreneur of the Year 2023 in the category of Social Innovation. I am exceedingly proud of the impact we are creating, and I can honestly say I love what we do. I see it as an absolute privilege to be able to give so much of my life's attention to fostering positive change in the world and I consider myself very lucky to have, over the years, gotten so much love in return, often going to bed at night with the beautiful feeling of a real sense of purpose. So, in effect, my story seems to confirm that sustainable entrepreneurship does lead to improved wellbeing. BUT…

When taking this book's title into consideration, it's important for me to share with you the other, maybe not so shiny, part of my story. A part that comprises the foundation of my work and kindles the flame at the heart of my motivation. It's a part that leaves me feeling vulnerable but is paramount to understand so that social and sustainable entrepreneurship, in itself, is not seen as a magic formula to finding meaning and purpose in life. Because, as always, "where there is light, there is shadow."

And I believe it's time to start embracing those not so "shiny" aspects of sustainable entrepreneurship so we can learn from them, deepen our understanding, heal, and become healthier—individually and as a collective. After all, that's what sustainable entrepreneurship is about.

My rollercoaster ride in this field has led me to many, but, irrefutably, one leading conclusion that I find particularly important in this context, to embark on the journey of sustainable entrepreneurship with a vision that aims to make the world a better place, means embarking on a journey of honesty within. Those journeys are synchronous, characterized by the breaking down and transforming of existing structures and systems in both inner and outer worlds.

But back to my story. Here is how it begins:

I grew up very privileged as a white woman in a loving family in the safe and beautiful country of Austria. We lived in a small town surrounded by hills, forests, and crystal-clear lakes. My younger sister and I, due to our parents' divorce, learned quite early that life wasn't always easy, but at the same time, we learned that a loving togetherness and a strong network of family and friends can soothe the sting of painful experiences.

However, at 18 years old, on a hot summer's day, my life took a drastic turn. Not long after getting my driver's license, I was behind the wheel when a sudden wasp sting caused a moment of distraction. The consequence was a serious accident, tragically resulting in a loss of innocent life and inflicting severe injuries to beloved others. This was the beginning of a profound personal struggle with grief and guilt, and a painful, maybe life-long journey toward deep self-forgiveness.

Years of depression followed, and I endured lasting psychological and physical side effects. "I don't deserve to have a good life anymore" was the belief system that controlled my being during my early 20s. My journey back to joy could be a book of its own, but that is not the point of focus here. The truth is, as scary as it is for me to share this horrific yet incredibly formative experience with you, the entirety of my story can't be told without it. Without me sharing this part of my becoming, I can't write authentically about my life path and learnings as a social innovator—especially when considering their application within the topics of wellbeing and social change. But, why is that so? I believe that we all carry experiences, beautiful and painful ones, which have shaped us into who we are right now.

Of course, every story is different, and the ways in which we learn how to cope with these fundamental experiences are largely individual. In my case, one way out of the pain and inner prison of guilt I had built for myself was to engage in the world and "do good." One could say that I wanted to give back to life. I wanted to do something meaningful, and I was "obliged" to "use" the second chance I was given at the age of 18 when I survived the accident.

3 What Drives You?

Subconsciously, what happened back then is quite obvious to me now. After many years of inner work, I now realize: I wanted to do good on the outside, not only but also, to patch over some very painful, dark places on the inside. One could argue there are worse coping mechanisms than founding a social enterprise, I'll give you that, but when it comes to sustainable entrepreneurship in connection to wellbeing, I believe it is crucial to become crystal clear about your own motivation and inner drivers. For me, I can see clearly now, with some time and distance from the actual event, the connection between the accident that occurred when I was 18, the deep pain, and the subsequent necessity of finding my way out of the brutal emotional prison I had confined myself to, and, ultimately, recovering my journey to joy (also as a social entrepreneur). The motivation in my early 20s was strongly influenced by somehow wanting to prove to myself, and to the world, that "I was a good person" after all. I wanted to escape the guilt by doing good. And while I checked the boxes, meeting my own criteria for being a good person by creating positive impact with my actions, I also constantly ignored and crossed my own boundaries. I would even go as far as calling it self-exploitation—something, in my opinion, completely misaligned with wellbeing or sustainable impact.

I was, of course, motivated by other drivers, too, like the values instilled in me by my loving and supportive family. I was fortunate to have experienced, first-hand, role models throughout my upbringing who cared, not only about themselves, but for those around them and also for others around the world. I always seemed to feel outraged and angry when witnessing inequality, either, myself, as a young woman when being compared to men, or even worse, when realizing that friends and strangers from diverse cultural backgrounds were unjustly treated differently to myself. I could go on and on with a longer list of my personal "whys," but what I can say, with certainty, is that your motivational drivers and core values must be compatible with the way that you impact the world. Everything else will break you down in the long run. You cannot achieve a state of wellbeing if you act externally to avoid looking within—no matter how "good" your outward actions are, sustainable entrepreneurship or otherwise.

To help you on your quest to understand more clearly what motivates you, I would like to offer some guiding questions. They are those I have found useful over the years to attain clarity, myself, helping me to acknowledge and modify unhealthy patterns.

- What is the real "why" of your engagement as a social innovator?

- Do you act out of love for yourself and ultimately the world or is fear the driving force for your doing? What are you afraid of?
- What are the values that are dear to you, that are at the core of your unique self?
- What was the "spark" that led you to your doing?
- Imagine you tell your (unborn) children, in 20 years from now, about your engagement in sustainable entrepreneurship: what will you tell them about why you started? What was your initial motivation? How do you feel about it? Would you consider it a sustainable motivation compatible with your core beliefs and values?

4 Deep Care for Yourself and for the World

I strongly believe that you can only care about the world as much as you care about yourself. What sounds like a sentence from the first page of my granny's kitchen calendar holds a lot of truth for me. It may even sound a bit selfish at first glance, but the kind of care it alludes to is a profound and deep, yet gentle and warm love for your inner core, no matter how dark or distorted your insides might feel at times. What it doesn't imply is the kind of care that is flippant or self-absorbed as in "You don't care about what's happening" and "You only care about yourself."

When you nurture the kind of care I mean, for the beauty and the dark inside of you, that care becomes a certain kind of feeling. It's a feeling from where your care for the world organically develops. By caring to acknowledge, accept and appreciate the beauty and dark within, you develop the same capacity to feel those sentiments for the world. All around you and out of love (instead of fear), you appreciate the beauty that exists. Attuning yourself to this type of care is where, I believe, real change is made possible. I believe that if sustainable entrepreneurship should lead to a greater wellbeing, then it is important that any action originates from a relatively healthy place—a place of real love for the world.

This isn't to say that you need a lifetime of therapy for you to start taking action in the world. In fact, the more you align yourself with your core values, seek clarity in understanding your motivation, and learn to love yourself throughout every stage of your sustainable entrepreneurial pursuit, the more the process will create opportunities for you to heal yourself. Though you might be presented with many seemingly impossible external challenges, you might actually discover through these situations that your own inner world presents the biggest challenges on your journey as a changemaker. And what a beautiful journey it is when you fully embrace it!

So yes, go out there and live by your values, stand up for your truth! I applaud anyone who dares to do so while choosing to stand up for something bigger than themselves. Just remember to move forward while also looking inward, for your sake and for the sake of the world. Though that sounds a bit like a cliché, it is part of my truth and has been proven through my experience over the past 16 years as a social entrepreneur and driver of change.

My story is one of self-forgiveness and of freeing myself from guilt, while accepting what happened and taking responsibility, where it is mine to take. Coming to peace with myself will likely be a lifelong process, and I don't feel that I am alone in this undertaking. We all carry trauma and stories that might have hurt us, yet shaped us. I encourage you to look at them, learn to love them, and, if you can't love them, at least acknowledge and accept them. You have to name them to tame them. You have to feel it to heal it.

Yes, sustainable entrepreneurship might bring you joy, but I am 100% certain that no good deed will save you from the inner work needed to heal yourself and find peace within. This work cannot be done by anyone else but you, but it cannot be done by you alone, either. And the good news is you are not alone. And while we can infer that there is a collective aspect that will lead to sustainable impact and wellbeing (I will get to that later), there are many ways for you to start your journey to inner peace, on your own accord, right now.

There are also many amazing professionals available to support you, just as there are many different paths to find your version of healing and peace. In my case, it was/is a mixture of learning to read my body and listening to its magnificent healing power. Learning how to breathe properly using vested techniques and allowing my breath to guide my doing. I practice listening to the silence within. I seek nature and other times professional coaching and therapy is what I need. Sometimes I read about trauma to understand it intellectually and, with loving support in safe spaces, I dive into it once again for a renewed sense of peace and healing. I tap into ancient healing techniques, harnessing their inherent magic, into the wisdom of yoga and into the power of music and movement. What nurtures me the most are deep connections with friends, family, and strangers and the opportunity to openly share what I'm going through—the mess and the beauty. This is how I exercise compassion through presence. There are so many ways to get started! I can only encourage you to discover what works best for you.

Here is a little exercise you might start with, even now, to uncover the roots of your journey:

To dig deeper into my granny's kitchen-calendar-sentence-collection, I present another quote, by nobody less than Mahatma Ghandi (go read about his life if you haven't already!): "Be the change you wish to see in the world". I urge you to take 20 min in silence and reflect on that sentence in connection with your engagement as a sustainable entrepreneur. After the 20 min have passed, take a piece of paper and a pen and give yourself another 5 min to write down what comes to you in reflection.

- Are you actually the change you wish to see?
- What is it that you want to change within yourself that you believe would be nice to see changed in the world?
- Who or what could support you in getting there?

- Who are people that you already know that could support you on your way to becoming that change? Do they know they are your elected supporters?
- Now, engage the world: go out and find people to initiate a conversation with based on your reflections and see what unfolds. Be authentic, show yourself, and dare to be vulnerable as an exercise in personal and collective growth.

5 Finding Your Star Collective

Being active in the field of sustainable entrepreneurship while actively improving my wellbeing has taught me so many incredible lessons. One of the most important has been learning how to open up to others and asking for help. You can't do that work alone. Let me repeat: you CANNOT do that alone.

To steer a social business through a world that is becoming ever more complex while navigating the complexities of your being and trying your best to make it well you need a real network of supporters and trusted allies, a team, and ecosystem to act in collaboration with.

In the latest social business I co-founded, my partners and I invested a lot of time and resources in establishing our relationship with support from professional externals, and every second of it paid off. Many start-ups and smaller businesses don't make it because of "founder clashes" and this happens for a reason. Complete openness and honesty are necessary to make the dynamics work. That requires you to expose your vulnerabilities while adopting an eager devotion to your common interest, during the ups but especially during the downs. I believe that is fundamental when it comes to sustainable business, where there is more at stake than "just" money and the framework hinges on aspects of your deep beliefs, your values, and your hopes for the world.

I am forever grateful to my co-founders for sustaining the common vision that guides our organization. I am grateful they step up and hold themselves accountable and step in to hold me accountable when I might lose my sense of direction. And I am grateful that when I am struggling through the hardships of my personal journey, I can count on them to support me unconditionally while helping me to stay on track. Our story has proven that many times.

We have managed to "dig through the mud" together and I am so proud of our accomplishments. Our tenacity has fueled the fire of our burning vision through many, sometimes hurricane-like storms which is one key reason why this sustainable enterprise of ours still continues to flourish.

However, to effect change in the world today, you must extend yourself beyond the support of your professional team and inner circle. I believe real change today is only possible through ecosystem collaboration. The challenges that our societies are facing in this globalized, intra-connected and fast-moving world can only be faced in networks that are based on trust and honest respect for each other. So, what does it take to become part of a collaborative ecosystem or how might you kick-start a collaborative initiative yourself?

For me, it takes the courage to be seen. To go out into the world and show myself authentically—flaws and all—and letting go of my perfectionism and my fear of not being good enough (and letting go of my worry that others will see that. Scary!). It takes talking about ideas that are not yet well-considered and releasing them to become someone else's in order to impact the world. It takes regularly stepping into someone else's shoes for the sake of finding common ground and trusting in the intrinsic wisdom of the collective. The following exercises might be helpful in this matter:

- *Your personal "star collective": In Karen Walrond's book,* A Lightmaker's Manifesto: How to Work for Change Without Losing Joy *(big recommendation by the way!* Walrond, 2021*), the author speaks of her tight support network as a "star collective"—a list of people you know (who themselves are aware of the list too) whom you can fully trust and count on. Based on that idea, I urge you to take a sheet of paper and a pen and spend a moment jotting down who is part of your personal star collective. Pay attention to the areas where you may be missing someone. These can be close friends, your grandmother, an experienced entrepreneur whose work you admire, or a trusted business consultant—most importantly, it's people that you feel comfortable sharing honestly and openly with and with whom you manage to be your most authentic self.*
- *Network map: Another effective exercise regarding this topic is to draw your personal network map. With paper and a pen, write your own name or the name of your sustainable enterprise in the middle. You can do this exercise for yourself, personally, or for your organization. Next, write down organizations or individuals that you consider important partners and collaborators. Who is part of your network? Then, when you can't think of any new names, try viewing the map from different perspectives.*
  - *Who could be supporters in terms of finding the right partners for your business?*
  - *Who is working on the same mission, shares the same or a similar vision with you?*
  - *What are goals you have in the long run and who would you need as partners to reach those goals?*

Add any more names corresponding to the new perspectives—you might be surprised to discover how many allies you already have. You might also identify gaps where it would be sensible for you to make more effort toward finding suitable collaborators.

6 From Reaction to Resonance

A friend of mine, who is also a business and life coach, and someone who is part of my star collective, asked me once during a time in which I was completely overwhelmed: "Do you think you are reacting or responding to what's going on in the world?"

At first, I did not understand what he actually meant, but today I believe this is one of the most important questions every sustainable entrepreneur should ask themselves on an ongoing basis—for the sake of the impact they want to create and for their own inner sanity and wellbeing.

Maybe you know the feeling of confusion, despair, sadness, and anger toward the immense inequality and suffering that goes on in the world (I know it, for sure). Taking action can sometimes seem insurmountable: What can I do about it? Can I even do anything at all? For a long time, I reacted in two ways to this overwhelming state I would find myself in by either shutting down completely, overcome by my emotions, or jumping in, headstrong and battle-ready, to resolve every worldly issue I could.

The first emotional response can sometimes be a necessary go-to when you're in need of stability amidst a bout of inner chaos, but I have come to realize it is not an appropriate option for me when considering my reactions to the world (and I believe you might feel the same; otherwise you would not be holding and reading this book). The second response, though it feels more productive, is actually a reaction trap and is great fuel for suffering achieving emotional burnout—which detracts from your ability to create a deep impact or embody healthy wellbeing. So, when there are, seemingly, only two options, I always ask myself what could possibly be a third. My answer to that here sounds fairly simple but is surprisingly difficult to do. It is: Listen. Listen deeply. And out of that, respond.

Only when you listen deeply, when you become quiet and hear what's happening around you and inside of you, do you start to understand that a lot of our (re)actions arise from fears and patterns, often developed in the past. When you acknowledge the fears that instigate your reactions, they begin to lose their power and influence over you. As your fears diminish, you create a new place within whence you start responding to the world. It is a place where you initiate your full creative potential and sustainable impact becomes possible. Recognizing reaction vs. response patterns will enhance your everyday leadership skills, which, incidentally, serves to deepen your impact and help you be well along the way.

As sustainable entrepreneurs, most of us are driven by a noble mission, but being human, we are all susceptible to moments where we act using our "lesser" qualities. This illustrates the contrast between leading with a sense of freedom and purpose through resonating with the world, versus operating from a place of fear and dependency. Do I adopt an internal or external perspective? Are my decisions motivated by fear or inspired by curiosity and, ultimately, love? Am I excessively influenced by external factors, or do I navigate based on my own internal compass?

The foundation of reactive behavior is quite straightforward. At some point in our (usually early) lives, we develop coping mechanisms as survival strategies, forming unconscious, habitual patterns. The patterns we develop originate from the interpretations we assign to significant experiences, often lacking a broader understanding at the time. These interpretations persist later in life when we encounter similar situations, unconsciously steering our actions. Despite our good intentions, these ingrained defense mechanisms activate, influencing our reactions and leading us to react impulsively rather than respond thoughtfully.[1]

So, I can only urge you to find moments of stillness every day to facilitate the listening process and become more conscious of your own reactive patterns. By listening to the world outside and inside as well, you will get to know yourself better. You will begin to understand your pain points and patterns in order to better access your creativity freely and find resonance with the world.

I recommend that you observe yourself in the upcoming week and take note daily of:

- *When are you reacting, triggered by some external factors and/or out of fear?*
- *When are you creating and responding freely? How does this make you feel? What effect does it have on the outcome of your doing?*

If you are curious to further explore the reactive and responsive patterns that influence your leadership, then check out the resources at leadershipcircle.com and take a free online assessment to develop your leadership profile. Here you can also access helpful videos and literature on the topic.

7 The Role of the Collective

I could go on forever about the importance of investigating your inner landscape and making peace with its ominous shadows to bring about positive change in the world. And while it is up to you to look at your inner landscape, inner development doesn't mean it's done solely at an individual level. It occurs more effectively at a collective level, instead. We grow and develop in relation to others, so it is the relation to others that has the capacity to bring healing to the whole.

If we now turn to the collective, which is in fact to all of us, what are those inner skills we need in order to move toward a more sustainable future? Presence, integrity and authenticity, complexity awareness, humility, empathy and compassion, mobilization and co-creation skills, trust, courage, perseverance, and optimism—to name just a few of the 23 skills currently identified in a co-creative research process by the Inner Development Goals (IDGs) initiative,[2] a non-profit, open-source and

[1] If you are interested in those issues, I recommend to check out the following website: https://leadershipcircle.com/leadership-assessment-tools/leadership-circle-profile/

[2] https://innerdevelopmentgoals.org/framework/

co-creative initiative and global movement focused on the inner skills we need to drive change in the outside world.

I have personally experienced the immense power of inner development in connection with sustainable impact. So, I can attest to the fact that enacting external, sustainable transformation is indeed possible if we also turn our focus inward, as individuals, but even more so as a collective. We, as humanity, need to address, focus on, and develop exactly that inner capacity needed to be able to deal with the increasingly complex environments we live in. And it is time to put the importance of inner development on political and organizational agendas. That cannot be carried only by individuals alone.

This is why I am a committed collaborator of the IDGs. You will find further references to this initiative in other chapters of this book (see Björkman et al., 2025; Lindencrona, 2025) and I encourage you to dig deeper into this matter. Please feel warmly invited to join the growing IDG movement and co-create with us.

When it comes to the daily realities of sustainable entrepreneurs, I also find it important to mention that the collective—that is, all of us—have a responsibility to support sustainable entrepreneurs on their journey to deeper and sustainable, even regenerative impact (which can only be achieved when personal wellbeing is in place). If we want sustainable and regenerative innovation to take us further as a society, we must create environments that foster creativity and (healthy) entrepreneurial action. Suggestions for the philanthropic sector in this matter have been made in this book by Björkman et al. (2025).

And while here is not the place to go too deep into a political agenda, I want to briefly mention what is usually left out of the dialogue about favoring ecosystems for social entrepreneurs: the myth of the "hero-preneur" and the danger of hyping individual entrepreneurs too much. My experience maintains itself as a thin line between a supportive ecosystem and an unhealthy hype that often pushes, especially early-stage entrepreneurs, to grow their impact beyond their comfort zones and even their capabilities. Remaining dedicated to your truth and growing at a rate within your comfort zone requires resolve, especially when outside factors are constantly pushing. I believe that stakeholders of impact ecosystems should be careful, to find balance between what is real, healthy support for impact entrepreneurs and what might be pushing entrepreneurs beyond their limits and wellbeing.

"There are times when you think 'Oh I can't do that, I can't be that.' But there is always a bigger you within you. So let that person out." Those are the words of Memory Banda, a young social entrepreneur and activist from Malawi fighting child marriage, as spoken in the movie *Bigger Than Us* (Vasseur, 2021). She superbly conveys what I wish to point out to end this chapter:

Embarking on the journey of bringing sustainable impact to the world is a compelling, noble, and beautiful decision to make—and one that is wholly necessary when observing the state of the world. There is no better time to take action!

Humans can change. You can change. And systems can change. Your journey will be heart-opening and incredibly challenging, filled with moments that test your resolve and push you beyond your comfort zone and into your growth zone, when approached consciously. This path is one that promises profound personal growth and the opportunity to make a meaningful difference.

Sustainable entrepreneurship is not a career choice; it is a call to adventure and depth—a quest that is as unpredictable as it is fulfilling, once you are receptive to your personal potential for growth and healing within it. This journey asks you to be bold, to challenge the status quo, to believe in the possibility of a better world and the potential that you, yourself, carry inside. It is a reminder that we are all connected, and through our collective efforts, no matter how daunting the challenge, change is within reach.

This journey is not only about the external work of building enterprises and initiating external change; it is deeply intertwined with the inner work of understanding oneself, embracing vulnerability, and expanding our capacity for empathy and love.

As we strive for sustainability, equity, and justice, we are invited to explore the depths of our being, to confront our shadows, and to illuminate our true potential. This alignment of purpose and self is what makes the journey, not just worthwhile, but deeply enriching. And, yes, this might lead to wellbeing.

"What in the world is going on?" I asked myself in April 2020, sitting on the floor of my Viennese apartment, with a racing heart and a longing for breath. "Letting go of fears, guilt and patterns that no longer serve or strengthen me. Growth. And healing," I would answer now, with immense gratitude for all the learning, lessons, and moments spent initiating and sustaining positive impact. Each a piece of the puzzle constructing a more just, balanced, gentle, and regenerative world—both outside and inside. And I would also say to myself, the me on the floor, "Dear, please find the courage to open up to others, you don't have to, you can't do it alone. Go find the necessary support and embrace the journey, for it is truly worth every step."

And I want to reach out and invite you, dear reader, with the same compassion addressed to my earlier self, to look boldly, dive deeply, and find (inner) peace for your wellbeing and for the world. Let's take these steps together.

References

Björkman, T., Garuti, C., Hillebrand, S., & Mühlböck, M. (2025). The philanthropic sector creating momentum for a sustainable transformation and the well-being of changemakers. In M. Mühlböck (Ed.), *Sustainable transformation and well-being. The role of business for individual and collective flourishing* (pp. 243–256). Springer.

Lindencrona, F. (2025). Eudaimonic well-being: A foundation for public administration systems fit to build a sustainable future. In M. Mühlböck (Ed.), *Sustainable transformation and well-being. The role of business for individual and collective flourishing* (pp. 219–241). Springer.

Vasseur, F. (Director), (2021). *Bigger than us* [Film]. All You Need is Prod, Big Mother Productions, Elzévir Films, France 2 Cinéma.

Walrond, K. (2021). *The Lightmakers Manifesto: How to work for change without losing your joy*, 1517 Media.

Hannah Lux is a social entrepreneur, transformation consultant, and leadership coach for drivers of change. She is a visionary whose work demonstrates that social, ecological, and economic goals can be achieved collectively within an organization and how harnessing this ability leads to a more sustainable (social and ecological), just, and balanced society. In her role as co-founder of Austrian social business Vollpension, Hannah and her colleagues bridge generational disconnect, fight poverty, and resolve isolation among elderly people (www.vollpension.wien). A deep practice of mediation, yoga, and inner work is the base for all of Hannah's being and doing. She strongly believes that the capacity for inner transformation within every one of us defines the capacity for our ability to act in transforming societies and systems as a whole. Therefore she is deeply invested and actively engaged in several initiatives bridging these inner and the outer dimensions (such as the global movement encompassing the Inner Development Goals (IDGs)). In addition to establishing several social innovation and inner work initiatives, Hannah has made profound ecosystem experience in her work at Impact Hub Vienna and through her master studies on Public Policy for political entrepreneurship at Humboldt-Viadrina School of Governance in Berlin. Hannah is also co-founder of SENA (Social Entrepreneurship Network Austria), a member of the supervisory board for creative industries of the Federal Ministry for Labour and Economy of Austria, and advisory board member of "AWS," the promotional bank of the Austrian federal government.

Part VI
Concluding Chapter

A Transdisciplinary Economy for Individual and Collective Well-Being: Starting a Scientific Journey in Economics and Management Science

Marisa Mühlböck and René Schmidpeter

Abstract While defining a unifying concept of well-being has always been a challenging endeavor, it has nevertheless been investigated in many science fields throughout history. The pursuit of well-being in economics and management science has mostly been based on theories and measurement scales that focus on abstraction and aggregation and have thus been able to properly depict neither the complexity of human beings nor that of the world surrounding them. Yet, today's social and ecological challenges that are closely tied to the topic of sustainability call for improvement in science in order to offer better solutions. In this chapter, we explore the introduction of transdisciplinarity in economics and management science as a suitable approach in this regard. As the very last article of this book, it draws on the inputs of the 27 contributors from different academic and practical fields and attempts to sketch cornerstones for a transdisciplinary context-specific research program in the future that puts well-being at its heart. We discuss what a suitable research philosophy might look like, play with initial ideas about a research framework, and introduce potential research questions.

Keywords Transdisciplinarity in economics · Well-being in economics · Sustainability in business · Complexity of human beings · Research philosophy in economics · Context-specific research · Transdisciplinary research · Interdisciplinary approaches in management · Theories of well-being · Integration of practical and academic fields · Dealing with complexity · Sustainable business transformation

---

M. Mühlböck (✉)
Independent Researcher, Vienna, Austria
e-mail: research@marisamuehlboeck.com

R. Schmidpeter
Bern University of Applied Sciences, Bern, Switzerland
e-mail: rene.schmidpeter@bfh.ch

M. Mühlböck (ed.), *Sustainable Transformation and Well-being*, CSR, Sustainability, Ethics & Governance,
https://doi.org/10.1007/978-3-031-75566-8_21

1 Introduction

Well-being is a concept that has recently been discussed with increasing interest within society and in the academic sphere. The contributions in this book have showcased the broad range of perspectives, ideas, and fields of applications that touch upon the notion of well-being in the context of sustainable business transformation. Yet, we are aware of the fact that this publication depicts only a small piece of the whole picture and that existing well-being concepts are still rather vague and diverse. We therefore support Alexandrova and Fabian's (2022) stance that it is essential to contextualize definitions and theories of well-being: Who or what are they looking at? In an academic context, which discipline do they stem from? Is the well-being of a person, society, or the planet the focus? If it's a person, what is their life situation? Is it a child or an adult, a person with disabilities, with a migratory background? What is the cultural context? Are we talking about well-being for today's or future generations?

A unifying character, though, might be found in the fact that life itself seems to strive for well-being. While well-being in philosophy has extensively been explored as a key concept in regard to a good life for all human subjects (Fletcher, 2016; Tesar & Peters, 2019; Tiberius, 2014), biology has investigated the optimal functioning of cells, organs, or organisms and introduced concepts such as *homeostasis* in this regard (Billman, 2020). We therefore see well-being as a suitable guiding concept to facilitate a transformation of business toward a sustainable future. The role of economics in this context can be seen as an enabler of well-being at all levels.

However, in economics, the concept of well-being has until now mainly been tackled under the umbrella of welfare economics (at the macroeconomic level), which is linked to concepts such as utility maximization or rational choice theory at the microeconomic level. Alexandrova and Fabian (2022) criticized the approach to focus on the satisfaction of preferences as too simple and limited, referring to the fact that "choices may be self-defeating, weak-willed, based on false information" (p. 9).

Nonetheless, a transition to a different paradigm seems to be under way with more and more economists opening up to insights from disciplines such as behavioral economics, which has to some extent closed the gap toward a more holistic understanding of economic agents (see chapter "Paying Tribute to the Complexity of Human Beings: A Business Narrative Harmonizing the Well-Being of Individuals and Society" in this book, Mühlböck & Schmidpeter, 2025). That paves the way toward more and more interdisciplinary perspectives. It broadens the horizon in a way that human beings evolve from ratio-driven and utility-maximizing *homines economici* to multifaceted active agents of their own lives. However, its critique of *homo economicus* has until now mostly been criticized empirically (see behavioral economics, e.g., Fehr & Fischbacher, 2002; Kahneman, 2003) or on ethical grounds. We therefore suggest challenging the prevailing reductionist approaches. What we believe is missing is a concept that better integrates the complexity of human beings, which is based on a sound ontological and epistemological foundation and produces a better proxy than *homo economicus.*

As insights from economic research serve as the basis for many political decisions, it is crucial to scrutinize the underlying assumptions and to constantly develop economic theories further so that a policy emerges that really increases the well-being of community agents and not just the normative economic indicators, such as GDP or profit. Yet, even with an interdisciplinary approach, one still risks not seeing the entire picture. As theories and concepts transition from one discipline to the next, new insights are gained but others lost at the same time. To avoid a movement from one limiting scenario to another, we advocate for approaches that generate knowledge beyond disciplines. The introduction of transdisciplinarity and an open discourse in economic research might offer solutions in this regard.

2 Introducing Transdisciplinarity into Economics and Management Science

A beyond disciplinarity or rather beyond interdisciplinary approach includes the challenge of integrating knowledge that is not merely collected in the academic sphere. In particular with an eye on sustainability challenges, Scholz and Steiner (2023) acknowledge that "from an epistemological perspective, practitioners' experiential knowledge is necessary for a holistic framing that allows for an adequate conceptual structuring of complexity and for a proper understanding of the role of contextualization in the practical significance/relevance of ill-defined or wicked societal problems" (p. 1504). They thus see the main goal of transdisciplinary processes in the value creation for both science and practice through effective and functional knowledge integration.

Schüz (2025a)[1] proposes a somewhat extended aim of transdisciplinarity in "the necessity of a reintegration of the human mind into the original wholeness of the universe." He suggests a holistic model of transdisciplinarity that preserves the knowledge from single disciplines and at the same time transcends transdisciplinarily to "a new level of knowledge" by

> (a) […] horizontally accessing the immanent world through scientific research as well as extra-scientific experiential knowledge – generated for instance through tradition, aesthetic or religious experience – in a complementary way (Science-Practice-Approach of Transdisciplinarity, cf. contribution by Gerald Steiner in this volume, cf. chap. 8);[2]
> (b) […] vertically plumbing the invisible depth dimensions of the transcendent world for example through graded philosophical reflection on its reasons, meaning or even unitary origin (Transcendental-Metaphysical Approach of Transdisciplinarity).

[1] This publication is currently in print. As page numbers might change, we do not indicate them here.

[2] Schüz refers to a chapter in his book (2025a).

He postulates that the second part of the model prevents the world from disintegrating "into random manifestations without a reason for meaning" and objectification "without deeper ethical justification" and further states that transdisciplinary science not only contributes to a more holistic view of the world and reality but "serves the individual to become aware of his or her embeddedness as part of the whole and to perceive the responsibility associated with it."

That view fits the paradigm of "all one," which Mühlböck and Schmidpeter (2025) have outlined in chapter "Paying Tribute to the Complexity of Human Beings: A Business Narrative Harmonizing the Well-Being of Individuals and Society" of this book as a basis for a new narrative for and about business. "All one" represents an ontological view of the world that corresponds to how we are all connected, starting with the recognition of the relational aspects of our mind, which has been agreed on by a multidisciplinary group of researchers who established the framework of interpersonal neurobiology (Siegel & Drulis, 2023). Based on their concept, we have furthermore drawn attention to the complexity of humans' mental life and suggested that mastering a harmonious way of differentiation and integration might be a key skill for responsible leadership. Thus, we have argued that such narrative elements outline ways by which business can increase individual and collective well-being.

Summing up, we believe that introducing transdisciplinarity into economics and management science is necessary when we aim for a sustainable business transformation that targets well-being at all levels. In the following sections, we reflect on first cornerstones for a research agenda in this sense. First, we discuss what a suitable research philosophy might look like. Second, we play with initial ideas about a research framework that reflects a transdisciplinary research approach. And third, we introduce research questions that result from the various inputs in this book.

2.1 A Research Philosophy for Transdisciplinarity

For any research endeavor, defining the ontological (how do we perceive reality) and epistemological (how do we gain knowledge about that reality) background is essential when starting a research agenda for which coherence is paramount.

In chapter "Paying Tribute to the Complexity of Human Beings: A Business Narrative Harmonizing the Well-Being of Individuals and Society" of this book, we have briefly touched on critical realism's potential to serve as a suitable philosophical underpinning for complexity science, since it suggests a stratified reality (realist ontology) and merges it with constructionist epistemology by acknowledging that understanding complexity needs some kind of filtering or abstraction and thus interpretation. As such, it promotes multidisciplinary research that seeks to unveil causative mechanisms and thus to draw a holistic picture of the world while taking into account multiple perspectives.

Walsh and Evans (2014) further describe it as a research philosophy that

> bridges a divide between positivist and interpretivist research approaches by marrying a realist ontology with a constructionist epistemology. It is particularly apposite for mixed methods research, enabling it to have integrity and coherence as it addresses multiple research questions by using a variety of research methods, resisting the critique of methodological incommensurability. (p. e5)

Drawing from such a rather open and holistic worldview and acknowledgment of different context-specific ways to gain knowledge, we perceive critical realism as a legitimate starting point for a discussion about a suitable research philosophy for a transdisciplinary economy that puts well-being at its center. A thorough analysis of its appropriateness may be the subject of a transdisciplinary debate and further research.

We would furthermore like to point to the fact that the definition of a common philosophy of science does not only provide a solid foundation for decision-making about research methodology. It also supports the development of a common language. That is in particular vital for transdisciplinary research projects, since on the one hand, multidisciplinary and non-academic stakeholders are involved in the research process and, on the other hand, the acceptance and thus implementation of the knowledge gained will be more likely. Scholz and Steiner (2023) argue that the equal process ownership of science and practice, which they see as an integral part of transdisciplinary processes, "promotes the generation of socially robust orientations (Scholz, 2011)," whereby one supporting factor for the successful integration of epistemics from science and practice is that such processes produce "knowledge that can be understood principally by all representatives of stakeholder groups (and thus includes the potential for consensus formation […])" (p. 1513).

2.2 *The Differentiation-Integration Principle: Exploring a Framework to Foster Transdisciplinarity in a Harmonic Way*

As a cornerstone for future transdisciplinary research endeavors in economics and management science, we suggest investigating in which way and to what extent the principle of differentiation and integration, which we identified as a mechanism to deal with complexity, could be applied (see chapter "Paying Tribute to the Complexity of Human Beings: A Business Narrative Harmonizing the Well-Being of Individuals and Society"). Figure 1 depicts different dimensions for the concept of well-being in this context.

Various scientific disciplines have been trying to explore the individual's inner world (what we think and feel) as well as the outer world (what influences us and what is influenced by us through our actions). In this way, different theories, models, and concepts have consequently emerged and insights have been generated through mostly "discipline-specific" processes and methods. Recently, more holistic and relational science fields beyond "object focused" research have been

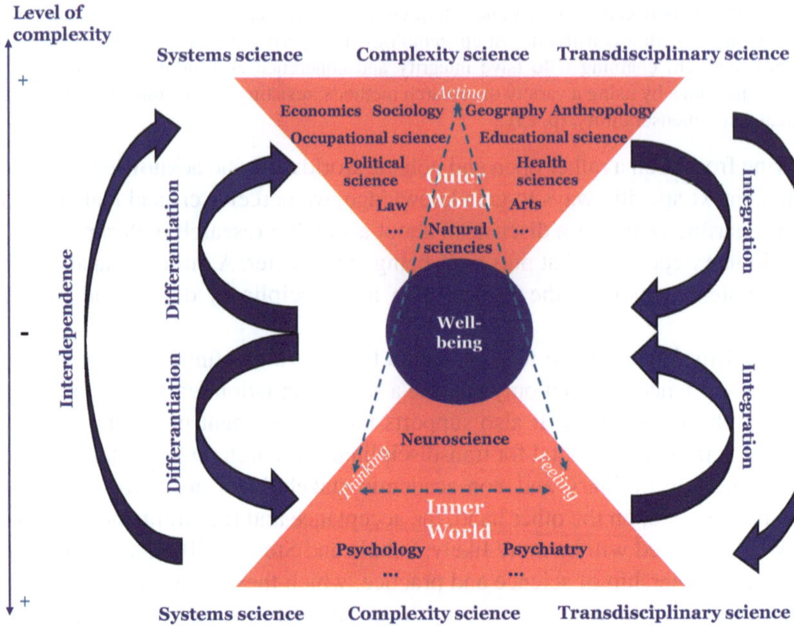

Fig. 1 Exploring well-being through differentiation and integration

developed, such as systems science, transdisciplinary science, or complexity science, which aim at offering answers and explanations to how to deal with the rising complexity in our world. To us, such plurality and abundance of insight generation contributes to an ever-expanding understanding of how the world works and how our actions might influence it. It represents a consequent differentiation under the motto of openness and acknowledgement of constant change. We suggest that the limits to such a promotion of heterogeneity should be defined by ethical deliberations in line with Schüz' (2025a) "transdisciplinary responsibility." With that concept, he promotes ethics "that teaches the art of getting along well with all forms of life," which he sees as a "precondition for well-being and happiness" and proposes "the destruction of the foundations of life" as "limits to the freedom of the human mind to create new realities with its cultural achievements."

Yet, that process of differentiation is only one side of the coin. The other represents the process of integration. Drawing from our past investigations, we postulate that without it the promotion and manifestation of well-being in the real world will be prohibited. Now, what could an approach look like that takes all the requirements outlined above into account?

As stated above, specific contexts impact the design of concepts. We have touched upon potential context-related questions that need to be answered when reflecting on the concept of well-being. Alexandrova (2017) proposes the use of mid-level theories of well-being, which requires that a theory of well-being be built for each context. We therefore suggest exploring how transdisciplinarity could be best practiced in a context-specific way. What is helpful is that both approaches seem to be mutually conducive, since transdisciplinarity calls for the inclusion of contextual experts from the practical sphere (in our case those could, for instance, be different types of entrepreneurs or business leaders or individuals in other contexts that are affected by economic actions). Figure 2 exemplarily illustrates such different contexts.

The respective contexts (in Fig. 2 illustrated as red boxes) serve as a frame for integration (which defines what factors of the subject/object of research need to be taken into consideration) and integration itself may take place through the application of the two approaches of transdisciplinarity introduced above, the science-practice approach and the transcendental-metaphysical approach. In other words, well-being in each context should be investigated transdisciplinarily by context-specifically integrating scientific and practical, philosophical, or even spiritual knowledge. Thus, a contextual knowledge universe for the concept of well-being emerges that integrates and thus honors the plurality of knowledge gained, be it scientific knowledge from the academic sphere, practical knowledge from contextual experts, or completely new types of knowledge that may be developed within the paradigm of "all one." What we consider essential in order to apply the differentiation-integration principle in a harmonic and well-being promoting way is that ethical knowledge in the sense of "transdisciplinary responsibility" be again taken into account.

Thus, a holistic framework emerges that promotes well-being not only because it may be the focus of a research project but also by its entire design.

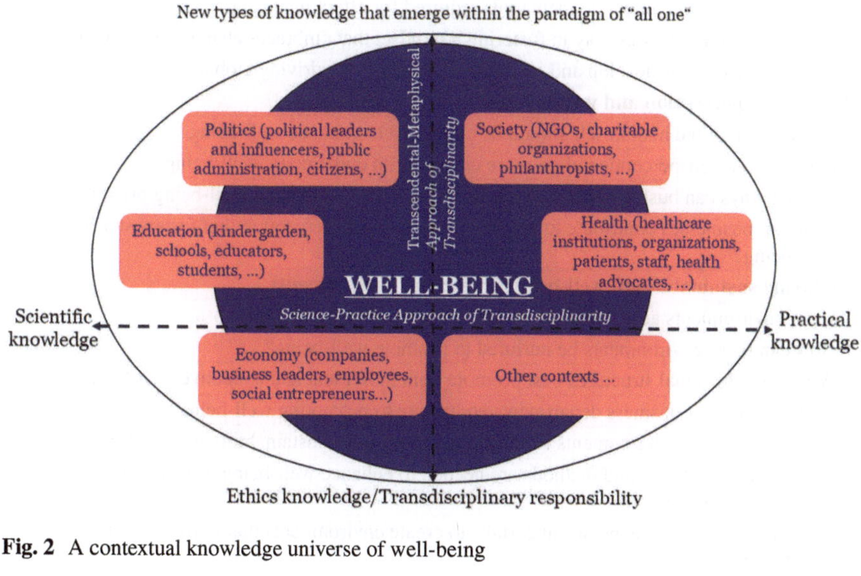

Fig. 2 A contextual knowledge universe of well-being

2.3 Emerging Questions

The 27 contributors to this book have shared various impulses and insights in the context of sustainable business transformation and well-being. Drawing from them, new questions emerge that pave the way for a future research program.

The list of questions in Table 1 and its structure are based on various sources and have been developed in what we, at this point, call a somewhat "experimental

Table 1 Emerging questions in the context of sustainable business transformation and well-being

Foundation for transformation
Do we have the right narrative to facilitate sustainable (business) transformation?
How can businesses reshape their narratives to align with the principles of well-being and sustainable development?
How suitable is the "all one" paradigm for driving such transformation?
How can businesses themselves shape narratives that drive transformative change?
To what extent can the "all one" paradigm guide businesses in their transformation efforts?
What are the prevailing business narratives across various sectors (business, politics, society), segments (i.e., family businesses, start-ups, social enterprises, listed companies, women-led businesses), and regions (and different cultural background)?
To what extent do those narratives align with the "all one" paradigm, and how can leadership be encouraged to embrace it?
Financial engineering and global challenges
Do we possess the necessary financial engineering to address and manage global challenges?
How can we effectively fund, manage, and hedge against global challenges?
Which financial frameworks do we need to foster a sustainable global financial system?
Mindset and tools for change
Do we possess the mindset to acknowledge our limitations and the necessity for new tools?
What are some fundamentally new tools required for effecting change on a global scale?
What role can businesses play in fostering a mindset that embraces change and innovation?
How can businesses develop and implement new tools for driving global change?
Human self-perception and well-being
What image of the human self should be cultivated to enhance well-being?
How can our self-perception align more closely with eudaimonic well-being?
In what ways can businesses align their practices with eudaimonic well-being principles?
How can businesses contribute to shaping a positive self-perception that fosters stakeholders' well-being?
Nurturing sustainable and well-being promoting environments
What environments are crucial for fostering well-being?
How can these environments be nurtured to promote well-being?
What organizational structures best promote both individual and collective well-being?
How do different business departments contribute to promoting well-being?
How can learning environments be developed to promote sustainability and well-being?
What tools, processes, and methods are needed to enhance well-being in workplaces and learning environments?
What initiatives can businesses undertake to create environments that nurture employee well-being?

(continued)

Table 1 (continued)

How can businesses extend their focus on well-being beyond the workplace to other societal "birth places"?
What tools, processes, and methods can businesses employ to create workplaces that prioritize employee well-being?
How can businesses integrate well-being considerations into their learning and development programs?
What strategies can businesses employ to identify and nurture emerging leaders who prioritize sustainability and well-being?
Economic policy and well-being
What type of economic policy is most conducive to individual and collective well-being?
How can businesses advocate for economic policies that prioritize profitability, sustainability, and well-being?
Business strategies and planning in the context of sustainable business development and well-being
What business strategies and planning processes can promote a harmonious integration for well-being at all levels?
Stakeholder engagement for sustainable development and well-being
How can we engage stakeholders from all sectors in discussions about sustainable development and well-being?
What role can businesses play in engaging stakeholders from all sectors in discussions and actions related to sustainable development and well-being?
What strategies can businesses use to foster collaboration and alignment among diverse stakeholders?
Inner development and well-being creation
Which aspects of the entrepreneur's inner world are conducive to promoting the creation of well-being in the outer world? What does integration (differentiation and linkage) mean in this regard?
What skills and capacities are necessary to align with the "all one" paradigm?
What is the current status of these skills across different sectors, segments, and regions?
What skills and capacities do businesses need to cultivate among their employees to embrace the "all one" paradigm?
How can businesses promote the development of these skills across different sectors, segments, and regions?
Which aspects of the entrepreneur's inner world contribute to creating well-being in the outer world?
How can integration be achieved to promote well-being on both personal and societal levels?
How can businesses support entrepreneurs in cultivating inner qualities that contribute to societal well-being?
In what ways can businesses integrate well-being principles into their entrepreneurial culture and practices?
Understanding and supporting business leaders
How can we better understand the thinking, feeling, and actions (and their interdependencies) of business leaders?
What leadership approaches can businesses employ to encourage the adoption of the "all one" paradigm within their organizations and beyond?
What initiatives can businesses implement to support leaders in promoting well-being at all levels of the organization?
Which implication does the "all one" paradigm have on the concept of responsible leadership?

(continued)

Table 1 (continued)

The role of arts in sustainable transformation
How can the arts contribute to sustainable business transformation and well-being?
How can businesses leverage the arts to drive sustainable transformation and promote well-being?
In what ways can businesses integrate artistic practices into their operations and engagement with stakeholders?
Ethical considerations
What ethical implications need to be considered in the pursuit of well-being and sustainable development?
What ethical dilemmas and trade-offs should businesses anticipate and address in their pursuit of sustainable business transformation and well-being?
How can "transdisciplinary responsibility" be implemented in the business context?

transdisciplinary process" involving our own knowledge as well as the knowledge from the contributors of this book and from artificial intelligence (AI).[3] Some questions have been actively posed in the respective chapters or were proposed over the course of follow-up conversations with various authors. Others have materialized through the process of reading and synthesizing the content. That concerns not only disciplines and fields that have already been included in this publication but also those that we consider relevant in the future, but could not be integrated at this point. That process resulted in a first rather unstructured list of questions. Next, we involved ChatGPT. We first asked the AI tool to structure the content. After a slight rearrangement of the structure done by ourselves, we asked ChatGPT to add questions that might be missing. From those suggestions only very few have been added to the following list, due to the fact that the majority of them were too general for us.

When asking ChatGPT to name disciplines that could help answer those questions, we received 36 propositions, from ethics and philosophy, anthropology, biology, and geopolitics to organizational behavior and health economics.[4] Fields such as complexity science or occupational science, which we also consider relevant in this context, were not part of that list. When asking the AI tool to suggest stakeholder groups that should be included in a transdisciplinary approach to promote sustainable business transformation and well-being, we received the categories and descriptions in Table 2.

Even after completing several run-throughs and using various input sources, we are aware of the fact that our list of questions makes no claims regarding proper

[3] Schmidpeter experimented with artificial intelligence in his essay "Crisis and transformation - a prerequisite for a new paradigm in the economy?", published in Schüz (2025a). Transdisciplinary thinking and acting - Fundamentals and perspectives. Springer.

[4] Two cycles were completed (Question 1: Which disciplines could help answer those questions? Question 2: Could you add more disciplines?)

Table 2 AI suggestions on stakeholder groups to be included in a transdisciplinary approach to sustainable business transformation and well-being (OpenAI, 2024)

Businesses and corporations
Engage representatives from small, medium, and large enterprises across different sectors to understand their challenges, perspectives, and innovative practices related to sustainability and well-being.
Academia and research institutions
Collaborate with researchers, scholars, and academic institutions from diverse disciplines to contribute expertise, conduct interdisciplinary research, and co-create knowledge that informs sustainable business strategies.
Government and policymakers
Involve policymakers, regulators, and government agencies at local, national, and international levels to develop supportive policy frameworks, incentives, and regulations for sustainable business practices and well-being promotion.
Civil society organizations
Partner with NGOs, advocacy groups, and community-based organizations working on environmental conservation, social justice, human rights, and public health to incorporate grassroots perspectives, mobilize collective action, and ensure inclusivity in decision-making processes.
International organizations
Collaborate with UN agencies, multilateral organizations, and global initiatives focused on sustainable development, such as the UN Global Compact, to align efforts, leverage resources, and scale impact on a global scale.
Labor unions and worker associations
Include representatives from labor unions, worker cooperatives, and employee associations to advocate for fair labor practices, worker rights, and equitable participation in decision-making processes within businesses and supply chains.
Investors and financial institutions
Engage investors, asset managers, and financial institutions to integrate environmental, social, and governance (ESG) criteria into investment decisions, drive capital toward sustainable businesses, and incentivize responsible corporate behavior.
Local communities and indigenous groups
Consult with local communities, indigenous peoples, and marginalized groups affected by business activities to respect their rights, incorporate traditional knowledge, and ensure that development initiatives are culturally sensitive and inclusive.
Consumers and customers
Solicit feedback from consumers, customers, and end-users to understand their preferences, values, and expectations regarding sustainable products, ethical business practices, and well-being outcomes.
Media and communication networks
Collaborate with journalists, media outlets, and communication networks to raise awareness, disseminate information, and amplify voices advocating for sustainable business practices and societal well-being.
Professional associations and industry groups
Partner with industry associations, trade unions, and professional bodies to facilitate knowledge exchange, promote best practices, and establish industry-wide standards and certifications for sustainability and well-being.
Youth and future generations
Involve young people, students, and future generations as stakeholders in decision-making processes, recognizing their stake in shaping a sustainable and inclusive future and empowering them as agents of change.

structural consistency or completeness. However, it may represent a first starting point for further investigation. The AI suggestions regarding disciplines to look out for might inspire further research projects such as a cross-disciplinary meta-analysis, and the proposals of stakeholders to include might serve as an impulse to reach out to those in order to discuss potential future collaborations.

3 Concluding Reflections

How could we steer economics and management science into a direction in which well-being at all levels is at the center? To us, that is the main purpose and thus the key question of a research agenda that serves the people and the planet. Technocratic factors such as GDP do not qualify to represent the unifying basis for political decisions that promote well-being, since they neither sufficiently mirror the complexity of human beings nor that of the world surrounding them. We therefore need to enter into a new debate on and clarify how we see humans when deciding which theories and which measurement scales to rely on in that regard. It might be easier to adhere to abstracting and aggregating concepts that are mainly based on the image of purely utility-optimizing agents. Instead, trusting in people's individual capacity of self-organizing their ever-changing multifaceted lives in a well-being-promoting way will require constantly investigating and providing resources that facilitate an environment for decision-making of complex humans in a complex world.

3.1 Dealing with Instead of Reducing Complexity

We therefore conclude that striving to reduce complexity will be the wrong answer to today's various social, ecological, and economic challenges. Consequently, the need to find a better proxy in economics and management science than *homo economicus* becomes evident. This book and its various contributions and in particular this closing chapter have pointed to transdisciplinarity as a potential approach to define such a better proxy and with it, in line with Alexandrova and Fabian (2022), "better theories, better methods and better ethics."

What we perceive useful in this regard, on top of integrating different knowledge from different disciplines, is to actively include knowledge beyond classical academic fields, such as ancient (see chapters "How Purpose Drives Sustainable Business Transformation and Well-Being: Practical Wisdom of Taoism"/Zu, 2025 and "Centennial Companies in Japan: Ethical Strategies for Sustainability and Well-Being"/Sato, 2025) and practical wisdom (see chapters "Teaching and Implementing Ethics for Well-Being in Business: A Field Report," "The Skull in Our Suitcase:

What We Can Learn from Zen in Challenging Moments," "From Paragliding in the Himalayas to Finding Purpose in Unleashing Women's Full Potential: A Personal Report," "If I Can Really Make an Impact, Then My Life Has Been a Success: A Personal Report on the Transformation of a Company and an Individual," and "Lessons from a Wasp Sting: A Plea for Inner Development in order to Change the World for the Better": Schüz, 2025b; Weetjens, 2025; Vermeulen, 2025; Welser, 2025; Lux, 2025). The arts and even spirituality might additionally serve as a transdisciplinary "vertical" input at the meta-level to help us build context-specific well-being promoting environments.

3.2 *Striving for Harmony*

The various contributions in this book have suggested that the pursuit of well-being challenges us to constantly recognize the opportunities of diversity and plurality that the world offers us and to integrate them in a harmonious way. They remind us that harmony is the consonance of different components. The uniqueness of musical pieces (see, e.g., Oliver Graber's (2025) very special discovery journey of Gustav Mahler's *Symphony of a Thousand* in chapter "The Zen of Music: On the Energetics of Music as a Source of Personal Well-Being") or of a delicious dish emerges only by creating a harmonious unity through diversity. Likewise, ancient wisdom traditions refer to the concept of harmony (e.g., Taoism, see chapter "How Purpose Drives Sustainable Business Transformation and Well-Being: Practical Wisdom of Taoism"/Zu, 2025, or Shintoism, see chapter "Centennial Companies in Japan: Ethical Strategies for Sustainability and Well-Being"/Sato, 2025) and teach us ways to successfully create a fruitful social co-existence.

Just by taking a quick look at the business world (a deep-dive may follow in future research endeavors), we find practical examples of the importance of harmony in the free market economy. As consumer needs change over time, companies provide solutions for the various demands and challenges. In such a process, even companies in the same sector that use the same resources, tools, and technologies should and must differentiate themselves in order to create value and thus survive. Just as the fastest horse will not always and everywhere be the "best," since in other living conditions being agile rather than fast may be the decisive factor, the market also requires differentiation. For example, the use of one technology may actually be about offering the fastest product, while another scenario may require more safety than speed. At the same time, the requirement to implement a business strategy that harmonically integrates various factors and parameters becomes evident. This can, for example, be illustrated by comparing the German and the Swiss railway systems (Süddeutsche Zeitung, 2023). Although the first provides many more high-speed trains than the latter, the German system struggles tremendously with delays due to infrastructural shortfalls. The benefit of speed will thus only be experienced by those few who only need to travel one leg without a transfer. All other customers are repeatedly left in frustration. By contrast, the Swiss system bets on capacity and trip frequency and is thus able to provide one of the most punctual railway systems in the world. Speed is thus only one factor out of many others that is harmoniously integrated within the whole system. Instead of minimizing the travel time for a few, all customers, and as such economy and society, benefit from a smooth trip. Truly entrepreneurial thinking must therefore be synonymous with recognizing and dealing with diversity and creating a product or service that optimally fits the context (integration). And truly sustainable entrepreneurial thinking ensures that a company's portfolio in that context is not only a market-fit, but also well-being-promoting.

Striving for harmony can therefore never be seen as a top-down, but as a bottom-up approach. It is a self-emerging and self-organizing organic process facilitated by a mindset that recognizes and embraces abundance instead of scarcity. Thus, even the debate about sustainability transforms from a one-dimensional development toward abandonment to one that seeks new ways of harmonizing with the abundance.

3.3 Mastering Adaptation and "the Art of Transformation"

Consequently, what complexity and diversity also demand is constant adaptation, which is an ability that has given us humans an evolutionary advantage throughout history, and as such to master "the art of transformation" (see chapter "The Art of Transformation: How We Can Simultaneously Adapt to and Change the World," Brunnhuber, 2025). In order to nourish that ability, we must keep on exposing ourselves to the unknown, even if that is the more difficult and uncomfortable way. That becomes an even more important task when we try to anticipate the future—because despite all our efforts to make predictions, in the end we cannot know what it will really be like and what will be demanded of us in such a process. The evolutionary biologist Ernst Mayr (1994) puts it this way: "Who could be sure that today he can define a certain number of ideal properties and that tomorrow it will not be a quite different […] set of properties that will lead to the desired harmonious society" (p. 19).[5]

Ultimately, sustainable business transformation challenges us to face up to the responsibility of the Anthropocene and align the economy as a system at a macro level, just as each individual company should transform itself with this aspiration. On that note, we are excited to further investigate how to integrate knowledge in a transdisciplinary way. We invite other researchers to join that scientific journey and help us shape a research program for a transdisciplinary economy for individual and collective well-being.

---

[5] This translation of the German quote was provided by Jeremy Gaines and published in Koch, S. (2012). Organisms. Verlag für modern Kunst Nürnberg (p. 29).

References

Alexandrova, A. (2017). *A philosophy for the science of well-being*. Oxford University Press.

Alexandrova, A., & Fabian, M. (2022). *The science of wellbeing*. John Templeton Foundation. https://www.templeton.org/wp-content/uploads/2022/02/Science_of_Wellbeing_JTF_1.pdf

Billman, G. E. (2020). Homeostasis: The underappreciated and far too often ignored central organizing principle of physiology. *Frontiers in Physiology, 11*, 200. https://doi.org/10.3389/fphys.2020.00200

Brunnhuber, S. (2025). The art of transformation: How we can simultaneously adapt to and change the world. In M. Mühlböck (Ed.), *Sustainable transformation and well-being. The role of business for individual and collective flourishing* (pp. 3–14). Springer.

Fehr, E., & Fischbacher, U. (2002). Why social preferences matter – The impact of non-selfish motives on competition, cooperation and incentives. *The Economic Journal, 112*, C1–C33.

Fletcher, G. (2016). *The philosophy of well-being: An introduction*. Routledge.

Graber, O. (2025). The Zen of music: On the energetics of music as a source of personal well-being. In M. Mühlböck (Ed.), *Sustainable transformation and well-being. The role of business for individual and collective flourishing* (pp. 117–128). Springer.

Kahneman, D. (2003). Maps of bounded rationality: Psychology for behavioral economics. *The American Economic Review, 93*, 1449–1475. https://doi.org/10.1257/000282803322655392

Lux, H. (2025). Lessons from a wasp sting: A plea for inner development in order to change the world for the better. In M. Mühlböck (Ed.), *Sustainable transformation and well-being. The role of business for individual and collective flourishing* (pp. 311–324). Springer.

Mayr, E. (1994). Evolution. In E. Mayr (Ed.), *Evolution. Die Entwicklung von den ersten Lebensspuren bis zum Menschen* (pp. 9–19). Spektrum der Wissenschaft.

Mühlböck, M., & Schmidpeter, R. (2025). Paying tribute to the complexity of human beings: A business narrative harmonizing the well-being of individuals and society. In M. Mühlböck (Ed.), *Sustainable transformation and well-being. The role of business for individual and collective flourishing* (pp. 69–101). Springer.

OpenAI. (2024). ChatGPT (13 May version) [Large language model]. https://chat.openai.com/chat

Sato, M. (2025). Centennial companies in Japan: Ethical strategies for sustainability and well-being. In M. Mühlböck (Ed.), *Sustainable transformation and well-being. The role of business for individual and collective flourishing* (pp. 45–66). Springer.

Scholz, R. W. (2011). *Environmental literacy in science and society: From knowledge to decisions*. Cambridge University Press.

Scholz, R. W., & Steiner, G. (2023). Process ownership in science – Practice collaborations: The special role of transdisciplinary processes in sustainable transitioning. *Sustainability Science (2023), 18*, 1501–1518. https://doi.org/10.1007/s11625-023-01291-7

Schüz, M. (2025a). Transdisciplinary management of polycrises – A philosophical approach. In M. Schüz (Ed.), *Transdisciplinary thinking and acting – Fundamentals and perspectives*. Springer..

Schüz, M. (2025b). Teaching and implementing ethics for well-being in business: A field report. In M. Mühlböck (Ed.), *Sustainable transformation and well-being. The role of business for individual and collective flourishing* (pp. 259–277). Springer.

Siegel, D. J., & Drulis, C. (2023). An interpersonal neurobiology perspective on the mind and mental health: Personal, public, and planetary well-being. *Annals of General Psychiatry, 22*, 5. https://doi.org/10.1186/s12991-023-00434-5

Süddeutsche Zeitung. (2023, December 21). Was die Deutsche Bahn von der Schweizer Bahn lernen kann. https://www.sueddeutsche.de/wirtschaft/deutsche-bahn-schweiz-tipps-verspaetung-puenktlichkeit-1.6322979

Tesar, M., & Peters, M. A. (2019). Heralding ideas of well-being: A philosophical perspective. *Educational Philosophy and Theory, 52*(9), 923–927. https://doi.org/10.1080/00131857.2019.1696731

Tiberius, V. (2014). Philosophical theories of well-being. In A. C. Michalos (Ed.), *Encyclopedia of quality of life and well-being research* (pp. 7110–7113). Springer Reference.

Vermeulen, V. (2025). From paragliding in the Himalayas to finding purpose in unleashing women's full potential: A personal report. In M. Mühlböck (Ed.), *Sustainable transformation and well-being. The role of business for individual and collective flourishing* (pp. 291–302). Springer.

Walsh, D., & Evans, K. (2014). Critical realism: An important theoretical perspective for midwifery research. *Midwifery, 30*(1), e1–e6. https://doi.org/10.1016/j.midw.2013.09.002

Weetjens, B. (2025). The skull in our suitcase: What we can learn from Zen in challenging moments. In M. Mühlböck (Ed.), *Sustainable transformation and well-being. The role of business for individual and collective flourishing* (pp. 281–289). Springer.

Welser, T. (2025). If I can really make an impact, then my life has been a success: A personal report on the transformation of a company and an individual. In M. Mühlböck (Ed.), *Sustainable transformation and well-being. The role of business for individual and collective flourishing* (pp. 303–310). Springer.

Zu, L. (2025). How purpose drives sustainable business transformation and wellbeing: Practical wisdom of Taoism. In M. Mühlböck (Ed.), *Sustainable transformation and well-being. The role of business for individual and collective flourishing* (pp. 29–43). Springer.

Marisa Mühlböck holds a doctorate in social and economic science. Over the course of her career, she has gained extensive insight into different fields of work, which has fostered her understanding of diverse organizational structures, thinking patterns, and social context. As a former social entrepreneur, she also knows the entrepreneurial life from practice and incorporates this experience into her research. As a Research Country Co-lead, she is involved in the Inner Development Goals (IDGs) Initiative, which is dedicated to researching and promoting inner development to achieve the UN Sustainable Development Goals. She also served as a founding board member of Verband für gemeinnütziges Stiften, Austria's first association representing and advocating for charitable foundations. Her current research interests are based on her theoretical background as well as on her practical experience and focus on sustainability in different areas, sustainable entrepreneurship, and well-being.

René Schmidpeter is a professor for Sustainable Management at the BFH in Bern and a scientific researcher at Parmenides Foundation in Munich. He also has been a guest professor/lecturer in Australia, England, China, India, and the USA. He is the series editor for Springer's CSR, Sustainability, Ethics and Governance books, editor of the Encyclopedia of Sustainable Management (ESM), and editor of the Dictionary of Corporate Social Responsibility (DCSR) as well as editor-in-chief of the International Journal of Corporate Social Responsibility (JCSR). His research and teaching activities focus on sustainable business transformation, international perspectives on sustainability, social innovation, and sustainable entrepreneurship with a special focus on the relationship between business and society. He has published widely and is the editor of the German management series on Corporate Social Responsibility at Springer Gabler and a scientific adviser to the FAZ-Institut and Academic Head of the annual Responsible Leadership Conference. Schmidpeter serves as an expert in many academic, economic, and political steering committees as well as on sustainability advisory boards. He has conducted research for and worked with the Federal Ministry of Education and Research, the European Union, the Bavarian Ministry of Social Affairs, the Anglo-German Foundation, the Bertelsmann Foundation, the Mercator Foundation, the Humboldt Foundation, the Federal State of Tyrol as well as the Deutsche Forschungsgesellschaft (DFG). He was co-founder of M3TRIX in Cologne and Tetranomics in Munich.

MIX
Papier aus verantwortungsvollen Quellen
Paper from responsible sources
FSC® C105338

If you have any concerns about our products,
you can contact us on
ProductSafety@springernature.com

In case Publisher is established outside the EU,
the EU authorized representative is:
Springer Nature Customer Service Center GmbH
Europaplatz 3, 69115 Heidelberg, Germany

Printed by Libri Plureos GmbH
in Hamburg, Germany